AF735278

REPUBLIC OF SOUTH AFRICA
REPUBLIEK VAN SUID-AFRIKA

DEPARTMENT OF AGRICULTURAL TECHNICAL SERVICES
DEPARTEMENT VAN LANDBOU-TEGNIESE DIENSTE

FLORA OF SOUTHERN AFRICA

VOL. 22

ISBN 0 621 02870 3

FLORA OF SOUTHERN AFRICA

which deals with the territories of

THE REPUBLIC OF SOUTH AFRICA, LESOTHO,
SWAZILAND AND SOUTH WEST AFRICA

VOLUME 22

Edited by
J. H. Ross

Botanical Research Institute,

Department of Agricultural Technical Services

1976

THE TERRITORIES DEALT WITH IN THIS FLORA

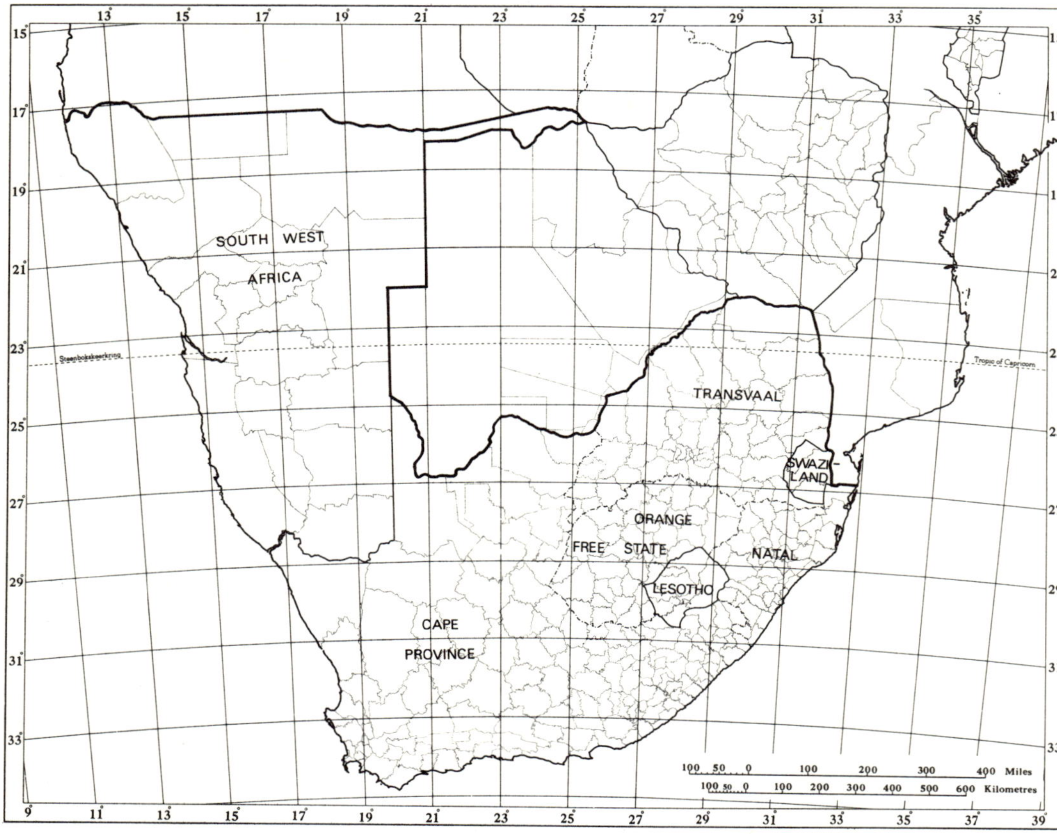

CONTENTS

INTRODUCTION

For a key to the families, the Flora should be used in conjunction with Phillips's Genera of South African Flowering Plants, ed. 2 (1951) and Dyer's Genera of Southern African Flowering Plants, Vol. 1 (1975) and Vol. 2 (in press), which are arranged on the lines of the Engler system. The genera are numbered according to the list published by De Dalla Torre and Harms in their Genera Siphonogamarum (1900–1907) in order to facilitate reference, though genera in the Flora are not necessarily arranged in this sequence.

As in previous volumes, generally accepted abbreviations are used for literature references, except in the following cases which appear frequently and are, therefore, considerably condensed:

C.F.A.	Conspectus Florae Angolensis
F.C.	Flora Capensis
F.C.B.	Flore du Congo et du Rwanda-Burundi
F.S.W.A.	Prodromus einer Flora von Südwestafrika
F.T.A.	Flora of Tropical Africa
F.T.E.A.	Flora of Tropical East Africa
F.W.T.A.	Flora of West Tropical Africa
F.Z.	Flora Zambesiaca
Phill., Gen. ed. 2	The Genera of South African Flowering Plants by E. P. Phillips, ed. 2 (1951)
Burtt Davy, Fl. Transv.	Manual of the Flowering Plants and Ferns of the Transvaal and Swaziland, Vol. 1 (1926) and Vol. 2 (1932).

As before, the abbreviation "l.c." is used for previously cited references even though "op. cit." or "tom. cit." would in certain cases be more correct.

In citing specimens the grid reference system has been used. The spelling of the names of some localities has been brought into line with the findings of the Committee on Standardisation of Place Names.

In the text, species which show evidence of becoming naturalized are treated in the same way as indigenous species. In the Index, synonyms are in italics while exotic species are signified by an asterisk.

A change in the existing Flora format is being introduced shortly and in future families or sections of large families will be published separately as they are completed.

PLAN OF CONTENTS OF VOLUMES

VOL. 1

Stangeriaceae
Zamiaceae
Podocarpaceae
Pinaceae
Cupressacea
Welwitschiaceae
Typhaceae
Zosteraceae
Potamogetonaceae
Ruppiaceae
Zanichelliaceae
Najadaceae
Aponogetonaceae
Juncaginaceae
Alismataceae
Hydrocharitaceae

VOL. 2

Poaceae

VOL. 3

Cyperaceae
Arecaceae
Araceae
Lemnaceae
Flagellariaceae

VOL. 4

Restionaceae
Mayacaceae
Xyridaceae
Eriocaulaceae
Commelinaceae
Pontederiaceae
Juncaceae

VOL. 5

Liliaceae

VOL. 6

Haemodoraceae
Amaryllidaceae
Velloziaceae
Dioscoreaceae

VOL. 7

Iridaceae

VOL. 8

Musaceae
Zingiberaceae
Burmanniaceae
Orchidaceae

VOL. 9

Piperaceae
Salicaceae
Myricaceae
Ulmaceae
Moraceae
Urticaceae
Proteaceae

VOL. 10

Loranthaceae
Santalaceae
Grubbiaceae
Olacaceae
Balanophoraceae
Aristolochiaceae
Rafflesiaceae
Hydnoraceae
Polygonaceae
Chenopodiaceae
Amaranthaceae
Nyctaginaceae

VOL. 11

Phytolaccaceae
Aizoaceae

VOL. 12

Portulacaceae
Basellaceae
Caryophyllaceae
Nymphaeaceae
Ceratophyllaceae
Ranunculaceae
Menispermaceae
Annonaceae
Monimiaceae
Lauraceae
Hernandiaceae
Papaveraceae
Fumariaceae

VOL. 13

Brassicaceae
Capparaceae
Resedaceae
Moringaceae
Droseraceae
Podostemaceae
Hydrostachyaceae

VOL. 14

Crassulaceae

VOL. 15

Saxifragaceae
Pittosporaceae
Cunoniaceae
Myrothamnaceae
Bruniaceae
Hamamelidaceae
Rosaceae
Connaraceae

VOL. 16

Fabaceae: (1) *Mimosoideae*

VOL. 17

Geraniaceae
Oxalidaceae

PLAN OF CONTENTS OF VOLUMES

OCHNACEAE

by P. C. V. du Toit and A. A. Obermeyer

Trees, shrubs or rarely herbs; bark rough or smooth, sometimes flaking; with cortical vascular bundles in stem; in *Brackenridgea* with a yellow pigment beneath bark. *Leaves* alternate, simple (rarely compound in genera outside Africa), discolorous, midrib prominent, nerves penninerved and with reticulate veining, margin entire or serrate; stipules free or united and intrapetiolar, entire or laciniate, early caducous or persistent. *Flowers* bisexual, regular (rarely zygomorphic in genera outside Africa), often fragrant, in terminal or axillary panicles, racemes or fascicled cymes, or solitary; pedicels articulated at or near base. *Sepals* usually 5 (3–10), free or nearly so, quincuncial, usually persistent and accrescent and often red coloured in fruit. *Petals* usually 5 (10), contorted in bud, fugaceous. *Stamens* few to many, free; (staminodes sometimes present in genera outside Africa); filaments free, persistent; anthers bilocular, basifixed, linear, opening by longitudinal slits or apical pores, usually caducous. *Carpels* (2) 5–15, superior, free, uni-ovulate, or fused, with 2–many biseriate ovules in each locule; styles as many as placentas, gynobasic or epigynous, completely united or free apically, stigmas terminal, usually somewhat enlarged. *Fruit:* drupes on an enlarged torus, berries or septicidal capsules; seeds 1–many, with or without endosperm.

Genera about 30, species about 250, pantropical, extending furthest south in subtropical and temperate southern Africa. Two genera and 13 species occur in our area.

Leaf margin serrulate with mucronate teeth or more or less entire; lateral veins not continuing upwards alongside margin; petals yellow or orange . . . . 1. **Ochna**

Leaf margin closely beset with small oblong-ovoid, white glands on the shallow teeth; lateral veins continuing upwards alongside margin; petals white or pink . . . . . . 2. **Brackenridgea**

5112 1. OCHNA

Ochna *L.*, Sp. Pl. 1: 513 (1753); Gen.Pl. ed. 5: 229 (1754); DC., Ann. Mus. Paris 17: 410 (1811); Prodr. 1: 735 (1824); Harv. in F.C. 1: 448 (1860); Oliv. in F.T.A. 1: 316 (1868); Phill. in Bothalia 1: 87 (1921); Gilg in Pflanzenfam. ed. 2,21: 53 (1925); Burtt Davy, Fl. Transv. 1: 237 (1926); Robson in F.Z. 2: 225 (1963); Schreiber in F.S.W.A. 42: 2 (1968). Type species: *O. jabotapita* L. from Sri Lanka.

Diporidium Bartl. & Wendl., Beitr. Bot. 2: 24 (1825).

Trees or shrubs or occasionally virgate, soboliferous shrublets; bark rough or smooth, sometimes flaking. *Leaves* semi-persistent, shed in spring or in times of drought, usually shortly petiolate, simple, ovate, obovate, oblong to linear, margin entire to serrate, midrib prominent on both sides, veins spreading, distinctly reticulate; stipules in S. Afr. spp. intrapetiolar, fused, bifid or bidentate. *Flowers* bisexual, regular, often fragrant, in terminal or axillary panicles or racemes or reduced to subumbellate, many- to few-flowered cymes or solitary, on short lateral spurs; bud scales or bracts distichous, caducous, leaving transverse scars on twigs; pedicels articulated near the base or in lower half, accrescent in S. Afr. spp. *Sepals* 5 in S. Afr. spp., quincuncial in bud, green or yellow, persistent, enlarging and usually turning red or pink in fruit, firm, enveloping bud and usually young fruit, but spreading at anthesis and when fruit is ripe. *Petals* 5 (in S. Afr. spp.), yellow, orange or rarely white, usually obovate, attenuate or clawed at the base, fugaceous. *Stamens* many (30–50), free, irregularly placed in a dense whorl on torus; filaments filiform, persistent, apex slightly capitate; anthers bilocular, opening by longitudinal slits or apical pores, basifixed, caducous. *Carpels* 5–15, free, attached basally or laterally to the receptacle; styles gynobasic, completely united or the upper ends free, recurved; stigmas apical, as many as carpels or fused and capitate. *Drupelets* 1–several, some usually aborting, kidney-shaped and attached centrally at the indented inner side or oblong-globose, attached basally to the swollen, often red receptacle, usually hard, black and shiny.

Species about 86, found in Africa south of the Sahara, Yemen, Madagascar, Mascarenes to Asia; tropics. Twelve species in Southern Africa, widespread but absent from the Winter-rainfall Region.

The fruit is distributed by birds who eat the oil-rich drupelets. This could explain the often erratic distribution of species. Of interest is the occurrence of 7 taxa in a small area in the N.E. Transvaal on the farm Cyprus, where, because of congenial conditions, *Ochna arborea* vars. *arborea* and *oconnorii, O. confusa, O. holstii, O. natalitia, O. pretoriensis* and *O. serrulata* were all well established.

Galls are found on *Ochna natalitia, O. serrulata, O. gamostigmata* and *O. barbosae.* Their shapes are typical for the respective species. The following species are soboliferous: *O. pulchra, O. confusa, O. natalitia, O. gamostigmata* and *O. cinnabarina.*

Flowers arranged in many-flowered inflorescences or in reduced few-flowered, subumbellate fascicles:
  Carpels and drupelets kidney-shaped, attached centrally to the receptacle on the indented side; bark flaking to expose patches of a different colour:
    Carpels 7 or 8; leaves elliptic c. 10 × 4 cm, firm; racemes in simple, many-flowered, pendulous bunches .... 1. **O. pulchra**
    Carpels 5; leaves narrowly oblong, c. 6 × 1,5 cm (larger in forest trees), thin; racemes compound, many-flowered, to short, subumbellate, few-flowered clusters .... 2. **O. arborea**
  Carpels and drupelets oblong-globose, attached basally to receptacle; bark not flaking, evenly coloured:
    Anthers opening by longitudinal slits; carpels 5–6:
      Usually arborescent with the leaves placed in a horizontal plane, appearing almost digitate on the ends of the branchlets; leaves acute basally, attenuate-acuminate distally; pedicels articulated above the base .... 3. **O. holstii**
      Usually soboliferous virgate dwarf shrubs with the bark of the woody stem-base not properly developed on one side; leaves long-attenuated basally, acute distally; pedicels articulated at the base .... 4. **O. confusa**
    Anthers opening by apical pores; carpels 6–12 .... 5. **O. natalitia**
Flowers solitary (rarely 2 placed close together) on short lateral spurs; carpels 5(6), oblong-globose, attached basally to torus:
  Sepals 1–1,5 cm long in fruit:
    Leaves green, flat and spreading; petioles 1–2 mm long, stout:
      Pedicels erect or spreading, 1–1,5 cm long, articulated at or near base:
        Young twigs with some scattered lenticels; leaves narrowly obovate; margin indistinctly to distinctly serrulate; Transvaal .... 6. **O. pretoriensis**
        Young twigs conspicuously and densely pale-mottled with lenticels; leaves narrowly elliptic or linear-acuminate; margin sharply and closely serrulate:
          Leaves narrowly elliptic; style with 5 free, recurved apical ends bearing small discoid stigmas; pedicels articulated 1–2 mm above the base; E. Cape, Natal, E. Transvaal .... 7. **O. serrulata**
          Leaves linear-acuminate; style ending in a capitate stigma; pedicels articulated at the base; Natal, E. Transvaal .... 8. **O. gamostigmata**
      Pedicels recurved, 2–3 cm long, articulated below centre with upper part often forming an angle with lower part .... 9. **O. inermis**
    Leaves glaucous, drooping and folded; petioles 4 mm long, thin; sprawling shrubs in granite rock fissures; N. Transvaal .... 10. **O. glauca**
  Sepals 1,5–2 cm long in fruit:
    Virgate, soboliferous shrubs up to 1 m high; young twigs densely lenticellate; sepals loosely arranged around maturing fruit; northern S.W. Africa .... 11. **O. cinnabarina**
    Trees or large shrubs; young twigs with few lenticels; sepals forming a closed, ovoid covering around maturing fruit, spreading later; coastal dunes of Natal .... 12. **O. barbosae**

1. **Ochna pulchra** *Hook.*, Ic. Pl. 6, t.588 (1843), as *pulchrum*; Harv. in F.C. 1: 449 (1860); Oliv. in F.T.A. 1: 317 (1868); Gilg in Bot. Jahrb. 33: 234 (1903); Phill. in Bothalia 1: 91 (1922); Burtt Davy, Fl. Transv. 1: 238 (1926); Verdoorn in Flow. Pl. Afr. 29: t.1139 (1952); Letty, Wild Flow. Transv. 217, t.109, f.1 (1962); Story in Mem. Bot. Surv. S. Afr. 30: 35, photo. 33 (1958); Robson in F.Z. 2: 228 (1963); Schreiber in F.S.W.A. 42: 2 (1968); Palmer & Pitman, Trees S. Afr. 3: 1503 (1973). Type: Transvaal, Magaliesberg, *Burke & Zeyher* 191 (K, holo.; PRE, photo.; SAM!).

O.rehmannii Szyszyl., Polypet. Disc. Rehm. 2: 28 (1888). Type: Transvaal, Pretoria, hills above Apies River, *Rehmann* 4341 (Z, holo.; K; PRE, photo.). *O. aschersoniana* Schinz in Verh. Bot. Ver. Prov. Brandenb. 29: 61 (1888). Type: S.W. Africa, Ovamboland, Otjiheveta, *Schinz* 1139 (Z, holo.; K; PRE, photo.). *O. fuscescens* Heine in Mitt. Bot. Staatssamml. München 1: 340 (1953). Syntypes: S.W. Africa, Waterberg Plateau, *Volk* 1135(M); farm Rotenfels, *Rehm* s.n.(M).

Trees up to 10 m tall or shrubs, soboliferous, occasionally dwarfed; trunk with rough broken bark below, smooth above, grey, flaking to expose cream patches. *Branches* spreading, brittle, rough, grey, new growth pale fawn. *Leaves* semi-persistent, turning yellow in winter, shed in spring, young leaves appearing with flowers, oblong to elliptic, c. 10 cm long and c. 2 cm wide but variable in size, apex obtuse, apiculate, base rounded or somewhat attenuate, margin pale, entire in lower half, shallowly serrulate above, midrib and veins prominent, firm and glossy; stipules intrapetiolar, paired, subulate, up to 15 mm long, rarely narrowly spatulate at the apex, early deciduous. *Flowers* fragrant, in simple, pendulous, clustered, many-flowered racemes terminating short annual branchlets; pedicels c. 1 cm at anthesis, up to 1,8 cm in fruit, articulated above the base. *Sepals* broadly ovate, 8 mm long, green to yellow, enlarged, reflexed, pink or red and white-edged in fruit. *Petals* soft yellow, obovate-attenuate, 12 mm long, fugaceous. *Stamens* numerous (c. 50) with filaments 2 mm long; anthers 1 mm long, biporose. *Carpels* 6–8, attached centrally on inside of torus; styles fused below, free and recurved above; stigmas apical, tumid. *Drupelets* kidney-shaped, rounded, 12 mm long, black or mottled with cream. FIG. 1: 1; 2: 1.

Recorded from the Transvaal, South West Africa, Angola, Rhodesia, Zambia and western Mozambique; forming colonies on rocky sandstone slopes or in sandy areas. Flowering in September–October.

New growth from stolons in narrow rock fissures may remain hemicryptophytic and has on occasion been mistaken for the poisonous *Dichapetalum cymosum.*

S.W.A.—1715 (Ondangua): Oshikango near Angola border, *Rodin 2615; 2662*. 1718 (Kuring-Kuru): 12 km E. of Kuring-Kuru, *Giess 9504*. 1722 (Chirundi): Western Caprivi Strip, Bwabwata Rest Camp, *Watt 18*. 1918 (Grootfontein): N.E. of Grootfontein, *Schoenfelder 220*. 1919 (Kanovlei): Okavango Territory, Kanovlei, *Barnard 4*. 1920 (Tsumkwe): Garu, 88 km S. of Runtu, *Maguire 1576*. 2012 (Waterberg): Waterberg Plateau, *Boss sub TRV 35033*.

TRANSVAAL.—2230 (Messina): Makonde, 24 km E. of Sibasa, *Pienaar 47*. 2231 (Pafuri): Punda Milia, *Van der Schijff 947; 996*. 2427 (Thabazimbi): Rankin's Pass, *Van Wyk 42*; near Rooiberg, *Codd 6154*. 2428 (Nylstroom): 13 km. N. of Nylstroom on road to Naboomspruit, *Vahrmeijer 134; Maguire 1354*; Mosdene, *Galpin M52*. 2429 (Zebediela): Potgietersrus, *Leendertz 1258*. 2526 (Marico): Zeerust, *Thode A1367*. 2527 (Rustenburg): Rustenburg, *Galpin 11633; 11635*; Magaliesberg, *Pegler 1032*; Hartebeestpoortdam, *Gerstner 6416*. 2528 (Pretoria): Meintjieskop, *Mogg 15310; Burtt Davy 5035*. 2529 (Witbank): Loskop Dam area, *Prosser 1885; Mogg 30579*. 2627 (Potchefstroom): Jack Scott Reserve, near Krugersdorp, *Wells 2338*.

Common names: Lekkerbreek; Monyelenyele (Tsw); Monamane, Mopha (NS); Musuma (V).

2. **Ochna arborea** *Burch. ex DC.*, Prodr. 1: 736 (1824). Type: E. Cape, Port Alfred, Kowie River, *Burchell*, Cat. 4012 (G, holo.; K; PRE, photo.).

Understorey forest trees up to 12 m tall, or smaller shrubs; trunk with light red wood, covered with smooth grey bark, flaking irregularly exposing brown patches; crown small, with ascending grey and brown mottled branches which are irregularly ridged with grey raised lenticels. *Leaves* semi-persistent or deciduous, shed in spring or during a period of drought; lamina deep green, variable, broadly to narrowly elliptic, 5–10 cm long, 2–3 cm broad, apex rounded to acute, base truncate, margin nearly smooth or with some shallow teeth to sharply serrate, venation distinctly reticulate, leathery; stipules intrapetiolar, fused. *Flowers* scented, arranged in compound, complex, many-flowered usually erect panicles on well developed, mostly leafless branchlets to a reduced subcorymbose fascicle with 4–8 flowers on a very short spur. *Sepals* 5, ovate-convex, c. 7 mm long, green; enlarged, dark red and reflexed at anthesis and in fruit. *Petals* 5, obovate-cuneate, 12 mm long, yellow. *Stamens* c. 30, with filaments about as long as biporose anthers. *Carpels* 5, reniform; styles completely

FIG. 1.—1, **Ochna pulchra,** flowering branch, × 1/3; 1a, old flower showing persistent calyx, torus and 1 drupelet attached ventrally, × 1 *(ex garden, Bot. Res. Inst.). 2,* **O. pretoriensis,** flowering branch, × 1; 2a, old flower showing persistent calyx, torus and 3 drupelets attached basally, × 1 *(ex garden, Bot. Res. Inst.)*; 2b, anther with locules opening through apical pores, × 10. 3, **O. confusa,** anther with longitudinal dehiscence, × 10. 4, **O. natalitia,** gall *(Moll* 3612) × 1. 5, **O. serrulata,** gall *(McClean & Ogilvie* sub PRE 39308), × 1.

united or free apically; stigmas capitate. *Drupelets* reniform, 10–15 mm long, black.

Transvaal, Swaziland, Natal to eastern and south-eastern Cape as far as George; also in Mozambique and Rhodesia. Flowering in spring.

Two varieties are recognized.

Leaves broadly elliptic, c. 4 × 1, 5 cm; marginal teeth often obscure; inflorescence few-flowered on short spurs; trees to shrublets found outside or inside forests ....(a) var. *arborea*

Leaves narrowly elliptic—acuminate, c. 8 × 2,5 cm; margin distinctly serrate; inflorescence many-flowered, paniculate; forest tree ....(b) var. *oconnorii*

(a) var. **arborea.**

Ochna arborea Burch. ex DC., Prodr. 1: 736 (1824); J. Smith in Bot. Mag. t.4519 (1850); Harv. in F.C. 1: 449 (1860); Sim, For. Fl. Cape Col. 163, t.29, f.1 (1907); Phill. in Bothalia 1: 92 (1922); Burtt Davy, Fl. Transv. 1: 238 (1926); Robson in F.Z. 2: 230 (1963); Palmer & Pitman, Trees S. Afr. 3: 1504 (1973). Type: E. Cape, Port Alfred Kowie River, *Burchell*, Cat. 4012 (G; holo.; K; PRE, photo.).

Diporidium arboreum (Burch. ex DC.) Wendl. in Bartl. & Wendl., Beitr. Bot. 2: 26 (1825). Type as above.

The variety *arborea* is distinguished by its shorter and relatively broader elliptic leaves, which usually have subentire margins. The inflorescence is few-flowered.

It appears to be the form adopted in less favourable conditions. The distribution is mainly southern, in forest scrub, dune forest and bushveld. FIG. 2: 2a.

TRANSVAAL.—2229 (Waterpoort): Wylliespoort, *Hafström & Acocks 1895*. 2430 (Pilgrims Rest): Cyprus, *Renny 137*; The Downs, *Du Toit 178*.

SWAZILAND.—2631 (Mbabane): 25 km S. of Stegi, *Compton 29282*.

NATAL.—2632 (Bela Vista): Ndumu Game Reserve, *Pooley 870*. 2731 (Louwsburg): Sokosoko forest, Nongoma, *Gerstner 4910*. 2732 (Ubombo): 21 km from Ingwavuma to Ndumu, *Moll & Strey 3710*; Mkuzi Game Reserve, *Ward 3604*. 2830 (Dundee): Qudeni Forest, *Edwards 2655*. 2831 (Nkandla): Eshowe, *Meebold 12952; Thode A1228*. 2832 (Mtubatuba): False Bay, *Ward 1634*. 2930 (Pietermaritzburg): Karkloof, Lions River, Ehlatini Forest, *Moll 2862*; Kloof near Pinetown, *Galpin 12080*. 2931 (Stanger): Durban, *Franks sub TRV 34099*. 3030 (Port Shepstone): Gibraltar, *Strey 10041*.

CAPE.—3129 (Port St. Johns): Mpingo Forest, Mtambala, *Fegen sub PRF 3089*. 3227 (Stutterheim): Kabaku Hills, *Acocks 8967;* Nahoon River, East London, *Galpin 3291*. 3228 (Butterworth): Kentani, *Pegler 1216;* near Komga, *Flanagan 359*. 3322 (Oudtshoorn): Ebb & Flow Nature Reserve, *Taylor 7984*. 3325 (Port Elizabeth): Van Stadens Pass, *De Winter 7618*. 3326 (Grahamstown): Bushmans River mouth, *Galpin 10667;* Port Alfred, *Salisbury 68*. 3423 (Knysna): Harkerville Forest, *Keet 527;* Storms River, *Keet 525*. 3424 (Humansdorp): Witelsbos, *Fourcade 845*.

Common names: Cape plane, redwood; rooihout; um-Tensema (X); umBomvane, umTelele (Z).

(b) var. **oconnorii** *(Phill.) Du Toit* in Bothalia 11: 518 (1975).

O. oconnorii Phill. in Bothalia 1: 92 (1922); Robson in F.Z. 2: 230 (1963). Lectotype: Transvaal, Woodbush forest, *O'Connor* sub PRE 1257 (PRE, lecto.; PRF 2198).

This variety is a more luxuriant forest form with longer, larger leaves and erect, compound, elongate inflorescences. There are intermediates. FIG. 2: 2b.

Recorded from the mist belt forests of Transvaal, Natal and eastern Cape; also in Rhodesia and Mozambique.

TRANSVAAL.—2230 (Messina): Entabeni Forest Reserve, *Obermeyer sub TRV 30346*. 2330 (Tzaneen): Westfalia Estate, *Scheepers 728;* Woodbush, *Botha 1; Grenfell 3*. 2430 (Pilgrims Rest): The Downs, *Du Toit 178;* Pilgrims Rest, *Burtt Davy 1409*. 2531 (Komatipoort): Barberton, *Oranje sub PRE 1264; Thorncroft 2029*.

NATAL.—2632 (Bela Vista): Ndumu Game Reserve, *Pooley 135c*. 2831 (Nkandla): Eshowe, *Moberly 9*. 2930 (Pietermaritzburg): York, Benvie, *Fisher 1038*.

CAPE.—2326 (Grahamstown): The Haven, *Gordon-Gray 592*.

Common names: Cape Plane, Rooihout, Rooi ysterhout, Morelle; Murambo, murambothavha (V).

3. **Ochna holstii** *Engl.* in Abh. Preuss. Akad. Wiss.: 69 (1894); Pflanzenw. Ost-Afr. C: 273 (1895); Gilg in Bot. Jahrb. 33: 234, 241 (1903); Phill. in Bothalia 1: 93 (1922); Burtt Davy, Fl. Transv. 1: 238 (1926); Robson in F.Z. 2: 240 (1963); Palmer & Pitman, Trees S. Afr. 3: 1507 (1973). Type: Tanzania, Usambara, *Holst* 2601 (B†; K; PRE, photo.).

Understorey trees up to about 8 m tall or shrubs with red wood and rough ridged, grey or brown bark which does not flake; branches spreading; young branchlets with scattered raised lenticels, glabrous or puberulous. *Leaves* tardily deciduous, immature at time of flowering, radiating in a horizontal plane from the tips of young shoots in a somewhat digitate pattern; blade narrowly elliptic, 5–10 cm long, 1,5–2,5 cm broad, apex attenuate-acuminate, base cuneate, regularly and closely serrulate, stiffly membranous, lateral veins curving upwards from midrib; stipules intrapetiolar, fused. *Flowers* usually in (2) 6–10-flowered racemes, on short lateral spurs below new leafy shoots (tropical forms with many-flowered elongate racemes); pedicels thin, up to 3 cm in fruit, articulated above base, glabrous or puberulous.

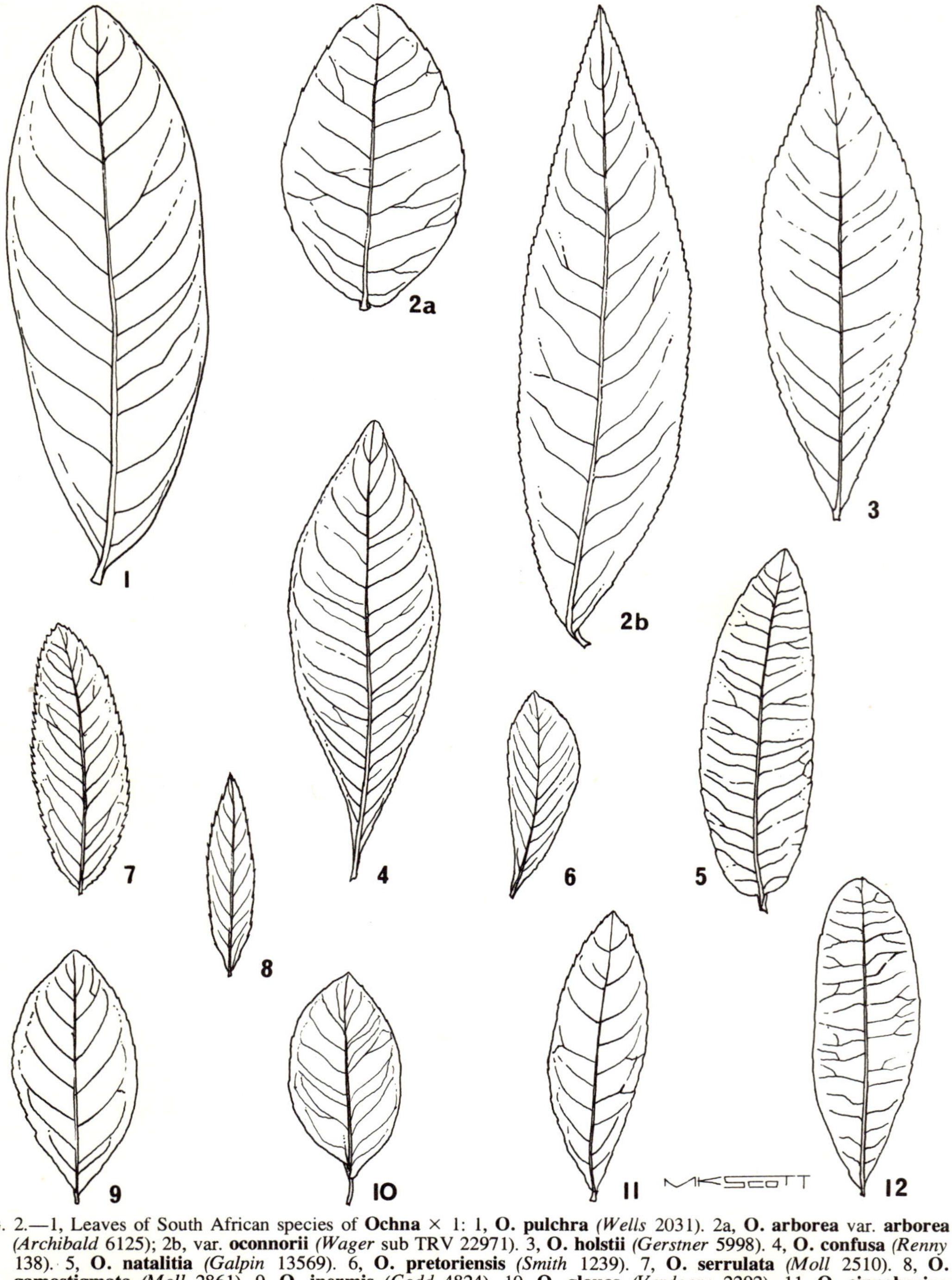

FIG. 2.—1, Leaves of South African species of **Ochna** × 1: 1, **O. pulchra** *(Wells* 2031). 2a, **O. arborea** var. **arborea** *(Archibald* 6125); 2b, var. **oconnorii** *(Wager* sub TRV 22971). 3, **O. holstii** *(Gerstner* 5998). 4, **O. confusa** *(Renny* 138). 5, **O. natalitia** *(Galpin* 13569). 6, **O. pretoriensis** *(Smith* 1239). 7, **O. serrulata** *(Moll* 2510). 8, **O. gamostigmata** *(Moll* 2861). 9, **O. inermis** *(Codd* 4824). 10, **O. glauca** *(Verdoorn* 2292). 11, **O. cinnabarina** *(Killick & Leistner* 3202). 12, **O. barbosae** *(Balsinhas* 1372).

Sepals narrowly elliptic, c. 9 mm long, obtuse, ivory at anthesis, up to 1,5 cm long, red, pink, purple or green and spreading in fruit, somewhat crinkled when dried. *Petals* obovate, about 1 cm long, clawed, pale to bright yellow, caducous. *Stamens* c. 40, with filaments 3 mm long; anthers 1,5 mm long, dehiscing by longitudinal slits. *Carpels* 5(–6), erect; style with a swollen, lobed stigma. *Drupelets* oblong-globose, c. 8 mm long, attached basally to swollen torus. FIG. 2: 3.

An understorey forest tree recorded from the Transvaal, Natal and eastern Cape; widespread and variable in its range to Rhodesia and tropical eastern Africa; inhabiting shady ravines in mistbelt forests, gallery forests along rivers or as a forest relic in open areas; flowering October – November.

TRANSVAAL.—2229 (Waterpoort): Soutpansberg above Louis Trichardt, *Hutchinson 2025; Story 5935.* 2230 (Messina): Soutpansberg, Phiphidi Waterfall near Sibasa, *Obermeyer sub TRV 30037;* Lake Funduzi, *Story 4850.* 2329 (Pietersburg): Lejuma Plateau, *Strey 7945; Nicholson 669.* 2330 (Tzaneen): Woodbush, *Botha sub PRF 2935.* 2428 (Nylstroom): near Palala, *Smuts & Gillett 3357.* 2430 (Pilgrims Rest): Cyprus, *Renny 186;* Mariepskop, *Van der Schijff 4754; 5918; 6237.* 2527 (Rustenburg): kloof in western Magaliesberg range, *Rose Innes 203;* near Beacon 6078, *Taylor sub PRE 28763.* 2530 (Lydenburg): Kaapse Hoop, Berlin Forest Reserve, *Hofmeyr sub PRE 28762.* 2531 (Komatipoort): Barberton, 5 km S.W. of Agnes Mine, *Story 5993; 5455.*

NATAL.—2731 (Louwsburg): Ngome Forest Reserve, *Gerstner 4496; 5152; Tustin sub PRF 3115.* 2830 (Dundee): Qudeni Forest Reserve, *Edwards 2667.* 2930 (Pietermaritzburg): Noodsberg, *Ward 938.* 3029 (Kokstad): Ingeli Forest Reserve, *Tustin sub PRF 2907.*

CAPE.—3127 (Lady Frere): Engcobo Mt., *Flanagan 2696.*

Common names: Rooi Ysterhout; Real Red Pear; isiBanku, Pambane (Z); Tshipfure (V); ngqelene (X).

4. **Ochna confusa** *Burtt Davy & Greenway* in Burtt Davy, Fl. Transv. 1: 238 (in clavis), 239 (1926); in Kew Bull. 1926: 239 (1926); Robson in F.Z. 2: 250 (1963). Syntypes: Transvaal, Barberton, *Rogers* 18264 (PRE); Pilgrims Rest, *Rogers* 23068 (PRE).

O. leptoclada sensu Phill. in Bothalia 1: 94 (1922) pro parte quoad specim. Transvaal.

Saxicolous, soboliferous shrubs or dwarf shrubs 0,5–1(–2) m tall, the virgate branches arising from a woody base; this basal sprawling stem lacking bark on one side (in Transvaal specimens seen). *Branches* and branchlets pale dirty grey, lenticellate, rough. *Leaves* deciduous, sessile, narrowly elliptic, 6–9 cm long, 1,5–2,5 cm broad, apex attenuate, acute or obtuse, base long-attenuate, margin serrulate, firm, with close lateral veins curving upwards; stipules intrapetiolar, elongate-deltoid. *Flowers* in sessile (1–) 2–6-flowered pseudumbels on short lateral spurs below new leaf growth; pedicels about 12 mm long in fruit, articulated at the base. *Sepals* c. 5 mm long, ovate at anthesis; 1 cm long, deep red, flat and spreading in fruit. *Petals* obovoid, clawed, c. 1 cm long, 7 mm broad, bright yellow. *Stamens* c. 25–30, with filaments 5 mm long; anthers 2 mm long, opening by longitudinal slits. *Carpels* 5, attached basally to receptacle; styles fused, terete; stigma capitate. *Drupelets* oblong-globose, c. 1 cm long, black. FIG. 1: 3; 2: 4.

Recorded from eastern Transvaal; also in Mozambique, Malawi, Rhodesia, southern Tanzania, Zambia. Inhabits rock crevices on hillsides in more open vegetation; flowering in spring.

TRANSVAAL.—2430 (Pilgrims Rest): farm Cyprus, *Renny 138;* Sikorora near Macoutsie River, *Van Dam sub TRV 22935;* Ohrigstad Nature Reserve, *Jacobsen 1242;* Graskop Falls, *Galpin s.n.* 2530 (Lydenburg): Waterval Boven, *Britten 4741; 4770.* 2531 (Komatipoort): Barberton, *Nel 52; Thode A1570.*

O. confusa shows a relationship with *Brackenridgea* in that its anthers open by longitudinal slits, and in its leaves, which have lateral veins curving upwards and marginal teeth tipped with glands.

5. **Ochna natalitia** *(Meisn.) Walp.*, Repert. 2: 826 (1843); Phill. in Bothalia 1: 93 (1922); Burtt Davy, Fl. Transv. 1: 239 (1926); Robson in F.Z. 2: 233 (1963); Palmer & Pitman, Trees S. Afr. 3: 1508 (1973). Type: Natal, Port Natal (Durban), *Krauss* 454 (K, holo.; PRE, photo.).

Diporidium natalitium Meisn. in Hook., J. Bot. Lond. 2: 58 (1843). Type as above.

O. atropurpurea var. *natalitia* (Meisn.) Harv. in F.C. 1: 448 (1860). *O. chilversii* Phill. in Bothalia 1: 90 (1922). Syntypes: E. Cape, Ingeli Forest Reserve, *Chilvers* sub PRF 1945 (PRE; PRF; K); at mouth of Umkwani River, *Tyson* 2619 (PRE).

Understorey forest trees up to 6 m tall with red wood and a dark rough bark not peeling off; shrubby in open vegetation or reduced to soboliferous shrublets when exposed to fire and frost, 0,5–2 m tall. *Branches* and branchlets dark brown or grey, lenticellate; young twigs pale fawn coloured, the epidermis peeling off in small thin membranous white flakes. *Galls* often present, globose with rosulate, ovate, obtuse bracts. *Leaves* tardily deciduous, narrowly oblong, variable in size, 5–12 (–14) cm long, 1,5–2,5 cm broad, apex and base generally obtuse, margin serrulate, rarely subentire, midrib distinct, lateral veins at right angles to midrib, leathery, shortly petiolate; stipules intrapetiolar, elongate-deltoid. *Flowers* fragrant, in compound, many- to few-flowered racemes or subumbellate, terminating short lateral spurs; pedicels up to 2,5 cm long, articulated in lower quarter. *Sepals* elliptic-convex, about 9 mm long, green, slightly larger in fruit, turning wine red. *Petals* broadly obovate, c. 15 mm long, clawed, yellow. *Stamens* c. 30, with filaments 3 mm long; anthers biporose, 2 mm long. *Carpels* 6–12, usually about 8, erect; styles fused, terete, the apical ends free, recurved, forming swollen irregular glandular stigmas. *Drupelets* oblong-globose, 10–15 mm long, attached basally, black, shiny. FIG. 1: 4; 2: 5.

Fairly common along the eastern Cape coastal belt from Knysna to Natal and northern Mozambique; further inland in Natal and Swaziland to eastern and northern Transvaal, in forests or coastal scrub, in sandy soil or amongst rocks; flowering in spring.

TRANSVAAL.—2230 (Messina): Makonde Mission Station, *Westphal sub TRV 29126*. 2231 (Pafuri): Kruger National Park, Punda Milia, *Rowland Jones 27*. 2327 (Ellisras): Soutpan, *Galpin 15150*. 2329 (Pietersburg): Louis Trichardt, *Breyer sub TRV 23390; 24385*. 2430 (Pilgrims Rest): Mariepskop, *Killick & Strey 2440*. 2431 (Acornhoek): near Skukuza, *Cholmondely s.n.* 2528 (Pretoria): Kameeldrift, 64 km N.E. of Pretoria, *Repton 3875*. 2529 (Witbank): Loskopdam, Tweefontein, *Theron 1512*. 2530 (Lydenburg): Nelspruit, *Breyer sub TRV 17701; Buitendag 323; 226*. 2531 (Komatipoort): Pretoriuskop, *Van der Schijff 6*.

SWAZILAND.—2531 (Komatipoort): Komassan River, Havelock Concession, *Saltmarshe sub Galpin 983*. 2631 (Mbabane): Hill N.E. of Mbabane, *Compton 26184;* Ntondozi, *Compton 29211*.

NATAL.—2632 (Bela Vista): Ndumu Game Reserve, *Pooley 135b*. 2731 (Louwsburg): Nongoma, *Gerstner 3642; 5105*. 2732 (Ubombo): Sihangwane Store, *Moll 4400;* Mkuzi Game Reserve, *Ward 3611*. 2831 (Nkandla): Eshowe, *Gerstner 2818;* Nkandla Forest, *Edwards 2663*. 2832 (Mtubatuba): 25 km N. of Mtubatuba, *Codd 2018*. 2930 (Pietermaritzburg): "The Start", near Howick, *Moll 3370*. 2931 (Stanger): Durban, *Marloth 4318; Medley Wood 66; 12702*. 3030 (Port Shepstone): near Southbroom, *Codd 9706*.

CAPE.—3129 (Port St. Johns): Ntsubane, *Strey 9006;* Notinsila Forest, *Fegen sub PRF 1816;* Manteku, Dakane Location, *Strey 10197*. 3325 (Uitenhage): Groendalkloof, below dam, *Mauve 4934*.

Common names: Coast Boxwood, Coast Redwood; umBomvu, umBomvane, umMilamatsheni (Z).

Ochna natalitia and *O. arborea* are often difficult to distinguish in the herbarium when the specimens consist of sterile branchlets only and offer no information on bark, etc. Galls are found only in *O. natalitia*. Young twigs of this species are fawn coloured. Mature leaves from forest trees of both species are usually indistinguishable for both become narrowly oblong with coarse marginal teeth, possess spreading lateral veins and a dense reticulation. More often the lamina of *O. natalitia* is narrowly oblong whereas that of *O. arborea* is ovate and lighter coloured below.

In the sandy area of Tongaland, in northern Natal, north of Manzengwenya in a strip a few km inland from the sea, a form of *Ochna natalitia* occurs, which is from 7,5 to 15 cm tall. These plants can cover an extensive area and it was found that the individual plants were interconnected by underground stems. At first thought to be a new species, investigation showed that morphologically they differ from the shrub or tree-like form only in the much smaller growth form and in having underground stems, a feature in common with a few other *Ochna* species, e.g. *O. pulchra, O. cinnabarina* and *O. confusa*. The anatomy has proved these underground interconnecting parts to be stems, in no way differing from the aerial stems.

6. **Ochna pretoriensis** *Phill.* in Flow. Pl. S. Afr. 2: t.70 (1922); in Bothalia 1: 95 (1922); Burtt Davy, Fl. Transv. 1: 239 (1926). Syntypes: Transvaal, Pretoria, *Fouché* sub PRE 1491 (PRE); *Phillips* sub PRE 1422 (PRE).

Shrubs or very rarely trees, much branched. *Stems* and branches with a rough brown bark, flaking in thin, small, longitudinal strips. *Leaves* tardily deciduous, obovate to narrowly obovate or oblong, usually about 3,5 cm long and 1–1,5 cm broad, apex acute to obtuse, base cuneate, margin shallowly serrulate with mucronate teeth, midrib occasionally arched with the convex half of lamina somewhat larger than other side, dark green, shiny; petiole 1–2 mm; intrapetiolar stipules elongate-deltoid, fused at the base. *Flowers* fragrant, solitary or rarely some 2–3-nate, on short spurs, appearing before the leaves; pedicels c. 8 mm at anthesis, up to 15 mm and erect or spreading in fruit; articulated at or near the base. *Sepals* elliptic, c. 5 mm long, obtuse, about 14 mm long and wine red in fruit, loosely enveloping drupelets. *Petals* broadly obovate, clawed, c. 14 mm long, rounded at the apex, yellow. *Stamens* c. 40, with the filaments 3 mm long; anthers biporose, 2 mm. *Carpels* 5, attached basally to torus; styles fused with free apical recurved tips; stigmas apical, discoid. *Drupelets* oblong-globose, c. 9 mm long, black. FIG. 1: 2; 2: 6.

Endemic, gregarious and widespread in the Transvaal with one record from near Mafeking; saxicolous; common on northern hill-slopes around Pretoria; flowering in spring.

TRANSVAAL.—2229 (Waterpoort): farm Soutpan 193, lower northern slopes of Soutpansberg, *Obermeyer, Schweickerdt & Verdoorn 312*. 2429 (Zebediela): granite ridge near Utrecht hills, *Acocks 8804*. 2527 (Rustenburg): Buffelspoort, *Turner 46*. 2528 (Pretoria): Hammanskraal,

Hutchinson & Mogg 2875; Magaliesberg above Wonderboom Nature Reserve, *Dyer 3131;* Meintjies Kop, *Pole Evans 353.* 2529 (Witbank): Olifants River Gorge, *Mogg 22422;* Doornkloof, *Du Plessis 168.* 2730 (Vryheid): Piet Retief, *Leipoldt sub PRE 30203.*

CAPE.—2525 (Mafeking): Blikplaas, Mafeking district, *Brueckner 472.*

Related to *O. inermis* but can be distinguished by the shorter usually erect pedicels articulated just above the base, and the somewhat smaller leaf.

7. **Ochna serrulata** *(Hochst.) Walp.*, Repert. 5: 400 (1846); Du Toit in Bothalia 11: 517 (1975). Type: Natal, Natal Bay (Durban), *Krauss* 473 (B†; K; PRE, photo.).

Diporidium serrulatum Hochst. in Flora 27, 1: 304 (1844). Type as above.

O. atropurpurea sensu Harv. in F.C. 1: 448 (1860); sensu Sim, For. Fl. Cape Col. 163, t.29, f.2 (1907); sensu Phill. in Bothalia 1: 94 (1922); sensu Stapf in Bot. Mag. t.9042 (1925); sensu Burtt Davy, Fl. Transv. 1: 239 (1926); sensu Van der Watt in Flow. Pl. Afr. 35: t.1392 (1962); sensu Palmer & Pitman, Trees S. Afr. 3: 1509 (1973), non DC. *O. multiflora* Williams in J.R. Hort. Soc. 5: 160 (1897), nom.; Gard. 22: 574, t.369 (1882); Gard. Chron. 40: 212, icon. suppl. (1906); non DC.

Decorative evergreen shrubs, 60–200 cm tall, or rarely a small tree. *Stems* with red wood; branches brown, densely verrucose with pale grey, raised rounded lenticels. *Galls* often present, forming a globose body consisting of numerous subulate bracts. *Leaves* partly shed at time of flowering, narrowly elliptic, 2,5–5 cm long, 1–1,5 cm broad, apex and base obtuse to acute, margin sharply serrulate with the teeth and apex mucronate, thinly coriaceous, shiny, dark green, midrib distinct, reticulate-veined; petiole 1 mm; intrapetiolar stipules linear, bidentate, margin toothed. *Flowers* usually solitary or rarely binate, on short lateral spurs; pedicels 1 cm long at anthesis, 1,5 cm in fruit, articulated 1–2 mm above base. *Sepals* ovate, 8 mm long at anthesis, green, up to 15 mm long and deep wine red in fruit, spreading or reflexed. *Petals* broadly obovate, cuneate, 1 cm long. *Stamens* c. 20, with filaments c. 3 mm long; biporose anthers about equal in length. *Carpels* 5, attached basally; style terete with short, recurved tips; stigmas apical, discoid. *Drupelets* oblong-ovoid, c. 1 cm long, black. FIG. 2: 7.

Recorded from the eastern Transvaal to Natal and eastern Cape as far west as George; common along forest edges or beside paths, in scrub forest or fynbos. Flowering in spring, but with some flowers throughout the year.

Schlechter collected specimens at Genadendal in 1896 but there are no further records from this area; it is possible that the plants were introduced at the Mission Station.

TRANSVAAL.—2430 (Pilgrims Rest): road from the Downs to Madeira, *Crundall sub PRE 32796.* 2730 (Vryheid): Mooihoek, near Piet Retief, *Devenish 1264.*

SWAZILAND.—2631 (Mbabane): Hlatikulu, *Galpin 9622.*

NATAL.—2731 (Louwsburg): Ingwavuma, *Ward 2374.* 2829 (Harrismith): Cathedral Peak Area, Inhlwasine River Valley, *Killick 1816;* Pieters, *Strey 9545.* 2831 (Nkandla): Melmoth, *Mogg 6142;* Eshowe, *Gerstner 2031.* 2929 (Underberg): Tabamhlope Research Station near Estcourt, *West 817; 531.* 2930 (Pietermaritzburg): Albert Falls, *Meebold 12948;* Thornville, *Moll 3418.* 2931 (Stanger): Groutville, *Moll 2510;* Durban, Natal Bay, *Gueinzius s.n.* 3029 (Kokstad): 10 km S. of Umzimkulu on Harding Road, *Story 655.* 3030 (Port Shepstone): Dumisa, *Rudatis 667;* Horseshoe farm, *Strey 5872.*

CAPE.—3128 (Umtata): Umtata commonage, *Miller B/952.* 3129 (Port St. Johns): Second Beach, *Bruce 448;* Prospect Siding, *Flanagan 322.* 3226 (Fort Beaufort): Katberg forest, *Staples sub PRF 2937.* 3227 (Stutterheim): Thomas River, *Compton 19288;* Keiskamma Hoek, *Story 2795;* Stutterheim, *Rogers 12726.* 3228 (Butterworth): Qora Mouth, *Meeuse 9695;* Komga, *Flanagan 322;* Kentani, *Pegler 194.* 3322 (Oudtshoorn): Wilderness, *Compton 15565.* 3323 (Willowmore): The Crags, *Compton 23593.* 3324 (Steytlerville): Melkhoutboom road, *Story 2608;* hills N. of Patensie, *Taylor 1283.* 3325 (Port Elizabeth): Enon, *Thode A1099;* Suurberg, Swartehoogte, *Zeyher sub SAM 38391.* 3326 (Grahamstown): Howiesonspoort, *MacOwan 1766;* Grahamstown, *Galpin 48.* 3419 (Caledon): Genadendal, *Schlechter 7749.* 3423 (Knysna): Knysna, *Keet 522; 590.*

The species is occasionally planted as an ornamental shrub in gardens in South Africa and in England. At Kirstenbosch it is used as a hedge.

Common names: Rooihout; iLitye (X); umBomvane (Z).

According to the Gardener's Chronicle (September 1906), it was first introduced to England in 1820 under the name of *Ochna multiflora,* a catalogue name (not of DC). It was lost to cultivation but re-introduced later. In Flowering Plants of Africa, t.1392 (1962), *Ochna serrulata* is illustrated (as *O. atropurpurea*) from shrubs flowering and fruiting freely in the garden of the Royal Horticultural Society at Wisley, England, in the early 1960s.

Suspected hybrids: Some specimens collected in Natal and the S.E. Transvaal appear to be intermediate between *Ochna natalitia* and *O. serrulata.* Like *O. natalitia* they bear (few-flowered) umbels, have larger leaves and are sometimes arborescent. On the other hand they possess the dense mottled covering of lenticels so typical of *O. serrulata* and, in fact, anatomically they agree with the latter and not with *O. natalitia.*

8. **Ochna gamostigmata** *Du Toit* in Bothalia 11: 517 (1975). Type: Transvaal, Barberton, Upper Moodies, *Galpin* 963 (PRE, lecto; NH; SAM).

O. atropurpurea DC. var. *angustifolia* Phill. in Bothalia 1: 95 (1921). Syntypes: Transvaal, Barberton, *Galpin* 963 (PRE; NH; SAM); Ermelo, Nooitgedacht, *Pott* 5096 (PRE); between Pilgrims Rest and Sabie, *Rogers* 23410; Swaziland: Havelock Concession, *Saltmarshe* in Herb. Galpin sub PRE 1270 (PRE).

Small shrubs or rarely trees up to 8 m tall, forming suckers. *Stems* and branches brown, densely covered with grey, raised, rounded lenticels. *Galls* sometimes present, forming globose bodies consisting of numerous linear-subulate bracts spirally arranged. *Leaves* shed at time of flowering, narrowly ovate- to linear-acuminate, 2–3 cm long and c. 8 mm broad, apex acuminate, rarely acute to obtuse, base rounded or acute, margin sharply serrulate with the teeth and apex mucronate, thinly coriaceous, shiny, midrib and lateral veins distinct; stipules intrapetiolar, fused, linear, bidentate. *Flowers* sweetly scented, solitary, on short spurs; pedicels 1 cm at anthesis, 2 cm long and recurved in fruit, articulated at the base. *Sepals* ovate, 7 mm, green, somewhat enlarged in fruit and turning brownish red. *Petals* obovate, 7 mm, yellow. *Stamens* 15–25, with filaments 1 mm long; anthers biporose, 3 mm long. *Carpels* 5, attached basally to receptacle; style terete; stigma entire, capitate, with short swollen lobes. *Drupelets* oblong-globose, 8 mm long, black. FIG. 2: 8.

Recorded from eastern Transvaal, Swaziland and Natal, usually at fairly high altitudes, in forested areas. Flowering in spring but with some flowers throughout the year.

TRANSVAAL.—2430 (Pilgrims Rest): 17 km E. of Graskop, *Codd & De Winter 3119*. 2531 (Komatipoort): Shiyalo-ngubu valley, *Pole Evans 4680*.

SWAZILAND.—2531 (Komatipoort): Mtutusi River, Havelock, *Compton 29126*.

NATAL.—2730 (Vryheid): Inkamana Farm, *Gerstner 3516*. 2830 (Dundee): Impati Hill, *Shirley 13*. 2831 (Nkandla): Eshowe, *Thode A1227*. 2929 (Estcourt): Griffin's Hill, *Acocks 10471*. 2930 (Pietermaritzburg): Karkloof forest, Farm Ehlatini, *Moll 2861; 3375; 3476;* 24 km N.W. of New Hanover, *Codd 1465*.

O. gamostigmata differs from *O. serrulata* in its smaller, usually shrubby habit, smaller and narrower, usually acuminate leaves, smaller flowers and capitate stigma. It grows at fairly high altitudes from Karkloof to Swaziland and the eastern Transvaal (Letaba district).

9. **Ochna inermis** *(Forsk.) Schweinf.* apud Penzig in Atti Congr. Bot. Intern. Genova 1892: 335 (1893); Gilg in Bot. Jahrb. 33: 247 (1904); Schweinf., Arab. Pfl. Aegypt. Alger. & Yemen 148 (1912); Robson in F.Z. 2: 237 (1963); Palmer & Pitman, Trees S. Afr. 3: 1510 (1973). Type: Yemen, *Forskål* 760 (C, holo.; PRE, photo.).

Euonymus inermis Forsk., Fl. Aegypt.-Arab. 204 (1775). Type as above.

O. rogersii Hutch., Botanist in S. Afr. 309, 317 (1946), nom. nud.

Spreading shrubs 1–2 m tall, rarely arborescent. *Stems* smooth, dark grey, with pale reddish wood; side branches lenticellate, numerous, short, bearing spurs which produce leaves and flowers. *Leaves* deciduous, being shed just before flowers appear (soon wilting after being picked), elliptic, 2–4 cm long and 1–1,5 cm broad, apex obtuse to acute, base acute or rounded, margin shallowly serrulate, the teeth mucronulate, shiny, distinctly reticulate; petiole 2 mm; stipules intrapetiolar, fused, bifid with apices filiform. *Flowers* sweet scented, solitary or rarely 2 close together, on short spurs; pedicels filiform, 1–1,5 cm long at anthesis, firm, wiry and up to 2 cm, red and recurved in fruit, articulated in lower quarter. *Sepals* ovate, convex, c. 5 mm long, green at anthesis, becoming c. 16 mm long and red in fruit, at first enclosing young drupelets, spreading when ripe. *Petals* yellow, orbicular, clawed, about 8 mm. *Stamens* c. 50, with filaments 3 mm long; anthers biporose, 1,5 mm long. *Carpels* 5, attached basally to receptacle; styles fused, slender, apices free, recurved, filiform with small discoid stigmas. *Drupelets* oblong-globose, 1 cm long, black. FIG. 2: 9.

The most widespread of all African *Ochna* species, extending from Yemen, Eritrea and Ethiopia along the eastern escarpment to Rhodesia, Botswana, Mozambique and Transvaal as far south as Komatipoort; inhabiting arid regions, saxicolous or in sandy soil. Flowering in spring.

TRANSVAAL.—2229 (Waterpoort): farm Hackthorne 608, Dongola Reserve, *Pole Evans 4436;* Wylliespoort, *Werdermann & Oberdieck 1944; Codd 4824; Story 5947*. 2230 (Messina): Messina, *Rogers 19398; Wild 7626*. 2231 (Pafuri): 27 km S.E. of Pafuri, *Mockford 1*. 2328 (Baltimore): Magoba's Kloof, *Maguire 1468*. 2331 (Phalaborwa): Kruger National Park, near Letaba River, *Galpin 15078*. 2427 (Thabazimbi): De Hoop trig. beacon, *Burtt Davy sub BOL 18207*. 2428 (Nylstroom): Near Nylstroom, on lower slopes of Krantzberg, *Prosser 1710;* Moorddrif, *Leendertz 2131;* Olifantspoort near Nylstroom, *Galpin 13796*. 2431 (Acornhoek): 10 km N. of Olifants River, *Codd 6195*. 2529 (Witbank): Loskopdam, *Theron 1558; 1885*. 2531 (Komatipoort): Kruger National Park, near Kemp's cottage, *Codd 5715;* Komati Poort, *Pole Evans sub PRE 16856*.

10. **Ochna glauca** *Verdoorn* in Bothalia 6: 232 (1951); Robson in F.Z. 2: 239 (1963). Type: Transvaal, Soutpansberg District, farm De Klundert, *Pole Evans* 4494 (PRE, holo.).

Spreading shrubs or small trees. *Branches* greyish-brown, smooth; side branchlets numerous, short, bearing knob-like spurs which produce leaves and flowers; lenticels sparse; young parts often covered with wax. *Leaves* deciduous, crowded on the apices of the branchlets, drooping, often folded, broadly elliptic, c. 3 cm long, 1,8 cm broad (in S. African material), obtuse at apex and base, midrib prominent, margin obscurely serrulate, thin, glaucous, metallic green; petiole thin, 4 mm long, usually recurved; stipules intrapetiolar, fused. *Flowers* appearing before leaves, solitary, apical, fairly small; pedicels 1 cm long, not accrescent in fruit but firmer, articulated at the base. *Sepals* elliptic, 5 mm long, green at anthesis, up to 1 cm long, reddish brown and reflexed in fruit. *Petals* ovate, clawed, 11 mm long, yellow. *Stamens* about 40, the filaments c. 4 mm long; anthers biporose, c. 2 mm long. *Carpels* 5, attached basally to torus; styles fused, apices recurved with slightly swollen stigmas. *Drupelets* oblong-globose, c. 8 mm long, brown. FIG. 2: 10.

Apparently restricted to the northern Transvaal and the Matopos in Rhodesia, on granite hills.

TRANSVAAL.—2228 (Maasstroom): Kremetartberg area, *Van Graan & Hardy 483*. 2229 (Waterpoort): farm De Klundert, *Pole Evans 4588; Verdoorn 2292; 2332*. 2328 (Baltimore): Steilloopbrug, *Van der Schijff 5330*.

11. **Ochna cinnabarina** *Engl. & Gilg* in Warb., Kunene-Samb. Exped. 305 (1903); Gilg in Bot. Jahrb. 33: 236 (1903); Exell & Mendonça in C.F.A. 1,2: 292 (1951); Robson in F.Z. 2: 236 (1963); Schreiber in F.S.W.A. 42: 2 (1968). Type: Angola, on the Kuito, below Longa, on sand hills, *Baum* 550 (B†; E, lecto.; PRE, photo.).

Small shrubs up to 1,5 m tall with virgate stems bearing short side branches, from a persistent woody base; lenticels pale, numerous on young brown twigs which become greyish and rough with age. *Leaves* deciduous, narrowly elliptic, 3–4 cm long and 1–1,5 cm broad, acute to obtuse at the apex and base, margin spinulose-serrate, leathery, discolorous, midrib distinct; stipules intrapetiolar, fused, linear. *Flowers* appearing before the leaves, solitary, apical on short lateral spurs; pedicels 1 cm long, erect at anthesis, up to 1,5 cm long and spreading in fruit, articulated below centre. *Sepals* ovate, 1 cm long, obtuse, flat, spreading, greenish-yellow, up to 2 cm long and scarlet in fruit, forming a loose cup around young fruit, spreading when drupelets are ripe. *Petals* obovate, clawed, c. 1,5 cm long, bright yellow. *Stamens* c. 50; filaments c. 6 mm long; anthers biporose, c. 3 mm long. *Carpels* 5, attached basally; styles fused, terete, apices free, recurved with capitate stigmas. *Drupelets* oblong-globose, c. 1 cm long, black. FIG. 2: 11.

Recorded from Botswana, northern South West Africa, southern Angola, Rhodesia and Zambia, in rock fissures, or on antheaps, in Kalahari sand. Flowering in November.

S.W.A.—1718 (Kuring-kuru): Omuramba Mpungu on road from Kuring-kuru to Tsintsabis on dunes of banks, *De Winter 3873*. 1723 (Singalamwe): Singalamwe area, *Killick & Leistner 3202*. 1724 (Katima Mulilo): near Katima Mulilo, *Pienaar & Vahrmeijer 191*. 1819 (Karakuwise): Cigarette, N.E. of Karakuwise, *Maguire 2271*. 1820 (Tarikora): Masari, *Giess 9519*. 1821 (Andara): 23 km W. of Andara Mission Station on road to Nyangana, *De Winter & Wiss 4225*.

12. **Ochna barbosae** *Robson* in Bol. Soc. Brot., Sér. 2, 36: 18 (1962); in F.Z. 2: 238 (1963). Type: Mozambique, Vila Luis, *Barbosa & Lemos* sub *Barbosa* 7895 (LISC, holo.; COI; K; LMJ).

A slender tree up to 8 m tall or a shrub with light brown, slightly rough bark, not flaking; branchlets with the epidermis peeling off in narrow membranous strips; lenticels many, scattered; woody ovoid galls often present. *Leaves* shed before flowering or shortly after, narrowly elliptic to elliptic, 3–5 cm long and 1,2–1,5 cm broad (in S. African material seen), apex rounded, base acute, margin indistinctly serrulate (appearing smooth), coriaceous, with distinct reticulate venation; petiole 1 mm; stipules intrapetiolar, fused, entire. *Flowers* scented, terminating short spurs, solitary; pedicels c. 4 mm long at anthesis, up to 1 (–2) cm in fruit, erect, articulated at the base. *Sepals* about 9 mm long at anthesis, about 2 cm long and pink or red in fruit, enclosing young drupelets, spreading when fruit is ripe. *Petals* yellow, obovate-cuneate, c. 12 mm long and 8 mm broad. *Stamens* c. 50, filaments c. 2,5 mm; anthers 1,5 mm, biporose. *Carpels* 5, attached basally to receptacle; styles fused, terete, with free apical, recurved stigmatic tips or completely fused with stigma apically, discoid, 5-lobed. *Drupelets* oblong-globose, 1 cm long, black. FIG. 2: 12.

Occurs in Natal, southern Mozambique, and south-eastern Rhodesia. A sand-loving tree or shrub inhabiting coastal dune forest, locally common where found; extending inland along low lying areas to south-eastern Rhodesia.

NATAL.—2632 (Bela Vista): 7 km W. of Muzi, *Moll & Nel 5590*. 2732 (Ubombo): Sihangwane Store, *Moll 4410;* Lake Sibayi, *Venter 5818;* Mbazwane–Mseleni Road, *Pooley 1512*. 3030 (Port Shepstone): Isipingo Beach, *Ward 1239; 3759; 5887*.

Hybrids: *Moll & Strey 3928* (2632 Bela Vista), near Banga Nek, and *Ward 3611* (2732 Ubombo), Mkuze Game Reserve,both collected in dune forest, northern Natal, appear to be hybrids between *O. barbosae* and *O. natalitia.*

5114 2. BRACKENRIDGEA

Brackenridgea *A. Gray,* Bot. U.S. Expl. Exp. Bot. 1: 361, t. 42 (1854); Walp., Ann. 4: 421 (1857); Gilg in Pflanzenfam. ed. 2,21: 74 (1925); Robson in F.Z. 2: 252 (1963); Schreiber in F.S.W.A. 42: 1 (1968). Type species: *B. nitida* A. Gray from Fiji.

Trees, shrubs or soboliferous dwarf shrubs, glabrous, forming a yellow pigment under the bark. *Leaves* simple, shortly petiolate, with the primary lateral nerves curving upwards, almost parallel to the margin, the margin closely serrulate with the teeth gland-tipped; petioles short; stipules deeply divided into subulate segments, striate, persistent at least on first year shoots in African species. *Flowers* variously arranged, usually in axillary fascicles or reduced to 1 flower; bracts resembling stipules. *Sepals* (4) 5, imbricate, white or pink, turning red and accrescent in fruit. *Petals* (4) 5, white or pale pink, fugaceous. *Stamens* 10–20, free; filaments persistent; anthers bilocular, basifixed, opening longitudinally, caducous. *Carpels* 5–10, attached basally to torus; style entire; stigma capitate. *Drupelets* globose, black attached basally to the red swollen torus; seeds without endosperm, embryo curved. FIG. 3.

Species about 12, widely distributed in S.E. Asia and Australia; 4 in tropical and subtropical Africa and Madagascar. One tropical to subtropical species recorded from South West Africa, usually in sandy soil.

Brackenridgea arenaria *(De Wild. & Dur.) Robson* in Bol. Soc. Brot., Sér. 2, 36: 37 (1962); in F.Z. 2: 254, t. 46, f.b (1963); Schreiber in F.S.W.A. 42: 2 (1968). Type: Zaire, Kisantu, *Gillett* 68 (BR, holo.).

Ochna arenaria De Wild. & Dur. in Bull. Herb. Boiss. 2, 1: 7 (1900). Type as above.

Small rhizomatous shrublets with virgate, grooved branches, up to 1 m tall (rarely somewhat taller); wood yellow below grey-brown peeling bark. *Leaves* deciduous, erect, close together, alternate, narrowly obovate, c. 8 cm long, 2 cm broad, usually obtuse above, apiculate, attenuate towards the base, forming a short petiole, margin closely and shallowly serrulate with the teeth bearing ovoid glands, firm, shiny, the lateral veins curved upwards along margin; stipules deeply dissected into subulate segments, striate, intrapetiolar, persisting on first year shoots. *Flowers* in short, axillary, sessile (1) 3–6-flowered fascicles; the bracts resembling the stipules; pedicels c. 8 mm long at anthesis, somewhat longer in fruit, articulated at the base. *Sepals* narrowly oblong, c. 5 mm at anthesis, up to 1 cm in fruit, pink or red, spreading or reflexed. *Petals* narrowly obovate, somewhat longer than sepals, white or pinkish, reflexed at anthesis. *Stamens* about 20; anthers opening by longitudinal slits. *Carpels* 5, rounded, inserted basally on the torus; style terete, stigma apical, small. FIG. 3.

Widespread in tropical and southern subtropical regions, as far south as northern South West Africa; in sandy soils.

S.W.A.:—1723 (Singalamwe): E. of base E.C.Z., *Pienaar & Vahrmeijer 207.* 1821 (Andara): Caprivi Strip, between Bagani Camp and Mahango, *De Winter & Wiss 4395;* 16 km from Bagani towards Angola border, *Watt 2.*

FIG. 3.—1, **Brackenridgea arenaria,** leafy branch, × 1 (*Pienaar & Vahrmeijer* 207); a, leaf-margin showing glands, × 4; b, base of leaf showing stipules, × 4; c, portion of flowering twig, × 1, (*Pocock* 1925); d, anther-locules with longitudinal dehiscence, × 10.

CLUSIACEAE

(nom. alt. Guttiferae)

by D. J. B. KILLICK and N. K. B. ROBSON*

Trees, shrubs or herbs, with resinous juice. *Leaves* opposite, sub-opposite or whorled, simple, exstipulate, glandular. *Flowers* actinomorphic, bisexual, dioecious or polygamous. *Sepals* 2–6, quincuncial or decussate. *Petals* same number as sepals, free, usually convolute, sometimes imbricate or decussate, alternating with sepals. *Androecium* basically of two whorls of stamen fascicles, the outer often sterile (fasciclodes) or absent, filaments variously united or free, inner fascicles variously united or free, very rarely each one reduced to a single stamen. *Ovary* superior, 1–many-locular, placentation usually axile, sometimes ± parietal, ovules 1–many; styles free, ± united or absent; stigmas equal in number to loculi. *Fruit* a septicidal (rarely loculicidal) capsule or berry or drupe. *Seeds* sometimes winged, carinate or arillate, without endosperm.

Almost exclusively tropical in distribution and composed of about 40 genera and 1000 species (Willis, 1966). 2 genera and 8 species occur in our area.

Flowers bisexual; sepals 5; styles (2) 3–5, elongate; ovules several–many per loculus ....1. **Hypericum**
Flowers polygamous or dioecious, rarely bisexual; sepals 4; styles absent and stigmas sessile; ovules 1 per loculus ....2. **Garcinia**

5168

1. HYPERICUM

Hypericum *L.*, Sp. Pl. 1: 443 (1753); Gen. Pl. ed. 5: 341 (1754); DC., Prodr. 1: 550 (1823); Sond. in F.C. 1: 117 (1860); Benth. & Hook.f., Gen. Pl. 1,1: 165 (1862); Oliv. in F.T.A. 1: 154 (1868); Keller in Pflanzenfam. ed. 2, 21: 175 (1929); Bredell in Bothalia 3: 574 (1939); Phill., Gen. ed. 2: 506 (1951); Milne-Redh. in F.T.E.A. Hypericaceae: 1 (1953); Robson in F.Z. 1: 379 (1961). Type species: *H. perforatum* L.

Trees, shrubs or herbs. *Leaves* opposite, sessile, amplexicaul or shortly petiolate, entire, with translucent glandular dots or linear striae and frequently with dark submarginal glandular dots. *Flowers* terminal, solitary or cymose, sometimes forming a corymb or panicle, bisexual. *Sepals* usually 5, equal, subequal or unequal, with translucent gland dots or striae, with or without dark often submarginal dots and sometimes with stalked marginal glands. *Petals* usually 5, usually yellow, often red-tinged, asymmetrical, with translucent or dark glandular dots or striae. *Stamens* usually very numerous, free or connate at the base into 3–5 distinct or indistinct groups; filaments thread-like; anthers with or without yellow or black gland. *Ovary* superior, sessile, ovoid or subround, 3–5-locular, with few to many ovules on parietal placentas in each loculus; styles 3–5, free or variously united; stigmas small, capitate. *Fruit* a septicidal capsule with few to many seeds, rarely indehiscent. *Seeds* minute, cylindric to ovoid, straight or slightly curved; testa thin ribbed, punctate or smooth; embryo straight or curved.

*Department of Botany, British Museum (Natural History), Cromwell Road, London SW7 5BD.

A large genus of about 400 species, widely dispersed, especially in the northern hemisphere. Six species in Southern Africa.

The generic name is derived from the Greek name for St. John's Wort, viz. hypereikon, i.e. *hyper* + *eikon* = above + image. The Greeks used plants of this genus to decorate religious images in order to ward off evil spirits, especially around midsummer's eve (St. John's eve).

Several species of *Hypericum* are cultivated as ornamentals in Southern Africa, namely *H. calycinum* L., *H. monogyrum* L. and *H. kalmianum* L. *H. perforatum* L. is a troublesome weed in the Stellenbosch area of the south-western Cape. Species of *Hypericum* contain the active principle hypericin, which renders the plants poisonous to stock through photosensitization.

Styles 5, wholly or partially united; shrubs:
- Flowers solitary at ends of branches; leaves 0,8–2,3 cm long ....1. *H. revolutum*
- Flowers in few–to many-flowered terminal corymbose cymes; leaves 4–8,5 cm long ....2. *H. roeperanum*

Styles 3–5, free; herbs:
- Stems quadrangular; sepals and petals without dark dots ....3. *H. lalandii*
- Stems terete; sepals usually with dark dots:
  - Sepals acute, subequal ....4. *H. aethiopicum*
  - Sepals obtuse to rounded, unequal:
    - Stems erect, 20–45 cm high; leaves sessile, rarely shortly petiolate ....5. *H. natalense*
    - Stems procumbent, 6–10 cm high; leaves shortly petiolate ....6. *H. wilmsii*

1. **Hypericum revolutum** *Vahl*, Symb. Bot. 1: 66 (1790); Christensen in Dansk Bot. Ark. 4,3: 39 (1922); Robson in Kew Bull. 14: 251 (1960); in F.Z. 1: 381 (1961); Moggi & Pisacchi in Webbia 22: 236, t.1 (1967). Lectotype: "Arabia felix, in montibus", *Forskål* 796 (C!).

H. kalmii Forsk., Fl. Aegypt.-Arab. CXVIII (1775), nomen nudum. *H. lanceolatum* Lam., Encycl. 4: 145 (1797); Oliv. in F.T.A. 1: 156 (1868); Bak.f. in J. Linn. Soc. (Bot.) 40: 26 (1911); Eyles in Trans. R. Soc. S. Afr. 5: 420 (1916); Good in J. Bot., Lond. 65: 330, t.582, fig. 1 (1927); Staner in Bull. Jard. Bot. Brux. 13: 74 (1934); Robyns, Fl. Parc Nat. Alb. 1: 620, t.62 (1948); Perrier in Fl. Madag. Hypericaceae: 3 (1951); Milne-Redh. in F.T.E.A. Hypericaceae: 4 (1953); in Mem. N.Y. Bot. Gard. 8: 221 (1953); Keay & Milne-Redh. in F.W.T.A. ed. 2,1: 287, t.109 (1954); Robson in Kew Bull. 12: 444 (1957). Type: Réunion ("Bourbon"), *Commerson* (LINN-SM, iso.!). *H. leucoptychodes* Steud. ex A. Rich., Tent. Fl. Abyss. 1: 96 (1847); Good, l.c. t.582, fig. 3–5 (1927); Norlindh in Bot. Notiser 1934: 100 (1934); Bredell in Bothalia 3: 580 (1939); Phill. in Flow. Pl. S. Afr. 20: t.787 (1940); Brenan, Checklist Tang. Terr. 249 (1949). Syntypes: Ethiopia, near Dschenausa, *Schimper* 834 (K!); Mt. Bachit, *Schimper* 1177 (K!). *H. lanuriense* De Wild., Pl. Bequaert. 5: 403 (1932); Robyns, l.c. 622 (1948). Type: Zaire, Ruwenzori, Ruaroli (Lanuri) Valley, *Bequaert* 4460 (BR, holo.; K, photo.!).

Shrub up to 3 m high (12 m in tropical Africa). *Stems* 4-angled when young, but becoming ± terete, often with scaly bark. *Leaves* sessile, narrowly elliptic, 0,8–2,5 cm long, 3–6 mm wide, apex acute, base cuneate, narrowing to a clasping base, venation sparsely pinnate, tertiary venation variously conspicuous, with translucent longitudinal glands or striae and few dark submarginal glands. *Flowers* solitary, terminal. *Sepals* unequal, ovate, 6–10 mm long, 3–5 mm wide, concave, with few dark submarginal glands and often numerous stalked marginal glands. *Petals* obovate-spathulate, 2–3 cm long, 0,8–1,8 cm wide, yellow, without dark submarginal gland dots. *Stamens* in 5 bundles of 20–30 each; filaments 8–15 mm long; anthers c. 1–1,3 mm long, with apical gland. *Ovary* ovoid, 4–6 mm long, 3–5 mm wide; styles 5, 4–8 mm long, united except for upper 1,5–4 mm. *Capsule* ovoid, 1–1,2 cm long, 0,7–1,1 cm wide, 5-valved.

Found on streambanks and the edge of forest in the Transvaal, Natal, Swaziland and the eastern Cape. Also occurring in tropical Africa, Fernando Po, Malagasy, the Comoro Islands and Réunion.

TRANSVAAL.—2230 (Messina): Entabeni Forest Station, *Codd 4206*. 2329 (Pietersburg): Houtboschdorp, *Van Vuuren 1203*. 2330 (Tzaneen): Westfalia Estates, Duiwelskloof, *Scheepers 412*. 2430 (Pilgrims Rest): Reitz's Grave, Mariepskop, *Van der Schijff 6803*. 2530 (Lydenburg): Duiwelskantoor, *Thode A1565*. 2627 (Potchefstroom): Witpoortjie, near Krugersdorp, *Grant 2687*.

SWAZILAND.—2531 (Komatipoort): near Havelock, *Compton 30123*. 2631 (Mbabane): Mbabane, *Rogers 11483; Miller S/122*.

NATAL.—2831 (Nkandla): Nkandla Forest, *Pentz 351;* Nhlazatshe, *Ward 3410*. 2930 (Pietermaritzburg): Illovo River, near Richmond, *Wood 1995* (NH); Pateni Estates, *Strey 7636*.

CAPE.—3029 (Kokstad): Malowe, *Tyson 3044*.

An attractive plant with large, yellow flowers, known as the Forest Primrose or Curry Bush, because of the curry-like odour given off by the leaves, particularly when crushed. According to Watt & Brandwijk (1962) the plant has been used as a source of balsam. In the Haenertsburg area of the Transvaal it is called "Nature's Fire Break" because, situated as it often is at the edge of forests, it is said to prevent fire from penetrating forest (*Hodgson* s.n. in PRE). However, it is doubtful whether the plant is particularly fire-resistant. Experimentally it has induced non-icteric photosensitization in sheep, due apparently to the presence of hypericin (Watt & Brandwijk, 1962).

2. **Hypericum roeperanum** *Schimp. ex A. Rich.*, Tent. Fl. Abyss. 1: 96 (1847); Keller in Pflanzenfam. ed. 2, 21: 176 (1925); Good in J. Bot., Lond. 65: 331, t.582, fig. 6 (1927); Staner in Bull. Jard. Bot. Brux. 13: 76 (1934); Exell & Mendonça in C.F.A. 1,1: 119 (1937); Bredell in Bothalia 3: 582 (1939); Brenan, Checklist Tang. Terr. 250 (1949); Suesseng. & Merxm. in Proc. & Trans. Rhod. Sci. Ass. 43: 88 (1951); Milne-Redh. in F.T.E.A. Hypericaceae:3 (1953); Robson in Kew Bull. 12: 444 (1957); in F.Z. 1: 380 (1961); Moggi & Pisacchi in Webbia 22: 244, t.2 (1967). Type: Ethiopia, "crescit ad montem Kubbi, prope Adoua et ad montem Aber prope Dschenausa in provincia Semiène", *Schimper* 866 (P, lecto.; BM!; FI; G; K!; LE; MO).

var. **roeperanum.**
Moggi & Pisacchi in Webbia 22: 247 (1967).

H. quartinianum A. Rich. var. *roeperanum* (Schimp. ex A. Rich.) Engl., Hochgebirgsfl. Trop. Afr. 307 (1892).

Shrub or small tree, up to 5 m high. *Stems* ± terete, smooth, glabrous. *Leaves* sessile, elliptic, 4–8,5 cm long, 1,2–3,4 cm wide, apex acute or obtuse (sometimes apiculate when obtuse), base cuneate, clasping, with several pairs of secondary veins and conspicuous reticulation of tertiary veins each reticulation containing a translucent glandular dot, dark glands on margin. *Flowers* in few- to many-flowered terminal corymbose cymes. *Sepals* unequal, ovate, 4–6 mm long, 3,5–4 mm wide, acute-obtuse, with submarginal dark gland dots and marginal stalked glands. *Petals* obovate-spathulate, 2–3 cm long, 1–1,2 cm wide, bright yellow, with few or no submarginal gland dots. *Stamens* in 5 bundles of about 45 each; filaments 1,2–1,4 cm long; anthers c. 0,75 mm long, with apical gland. *Ovary* ovoid, 5–7 mm long; styles 5, united to apex, 8–10 mm long. *Capsule* ovoid, 1–1,8 cm long, 0,6–1 cm wide, 5-valved.

Widely distributed in Africa from Sudan and Ethiopia southwards to the north-eastern Transvaal and Mozambique. Apparently restricted in South Africa to the Mariepskop area. Usually occurring on streambanks, in woodland or at the edge of forest.

TRANSVAAL.—2430 (Pilgrims Rest): Mariepskop area, *Killick & Strey 2424; Van der Schijff 4975;* Blyde River Road, *Van der Schijff 5529.*

Differing from *H. revolutum* Vahl in that the flowers are in few- to many-flowered terminal corymbose cymes instead of solitary and in the larger leaves with densely reticulate venation. Var. *schimperi* (Hochst. ex A. Rich.) Moggi & Pisacchi is restricted to Ethiopia.

3. **Hypericum lalandii** *Choisy* in DC., Prodr. 1: 550 (1824); Oliv. in F.T.A. 1: 155 (1868); Gibbs in J. Linn. Soc. (Bot.) 37: 430 (1906); Bak.f. in J. Linn. Soc. (Bot.) 40: 26 (1911); Eyles in Trans. R. Soc. S. Afr. 5: 420 (1916); Keller in Bot. Jahrb. 58: 197 (1923); in Pflanzenfam. ed. 2, 21: 181 (1925); Burtt Davy, Fl. Transv. 1: 251 (1926); Staner in Bull. Jard. Bot. Brux. 13: 70 (1934); Norlindh in Bot. Notiser 1934: 102 (1934); Gomes e Sousa in Bol. Soc. Estud. Col. Mocamb. 26: 42 (1935); Exell & Mendonça in C.F.A. 1,1: 120 (1937); Bredell in Bothalia 3: 575 (1939); Adamson in Fl. Cape Penins. 588 (1950); Suesseng. & Merxm. in Proc. & Trans. Rhod. Sci. Ass. 43: 88 (1951); Perrier in Fl. Madag. Hypericaceae: 14, t.1, figs. 7 & 8 (1951); Milne-Redh. in F.T.E.A. Hypericaceae: 7 (1953); Robson in F.Z. 1: 385 (1961); in Garcia de Orta 1: 85 (1973). Type: Cape, *Laland* s.n. (P, holo.; PRE, photo.).

H. lalandii Choisy var. *lanceolata* Sond. in F.C. 1: 118 (1860). Type: Port Natal, *Drège* d (S!). —var. *latifolia* Sond., l.c. Type: Uitenhage, *Zeyher* s.n. (S!). —var. *macropetala* Sond., l.c. Type: *Drège* c (S, holo.!; BM!; K!). —var. *lanceolatum* Keller in Bull. Herb. Boiss. Ser. 2,8: 187 (1908). Syntypes: Natal, *Gueinzius* s.n. (B†); *Gerrard* 190 (K!; PRE, photo.; W!). —var. *transvaalense* Bredell in Bothalia 3: 577 (1939). Type: Transvaal, *Rehmann* 6608 (PRE, lecto.!; BM!).

Herb, 4–40 cm high. *Stems* from a woody underground rootstock, erect or decumbent, rarely procumbent, branched or unbranched, 4-ribbed, glabrous, glandular or eglandular. *Leaves* sessile, ascending; blade very variable in shape and size, upper and middle leaves usually narrowly ovate, narrowly elliptic or linear, basal leaves broader and shorter, 3–36 mm long, 0,5–6 mm wide, apex acute or obtuse, base clasping, venation not conspicuous, pellucid-glandular punctate. *Flowers* up to c. 50 in a loose dichasial cyme or solitary. *Sepals* ± equal, narrowly ovate, 5,5–7 mm long, 1,2–1,5 mm wide, long-acute, with translucent longitudinal veins and gland dots. *Petals* 1–2 times as long as sepals, yellow or orange, sometimes marked with red, distinctly veined.

Stamens 40–60, irregularly arranged; filaments 4–6 mm long; anthers c. 0,3 mm long. *Ovary* ovoid, 1,5–3,5 mm long, 1–1,5 mm wide, 1-locular; styles 3, free, 1,5–2 mm long; stigmas capitate. *Fruit* capsular, (2) 3–4-valved.

Widely distributed in Africa from Nigeria and the Sudan southwards to the Cape Province. Also in Madagascar.

S.W.A.—2017 (Waterberg): Waterberg Plateau, *Boss sub TRV 34997;* Okosongomingo, *Dinter 1758* (SAM).

TRANSVAAL.—2230 (Messina): Entabeni, *Obermeyer sub TRV 30071.* 2327 (Ellisras): Ellisras, *Bayliss 2393* (NBG). 2328 (Baltimore): on Mohlakeng Plateau, Blauwberg, *Codd & Dyer 8992.* 2329 (Pietersburg): Houtbosch, *Rehmann 6340.* 2330 (Tzaneen): Westfalia, *Scheepers 1093; 1116.* 2429 (Nylstroom): Mosdene Farm, Naboomspruit, *Germishuysen 49.* 2430 (Pilgrims Rest): Mariepskop, *Van der Schijff 6021.* 2526 (Zeerust): Zwartruggens, *Fouche 12.* 2527 (Rustenburg): Rustenburg Nature Reserve, *Jacobsen 804.* 2528 (Pretoria): Wonderboom, *Smith 2291.* 2529 (Witbank): Loskop Dam—Donkerhoek, *Theron 1603.* 2530 (Lydenburg): Rosehaugh, *Mogg 13962.* 2531 (Komatipoort): Numbi, *Van der Schijff 4158.* 2627 (Potchefstroom): Buffelshoek, Potchefstroom, *Louw 1527.* 2628 (Johannesburg): Heidelberg, *Leendertz 1030.* 2629 (Bethal): Nooitgedacht, Ermelo, *Henrici 1348.* 2630 (Carolina): Bereton Park Outspan, near Sheepmoor, *De Winter 7520.* 2729 (Volksrust): Amersfoort, *Sidey 3504.* 2730 (Vryheid): Oshoek, Wakkerstroom, *Devenish 1094.*

O.F.S.—2727 (Kroonstad): Bothaville, *Goossens 1216.* 2827 (Senekal): Willem Pretorius Game Reserve, *Leistner 2969.* 2828 (Bethlehem): Bethlehem, *Phillips 3215.* 2829 (Harrismith): Harrismith, *Sankey 21* (K).

SWAZILAND.—2631 (Mbabane): 5 km N.E. of Forbes's Reef, *Schlieben 9531;* Poliniane River, *Compton 26491;* Nduma, *Compton 27269.*

NATAL.—2632 (Bela Vista): Maputa Store, *Moll & Strey 3875.* 2729 (Volksrust): "Glen Atholl", near Charlestown & Volksrust, *Smith 5697.* 2730 (Vryheid): Tweekloof (Altemooi), *Thode A 1145.* 2731 (Louwsburg): Nongoma Townlands, *Gerstner 4612.* 2732 (Ubombo): Sibayi, *Vahrmeijer 554.* 2828 (Bethlehem): Royal Natal National Park, *Galpin 10177.* 2829 (Harrismith): Cathedral Peak Forest Station, *Killick 1173.* 2830 (Dundee): Qudeni, *Gerstner 1950.* 2831 (Nkandla): Melmoth, *Mogg 6052.* 2832 (Mtubatuba): St. Lucia Park, *Dutton 35.* 2929 (Underberg): between Loteni and Giants Castle, *Killick 3915.* 2930 (Pietermaritzburg): "Ehlatini Farm", Karkloof, *Moll 3460.* 2931 (Stanger): Wentworth, Durban, *Ward 6463.* 3029 (Kokstad): Weza, Ingeli, *Strey 6316.* 3030 (Port Shepstone): Uvongo, *Strey 4889.*

LESOTHO.—2828 (Bethlehem): Leribe, *Dieterlen 678.* 2927 (Maseru): Mazenod, *Jacot Guillarmod 805.* 2929 (Underberg): Sehlabathebe, *Jacot Guillarmod, Getliffe & Mzamane 86.*

CAPE.—2822 (Glen Lyon): Witsand, Hay Division, *Acocks 2173.* 3029 (Kokstad): Clydesdale, *Tyson 2097.* 3128 (Umtata): Baziya Forest Station, *Mauve 4862.* 3129 (Port St. Johns): Ndindini, *Strey 10109.* 3227 (Stutterheim): Evelyn Valley, *Leighton 2672.* 3228 (Butterworth): Kentani, *Pegler 117, pro parte.* 3322 (Oudtshoorn): George, *Marloth 2545.* 3225 (Port Elizabeth): 11 km from Ann's Villa on Zuurberg Pass, *Acocks 20298.* 3326 (Grahamstown): near Dassie Krantz, *Britten 627.* 3327 (East London): near East London, *Acocks 21848.* 3418 (Simonstown): Sir Lowry's Pass, *Marloth 4866.* 3420 (Bredasdorp): Swellendam, *Compton 584* (NBG). 3422 (Mossel Bay): Vrijers Berg, *Muir 2043.* 3423 (Knysna): Harkerville, *Breyer sub TRV 23914.*

An extremely variable species in habit, leaf shape and size, flower size and colour and density of punctate glands.

4. **Hypericum aethiopicum** *Thunb.*, Prodr. 2: 138 (1800). Types: Cape Province, *Thunberg* α (UPS, lecto.!); *Thunberg* β (UPS!).

Perennial herb, 10–30 cm high. *Stems* erect or decumbent, unbranched, arising from an underground crown, terete or sometimes ± flattened or 2-lined above, eglandular or spotted with dark glands. *Leaves* sessile, ovate to broadly ovate, 0,8–2,3 cm long, 0,4–1,5 cm wide, apex obtuse to rounded, base cordate-amplexicaul, margins often revolute, with numerous translucent punctate glands and dark marginal glands. *Flowers* in lax or compact terminal few- to many-flowered cymes, rarely solitary; pedicels 2–4 mm long, except in fruit. *Sepals* subequal, ovate, 5–8 mm long, 1,5–2 mm wide, apex sharply acute or acuminate, with translucent gland dots and striae and with or without dark gland dots both marginal and on lamina, with or without stalked marginal glands. *Petals* 10–13 mm long, 2–4 times as long as sepals, primrose yellow, usually red-tinged, with dark gland dots and translucent gland dots or striae, distinctly veined. *Stamens* c. 50–70, irregularly arranged or in 3 or 4 indistinct groups, with filaments united at extreme base; anthers 0,4–0,5 mm long, often with black or orange glands at apex of connective. *Ovary* ovoid, 2,5–4 mm long, 1,5–2 mm wide, 3 (4)-locular; styles 3, free, 4,5–6,5 mm long; stigmas ± capitate. *Fruit* capsular, ovoid, 6 mm long, 3 mm wide, 3–4-valved, erect.

A grassland herb occurring in Rhodesia, Mozambique, Angola, Lesotho, Swaziland and South Africa.

Two subspecies are recognized and may be keyed out as follows:

Stems eglandular, sepals and bracts with marginal stalked glands ....(a) subsp. *aethiopicum*

Stems glandular, sepals and bracts without marginal stalked glands ....(b) subsp. *sonderi*

(a) subsp. **aethiopicum.**

Robson in Kew Bull. 12: 439 (1957); in F.Z. 1: 382 (1961).

H. aethiopicum Thunb., Prodr. 2: 138 (1800); Sond. in F.C. 1: 117 (1860), pro parte.

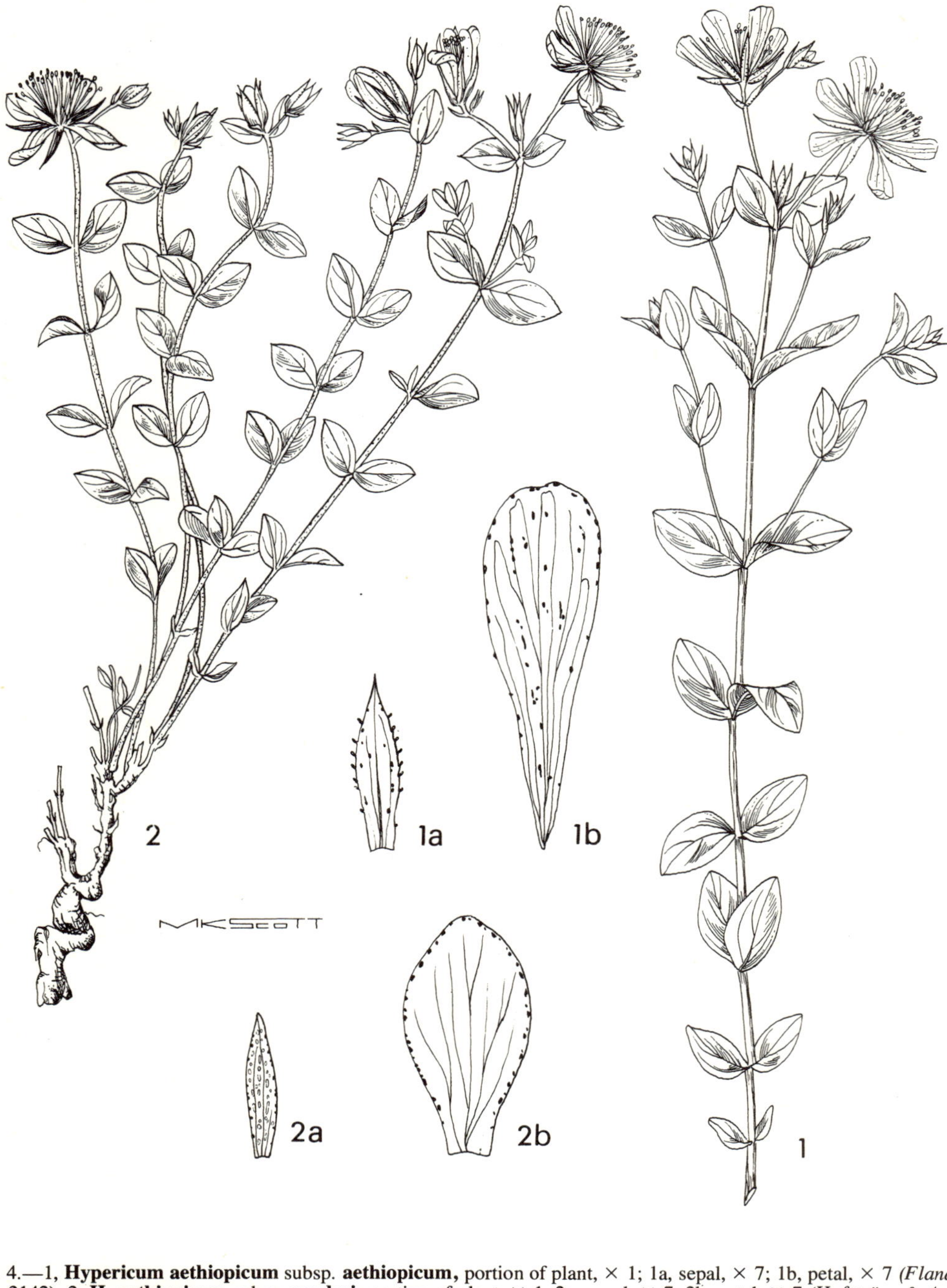

FIG. 4.—1, **Hypericum aethiopicum** subsp. **aethiopicum,** portion of plant, × 1; 1a, sepal, × 7; 1b, petal, × 7 *(Flanagan* 2142). 2, **H. aethiopicum** subsp. **sonderi,** portion of plant, × 1; 2a, sepal, × 7; 2b, petal, × 7 *(Hafström & Acocks* 2195).

This subspecies is confined to the Cape Province between the Korinte River near Riversdale and the mountains of Griqualand East. FIG. 4: 1.

CAPE.—3029 (Kokstad): mountain side, Griqualand East, *Tyson 1376*, pro parte. 3321 (Ladismith): Korinte River, Riversdale, *Muir 855*. 3226 (Fort Beaufort): Hogsback Forest Reserve, *Dahlstrand 1810*. 3227 (Stutterheim): Stutterheim Commonage, *Acocks 9544*. 3322 (Oudtshoorn): Kamanassie Hills near George, *Prior s.n.* (K). 3325 (Port Elizabeth): Zuurberg Sanatorium, *Long 225* (K). 3326 (Grahamstown): Kolsrand, *Johnson 1085;* Howiesons Poort, *MacOwan 397* (BM; K). 3423 (Knysna): Knysna, *Schonland 3498*.

(b) subsp. **sonderi** *(Bredell) Robson* in Kew Bull. 12: 440 (1957); in F.Z. 1: 382 (1961). Type: Natal, near Camperdown, *Schlechter* 3270 (K, lecto.!; G!; PRE!).

H. aethiopicum Thunb. var. *glaucescens* Sond. in F.C. 1: 118 (1860); Burtt Davy in Fl. Transv. 1: 251 (1926). Syntypes: Transvaal, Apies River and Magaliesberg, *Zeyher* s.n. (*Burke* s.n. from Magaliesberg in BM! and K! may be type material—also *Zeyher* 149 in BM!). *H. aethiopicum* sensu Sond., l.c., pro parte; sensu Bak.f. in J. Linn. Soc. (Bot.) 40: 26 (1911); Eyles in Trans. R. Soc. S. Afr. 5: 420 (1916); Norlindh in Bot. Notiser 1934: 101 (1934); Exell & Mendonça in C.F.A. 1,1: 120 (1937). *H. sonderi* Bredell in Bothalia 3: 578 (1939); Verdoorn in Flow. Pl. S. Afr. 23: t. 897 (1943). —var. *transvaalense* Bredell, l.c. 579. Syntypes: Transvaal, Woodbush, *Mogg 13996* (PRE!); Shiluvane, *Junod 4290* (PRE!); without precise locality, *Wager sub TRV 7223* (PRE!), pro parte.

Found in the Transvaal, Orange Free State, Swaziland, Natal, Lesotho, the eastern Cape and south tropical Africa. FIG. 4: 2.

TRANSVAAL.—2230 (Messina): near Entabeni, *Hutchinson & Gillett 4238* (BM; K). 2330 (Tzaneen): Pigeon Hole, *McCullum 17;* Woodbush, *Mogg 14711*. 2427 (Thabazimbi): below Waterberg, *Hutchinson 2291*. 2430 (Pilgrims Rest): The Downs, *Junod 4290*. 2527 (Rustenburg): 2,8 km N.E. of Derby Station, *Acocks 21971*. 2528 (Pretoria): Apies River, *Smith 207*. 2530 (Lydenburg): Rosehaugh, *Mogg 13960*. 2531 (Komatipoort): Saddleback Mountain, Barberton, *Galpin 1028*. 2626 (Klerksdorp): Goedgedacht, Ventersdorp District, *Sutton 502*. 2627 (Potchefstroom): "Uitkomst", near Skeerpoort, *Codd 10097*. 2628 (Johannesburg): Johannesburg, *Bredenkamp 742*. 2629 (Bethal): near Amsterdam, *Van der Merwe 1260*. 2730 (Vryheid): Piet Retief, *Galpin 9638*.

O.F.S.—2828 (Bethlehem): "Wyndford", Fouriesburg, *Gemmell sub BLFU 5812*. 2829 (Harrismith): "Grootvlei", near Swinburne, *Jacobsz 33*. 2927 (Maseru): Ladybrand, *Werdermann & Oberdieck 1566*.

SWAZILAND.—2531 (Komatipoort): Havelock Mine, *Miller 3074*. 2631 (Mbabane): Verdun, Hlatikulu, *Compton 26322*.

NATAL.—2730 (Vryheid): Tweekloof (Altemooi), *Thode A1146*. 2731 (Louwsburg): Nongoma Townlands, *Gerstner 4611*. 2828 (Bethlehem): Royal Natal National Park, *Codd 2784*. 2829 (Harrismith): Cathedral Peak, *Goodier 350*. 2830 (Dundee): hill above Dundee Reservoir, *Edwards 1088*. 2831 (Nkandla): Empangeni Village, *Venter 3251*. 2832 (Mtubatuba): Ezinkakeni, Hluhluwe Game Reserve, *Hitchens 745*. 2929 (Underberg): Cathkin Park, *Howlett 94*. 2930 (Pietermaritzburg): Balgowan, *Mogg 3516*. 3030 (Port Shepstone): Ifafa, *Rudatis 246* (BM; K); Marina Beach, Port Shepstone, *Strey 5950*.

LESOTHO.—2828 (Bethlehem): 5 km S. of Buthe-Buthe, *Galpin s.n.* 2927 (Maseru): bank of the Catai River, Mafeteng Distr., *Dieterlen 1293*.

CAPE.—3029 (Kokstad): 1 km from Bizana on Kokstad road, *Story 597*.

There is a certain amount of overlap between subsp. *aethiopicum* and subsp. *sonderi:* occasionally the odd specimen of subsp. *aethiopicum* has glandular stems, while specimens of subsp. *sonderi* sometimes have eglandular or nearly eglandular stems. According to Robson (l.c. 439), some specimens of subsp. *sonderi* from the Melsetter region of Rhodesia have sepals bearing stalked glands. This incomplete correlation between stem and sepal characters led Robson to treat these two taxa as subspecies rather than species.

According to a note on *Wood* 225 in Herb. Kew, the leaves of the plant are used to make a lotion for inflamed or diseased eyes; and the Zulu name is *Insukubile,* which means "two days".

5. **Hypericum natalense** *Wood & Evans* in J. Bot. 35: 487 (1897); Burtt Davy, Fl. Transv. 1: 25 (1926); Keller in Pflanzenfam. 21: 177 (1929); Bredell in Bothalia 3: 579 (1939); Robson in Kew Bull. 12: 440 (1957). Type: Natal, near bank of Mooi River, 1 200–1 500 m, *Wood* 4034 (NH, holo.!; BM!; BOL!; G!; K!; PRE, photo.).

H. woodii Keller in Bot. Jahrb. 58: 193 (1923), nom. superfl. Type: as for *H. natalense* Wood & Evans. *H. natalense* Wood & Evans var. *petiolatum* Bredell in Bothalia 3: 580 (1939). Type: Natal, Camperdown District, *Franks* sub NH 12968 (NH, holo.!; PRE, photo.).

Herb up to 45 cm high. *Stems* 1–many, from a persistent underground rootstock, erect, much-branched, round and smooth, glabrous. *Leaves* sessile or rarely shortly petiolate, elliptic, broadly elliptic or obovate, 0,9–1,9 cm long, 0,5–1,2 cm wide, apex obtuse to rounded, 3–7-nerved, tertiary venation fairly conspicuous, punctate with submarginal dark dots. *Flowers* terminal with flowering branches from uppermost pair of nodes. *Sepals* unequal, elliptic, obovate or spathulate, 4,5–8 mm long, 1,5–4 mm wide, apex rounded, slightly apiculate, distinctly veined, densely yellow-punctate with few dark submarginal gland dots. *Petals* elliptic, oblong or spathulate, 6,5–7 mm long, 1,7–2,5 mm wide, yellow, with 0–few yellow gland dots and occasional dark dots near apex. *Stamens* indefinite, 3–5 mm long; anthers 0,3 mm long, each with a dark gland at end of connective. *Ovary* ovoid, 2,5–3 mm long, 2 mm wide; styles usually 5, free, 2–2,5 mm long; stigmas capitate. *Capsule* ovoid, 5–6 mm long, 3 mm wide.

A plant of damp places occurring in the eastern Transvaal, Swaziland, Natal Midlands and the eastern Cape.

TRANSVAAL.—2430 (Pilgrims Rest): Mac Mac, *Mudd s.n.*

SWAZILAND.—2631 (Mbabane): Poliniane River, *Compton 26489* (NBG).

NATAL.—2831 (Nkandla): Nkandla Bush, *Van der Merwe 2453.* 2929 (Underberg): Ntabamhlope Research Station, *West 634.* 2930 (Pietermaritzburg): Balgowan, *Mogg 3809;* Howick Falls, *Rogers 5212;* Mooi River, *Wood 4034; 6251* (K).

CAPE.—3029 (Kokstad): 6 km W. of Bizana, *Acocks 12228.* 3128 (Umtata): Baziya, *Baur 582* p.p. (K). 3225 (Somerset East): *Bolus 319* (K). 3228 (Butterworth): Kentani, *Pegler 117.*

Apparently Keller overlooked *H. natalense* Wood & Evans (1897) and described *H. woodii* (1923) basing it on the same Wood gathering. The latter is therefore a superfluous name.

6. **Hypericum wilmsii** *Keller* in Bull. Herb. Boiss. Sér. 2,8: 179 (1908) in clav.; Bredell in Bothalia 3: 579 (1939); Robson in Kew Bull. 12: 440 (1957); in F.Z. 1: 383 (1961). Type: Transvaal, near Lydenburg, *Wilms* 136 (W, holo.; PRE, fragment!).

Perennial herb, 6–20 cm high. *Stems* several to many, arising from a persistent taproot, procumbent, branched, terete, or somewhat flattened or 2-lined above, dark gland-dotted or eglandular. *Leaves* shortly petiolate, rarely almost sessile; blade elliptic, ovate or obovate, 6–10 mm long, 2–6 mm wide, apex rounded, base cuneate or rounded, nerves 3- or 4-paired, tertiary venation inconspicuous, with translucent yellow gland dots and submarginal dark dots; petioles 1–2 mm long. *Pedicels* 3–10 mm long. *Sepals* unequal, oblong, 4,5–6 mm long, 1–2 mm wide, apex obtuse to rounded, sometimes apiculate, distinctly veined, yellow gland-dotted, and with 0–few dark dots along margin. *Petals* oblong, 5–8 mm long, 2–3 mm wide, dark gland-dotted along margin, yellow or orange. *Stamens* 18–30, in 3 or 4 groups, cohering at base; filaments c. 5 mm long; anthers 0,2–0,3 mm long, each with dark gland at end of connective. *Ovary* ovoid, 3–4 mm long, 3- or 4-locular; styles 3 or 4, c. 2 mm long. *Fruit* capsular, 3- or 4-valved, erect.

Occurring in moist areas and on mountain slopes in the Transvaal, Orange Free State, Lesotho and the Cape. Also in Rhodesia and Malagasy.

TRANSVAAL.—2530 (Lydenburg): Lydenburg, *Wilms 136.*

O.F.S.—2926 (Bloemfontein): Thaba 'Nchu, *Roberts 2350; 2913.*

LESOTHO.—2927 (Maseru): Ha-mvya-pela Mt., Likhoele, Mafeteng District, *Dieterlen 1222;* Catai River, Mafeteng District, *Dieterlen 1293.*

CAPE.—3026 (Aliwal North): Doctor's Drift, *Gerstner 137;* Elandshoek, *Bolus 153.* 3126 (Queenstown): Queenstown, *Galpin 1629.* 3223 (Rietbron): Palmietfontein, *Acocks 21770.* 3225 (Somerset East): Cradock, *Brynard 291.*

Differing from *H. natalense* in its shorter, procumbent stems and shortly petiolate leaves. *H. nigropunctatum* Norlindh, described from Rhodesia, is a synonym.

5199 2. GARCINIA

Garcinia *L.*, Sp. Pl. 1: 443 (1753); Gen. Pl. ed. 5: 202 (1754); DC., Prodr. 1: 560 (1824); Benth. & Hook.f., Gen. Pl. 1,1: 174 (1862); Harv., Gen. Pl. ed. 2: 26 (1868); Oliv. in F.T.A. 1: 164 (1868); Keller in Pflanzenfam. ed. 2, 21: 211 (1929); Phill., Gen. ed. 2: 506 (1951); Robson in F.Z. 1: 396 (1961). Type species: *G. mangostana* L.

Trees or shrubs, usually with yellow juice, sometimes with leafy branches modified into thorns. *Leaves* opposite or in whorls of 3, sometimes large, usually entire, coriaceous, venation usually prominent, often with translucent glandular canals and brownish resin canals. *Flowers* terminal or axillary, solitary or in few- to many-flowered cymes, dioecious or polygamous, rarely bisexual. *Sepals* 4 (5), decussate. *Petals* 4 (5), longer than the sepals. *Male flowers*: *Stamens* indefinite, free or all connate in (3) 4 or 5 (–9) fascicles (bundles) with each fascicle forming a solid staminal column bearing free or partially fused filaments above, sometimes with a whorl of sterile stamen-fascicles (fasciclodes) alternating with the stamen-fascicles or forming a cushion in which the stamens are inserted; anthers opening by a circular rim or longitudinal slits. *Female or bisexual flowers*: usually with 4 (5) stamen- or staminode-fascicles, similar to those of the male flowers, but smaller and with fewer members, sometimes with fasciclodes, free or fused together in a ring at the base of the ovary. *Ovary* 2–many-locular with 1 ovule per loculus; styles absent; stigma sessile, peltate, lobed, smooth or tuberculate. *Fruit* baccate. *Seeds* 1–4, large, arillate; embryo with minute cotyledons.

A large genus of about 400 species chiefly tropical, especially Asia, and South Africa. The genus was named in honour of Lawrence Garcin, an Englishman, who travelled and collected in India in the 18th century.

Some species of the genus yield commercial products, e.g. *G. mangostana* L.—mangosteen fruits and *G. morella* Desr.—gamboge, but none of these is apparently cultivated in Southern Africa.

Flowers terminal at ends of branches; staminal or staminodial fascicles present; leaves opposite; petioles 8–18 mm long .... 1. *G. gerrardii*

Flowers in fascicles in axils of older leaves on old wood; staminal or staminodial fascicles absent; leaves 3-whorled; petioles 4–8 mm long .... 2. *G. livingstonei*

1. **Garcinia gerrardii** *Harv. ex Sim*, For. Fl. Cape Col. 141, t.13 (14 May 1907); Henkel, Woody Pl. Natal 126 (1934); Robson in F.Z. 1: 398 (1961); Codd in Bothalia 8: 174 (1964); Bamps in Bull. Jard. Bot. Nat. Belg. 40: 282 (1970). Type: Natal, Umvoti District, *Gerrard* 1181 (BM!; K!).

G. gerrardii Harv., Gen. S. Afr. Pl. ed. 2: 26 (1868), nom. illegit. *G. natalensis* Schltr. in Bot. Jahrb. 40: 93 (21 May 1907). Type: Natal, "in bewaldeten Schluchten am Umtwalumi bei Fairfield", *Rudatis* 150 (not traced). *G. transvaalensis* Burtt Davy in Kew Bull. 1924: 229 (1924); Fl. Transv. 1: 252 (1926). Type: Transvaal, Rimers Creek, Barberton, *Galpin* 1202 (BOL!; K!; NH!; PRE!).

Shrub or tree up to 13 m high; stems acutely angled or winged by decurrent petioles, with yellow or reddish brown latex; bark light grey. *Leaves* opposite; blade elliptic to broadly elliptic, 7–13 cm long, 3–8 cm wide, apex acute, mucronate with a recurved point, base broadly cuneate to rounded, midrib projecting below, lateral veins 9–13, ½-angled, irregular, curved, looping near margin, margin subrevolute and callose, coriaceous, glossy on both surfaces, dark green above, paler below; petiole 0,8–1,8 cm long, furrowed, clasping. *Flowers* 3–9, terminal at ends of branchlets, subtended at base by small ovate-triangular bracts. *Male flowers*: *Sepals* 4, unequal, opposite and decussate, ovate-triangular, 1,5–2 mm long, 2 mm wide, keeled, mucronate (with brown tip), pale green with brown mottling (teste *Scheepers* 1238). *Petals* broadly ovate to subrotund, 6–8 mm long, 6–7 mm wide, concave, white. *Staminal fascicles* 4, 5,5 mm long, 1,5 mm wide, alternating with united rugose globose fasciclodes; stamens 10–16; filaments free or partially united, 0,5–1 mm long; anthers 0,25 mm long, 0,5 mm wide. *Female flowers*: *Staminodial fascicles* c. 1,5 mm long, 0,3 mm wide, alternating with rugose globose fasciclodes; filaments united almost to apex. *Ovary* globose, 3- or 4-locular, 2,6 mm long, 2,4 mm wide, surmounted by fleshy 3- or 4-lobed stigma. *Berry* subglobose, 1,8–3,25 cm diam., turning orange when ripe; seeds 1 or 2, subglobose, 1,4–1,5 cm diam.

A constituent of forest, sometimes on streambanks, in the eastern Transvaal, Swaziland, Natal and the eastern Cape. Said to be particularly common in the Egossa Forest in Pondoland and in the Ngoye Forest in Zululand (Palmer & Pitman, Trees S. Afr., 1973).

TRANSVAAL.—2531 (Komatipoort): 26 km from Barberton, *Scheepers 1238*; Rimers Creek, *Galpin 1202*.

SWAZILAND.—2531 (Komatipoort): Devils Bridge, *Dlamini s.n.; Compton 31496*.

NATAL.—2830 (Dundee): Qudeni Forest, *Edwards 2660*. 2831 (Nkandla): Eshowe, *Gerstner 2541;* Mtunzini District, *Wells & Edwards 29*. 2930 (Pietermaritzburg): Kloof Nature Reserve, *Moll 3256;* Inanda, *Wood 1082* (BM; K). 2931 (Stanger): Hawaan Forest, *Moll 3199*. 3029 (Kokstad): Bizana, *Miller 4420*. 3030 (Port Shepstone): Tributary of Uvongo River, *Cooper 25*. 3129 (Port St. Johns): Umsikaba, *Strey 1012*. 3130 (Port Edward): edge of Umtamvuna River Gorge forest, *Cooper 103*.

Known as the Forest Mangosteen, um Bindi (X) and isi Binda (Z). Sim (l.c., 141) states that the saplings are used as whipsticks. The fruits are edible.

2. **Garcinia livingstonei** *T. Anders.* in J. Linn. Soc. (Bot.) 9: 263 (1866); Oliv. in F.T.A. 1: 165 (1868); Gibbs in J. Linn. Soc. (Bot.) 37: 430 (1906); Sim, For. Fl. P. E. Afr. 15, t.4 (1909); Bak.f. in J. Linn. Soc. (Bot.) 40: 26 (1911); Eyles in Trans. R. Soc. S. Afr. 5: 421 (1916); Henkel, Woody Pl. Natal 165 (1934); Brenan, Checklist Tang. Terr. 241 (1949); Keay in F.W.T.A. ed. 2,1: 294 (1954); Robson in F.Z. 1: 400 (1961). Type: specim. cult. hort. bot. Calcutta ex Mozambique, "In rupibus schistosis prope flumen Zambesi", *Kirk* (CAL, holo.).

G. angolensis Vesque in A.DC., Monogr. Phan. 8: 335 (1893); R.E. Fr., Schwed. Rhod.-Kongo-Exped. 1: 151 (1914); Burtt Davy, Fl. Transv. 1: 252 (1926); Staner in Bull. Jard. Bot. Brux. 13: 120 (1934); Exell & Mendonça in C.F.A. 1,1: 127 (1937). Type: Angola, *Welwitsch* 1047/48 (K, iso!). *G. baikieana* Vesque, l.c. 336 (1893). Type: Nigeria, *Barter* 848 (K!; P). *G. ferrandii* Chiov. in Stefanini & Paoli, Result. Sci. Miss. Stefanini-Paoli Somal. Ital. 1: 26 (1916). Syntypes: Somalia, *Paoli* 847; 1012 (FI?).

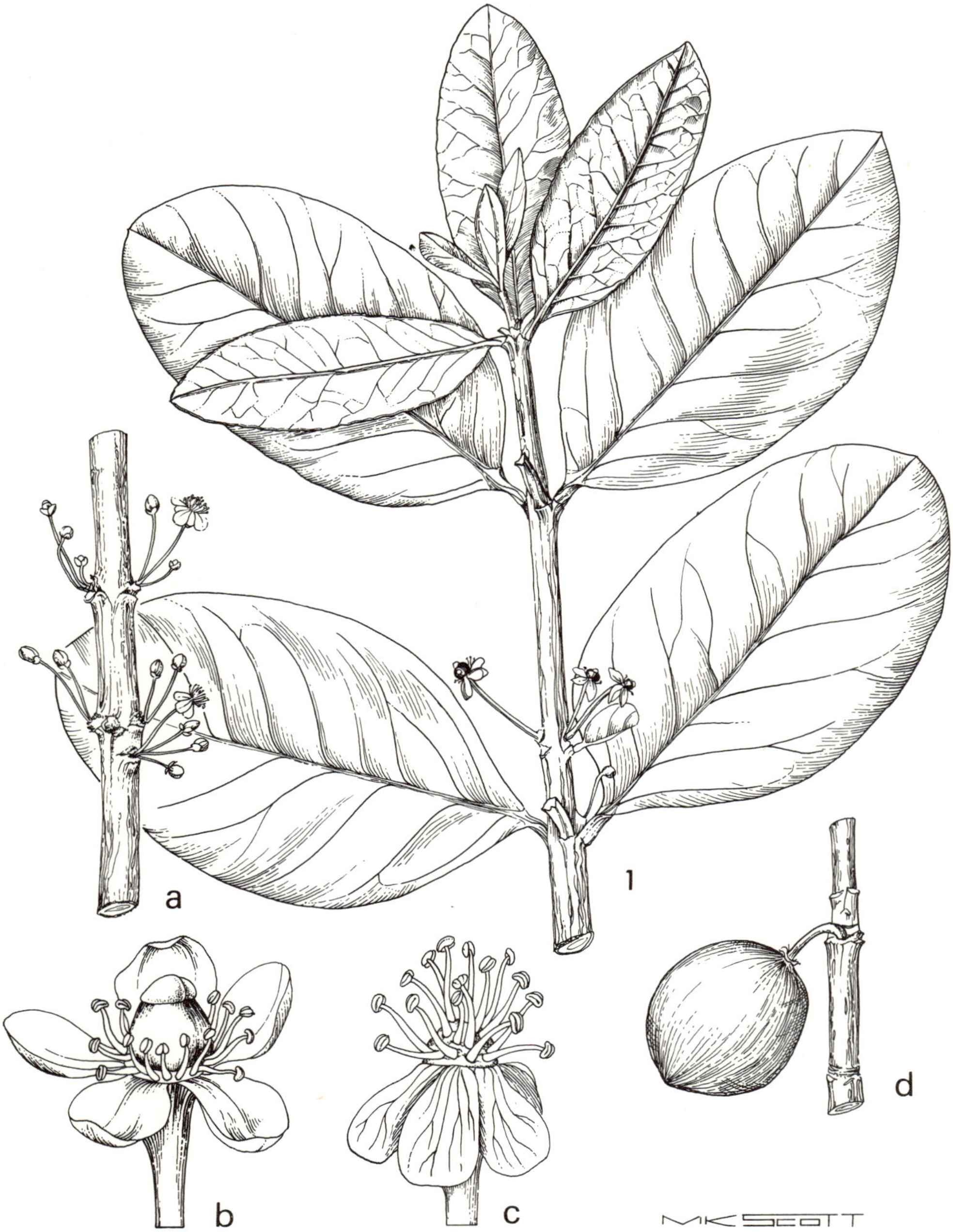

FIG. 5.—1, **Garcinia livingstonei,** branch with bisexual flowers, × 2/3 *(Harrison* 144); a, branch with male flowers, × 2/3 *(Stephen, Van Graan & Schwabe* 1020); b, bisexual flower, × 10 *(Harrison* 144); c, male flower, × 10 *(Stephen, Van Graan & Schwabe* 1020); d, fruit, × 2/3 *(Stephen* 27).

Shrub or tree up to 12 m high, narrowly triangular when young, becoming bushy and virgate; branchlets 3-whorled, thick, flattened or angled towards apex; bark moderately smooth, grey with shallow furrows producing a chequered effect (teste *De Winter & Vahrmeijer* 8572); sap yellow, thick. *Leaves* 3-whorled, rarely 4- or opposite; blade elliptic, ovate or obovate to broadly so, 6–11 cm long, 3–5,5 cm wide, apex acute to rounded, mucronate, base broadly cuneate, rounded or cordate, midrib projecting more on lower side, lateral veins 10–12 (–20), terminating in margin, tertiary venation variously conspicuous, margin entire or crenate, thickened, coriaceous, young leaves bright red; petiole 4–8 mm long, channelled above, transversely rugose. *Flowers* polygamous, in fascicles of 5–15 or more, in axils of older leaves on old wood; pedicels 0,8–1,3 cm long, varying in thickness. *Sepals* 4, unequal, in 2 opposite and decussate pairs, broadly elliptic to orbicular or triangular to transversely ovate, 1–3,5 mm long, 2–3 mm wide, concave. *Petals* 3–7 (9), ovate, obovate or orbicular, 4,5–5 mm long, 3–4 mm wide, concave, greenish white to cream or pale yellow with orange or reddish longitudinal glandular lines. *Male flowers* with numerous free stamens inserted on a fleshy cushion formed by the united fasciclodes. *Bisexual* (and female?) *flowers* with fewer stamens inserted in a fleshy fasciclodal ring below the ovary; ovary ovoid to globose, 2(3)-locular, 1,5 mm long, 2 mm wide, surmounted by fleshy bilobed stigma. *Berry* subglobose, 1–4 cm diam., yellowish red or orange, 1–2(3)-seeded. *Seeds* plano-ovoid, 1–2 cm long. FIG. 5.

Found in scrub, open woodland and forest, often riverine, in the Transvaal, Natal, Swaziland, Botswana, Okavango, Caprivi Strip and tropical Africa.

S.W.A.—1723 (Singalamwe): E. of Kwando River, *Curson 936.* 1724 (Katima Mulilo): Ngoma, *Killick & Leistner 3322.* 1821 (Andara): Andara Mission, *De Winter 4238.*

TRANSVAAL.—2229 (Waterpoort): "Trekpad", Vivo, *Vahrmeijer 1786.* 2231 (Pafuri): Pafuri Camp, *Codd & Dyer 4625.* 2329 (Pietersburg): Louis Trichardt, *Stayt 4.* 2531 (Komatipoort): Malelane, *Codd 4362; Van der Schiff 3923.*

SWAZILAND.—2631 (Mbabane): Usutu Pont, Hlatikulu, *Dlamini s.n.* 2731 (Louwsburg): Hluti, *Pole Evans 3392(1).*

NATAL.—2632 (Bela Vista): Maputa Region, *Vahrmeijer 1204;* Maputa to Kosi Nature Reserve, *Edwards 2591.* 2731 (Louwsburg): banks of Black Umfolozi River, Nongoma, *Codd 1936.* 2732 (Ubombo): 13 km N.E. of Mseleni Mission Station, *De Winter & Vahrmeijer 8572.* 2832 (Mtubatuba): Hluhluwe Game Reserve, *Ward 2948;* Palm Ridge Farm, *Harrison 144.*

An extremely variable species, but readily distinguishable from *G. gerrardii* (see key). According to a note on *De Winter* 4238, *G. livingstonei* is easily distinguished from other trees in the Okavango by its narrowly triangular shape and blue-green appearance. Palmer & Pitman (1973) state that "the general effect when the tree is bare of leaves is of a light-creamy grey or dark grey shape, all angles and points". The flowers are said to be sweetly scented and the fruits, which have a pleasant acid taste, are relished by humans and animals. The wood is light-coloured and semi-hard and is used for fencing posts and boxes and sometimes furniture. The common names are Livingstone Garcinia, African Mangosteen and uPhimbi (Z).

ELATINACEAE

by A. A. OBERMEYER

Annual herbs or low shrublets with the branches either deciduous from a persistent woody crown, or perennial, often aromatic. *Leaves* decussate (rarely verticillate), simple, sessile or subsessile, serrate or rarely entire; stipules paired, interpetiolar, small, deciduous or persistent. *Flowers* axillary, solitary or in dichasia, pedicellate or subsessile, bisexual, actinomorphic, persistent; bracts resembling reduced stipules or 0. *Sepals* 2–5, free or fused at the base, imbricate with the outermost somewhat longer than the others. *Petals* 3–5, free, imbricate, shorter or longer than the sepals. *Stamens* diplostemonous or the inner whorl absent; filaments of outer whorl often expanded below, those of inner whorl shorter, not, or somewhat expanded, anthers bilocular, opening by longitudinal slits, introrse, versatile, dorsifixed. *Gynoecium* syncarpous, superior, 3–5-locular, placenta axile, styles 3–5, free with apical, usually swollen stigmas, ovules numerous, attached to a convex placenta with the axis protruding and subulate above. *Capsule* septifragal;

seeds subcylindrical, obtuse, straight or somewhat reniform to horseshoe-shaped, smooth or reticulate, exendospermous.

Genera 2 with about 30 species. Cosmopolitan, inhabiting damp areas such as marshes, riverbanks or temporary pools; a few *Bergia* species are found in grassland communities.

Flowers 5-merous; sepals with a prominent, firm midrib and membranous margin; ovary ovoid or globose, attenuate into short or long styles; herbs or shrublets ....1. **Bergia**
Flowers 3 (rarely 2)-merous; sepals without a prominent midrib, texture homogeneous; ovary depressed-globose with minute styles; small, creeping, soft herbs, usually hygrophytic ....2. **Elatine**

5230 1. **BERGIA**

Bergia *L.*, Mant. Pl. Alt. 152, No. 1309 (1771); Harv. in F.C. 1: 115 (1860); Benth. & Hook., Gen. Pl. 1: 162 (1862); Niedenzu in Pflanzenfam. 3, 6: 281 (1895); ed. 2, 21: 273 (1925); Phill., Gen. ed. 2: 507 (1951); Hutch., Fam. Flow. Pl. ed. 2, 1: 428 (1959). Type species: *B. capensis* L.

Erect shrublets or with annual repent, decumbent or ascending branches from a woody crown, or annuals, glabrous, pilose or glandular-pubescent, often aromatic. *Roots* of the hygrophilous plants usually with a spongy outer covering. *Stems* and branches woody or herbaceous. *Leaves* decussate or rarely verticillate; lamina linear to ovate or obovate, margin entire or serrate, the teeth usually gland-tipped, pointing upwards. *Flowers* actinomorphous, small. *Sepals* 3–5 (5 in South African species), midrib prominent, margin membranous. *Petals* 3–5 (5 in South African species), delicate, white, pink or mauve. *Stamens* 10 (in South African species, except for *B. ammannioides* which has 5). *Gynoecium* 5-locular (in South African species) bearing 5 short or long, patent styles with apical, usually capitate, stigmas (in older flowers these curve outwards and touch the stamens; probably self-pollinated), ovules many. *Capsule* with the 5 valves breaking away from the hard thin dissepiments, the placentas bulging or recurved to horn-shaped, covered with the short hardened funicles; above the axis is prolonged into a subulate tip. *Seed* very small, subcylindric to somewhat reniform, usually obtuse above and below, tesselated or smooth, yellow to brown or black.

A cosmopolitan genus with about 25 species; 10 of these occur in Southern Africa. Those species which are hygrophilous may well be distributed by birds or floodwaters for the seeds are very small.

The genus was named in honour of Petrus Jonas Bergius a professor in Stockholm and author of Descriptiones plantarum ex Capite Bonae Spei, 1767, an early flora on Cape plants.

An anatomical study of this genus might prove interesting.

Annuals:
Flowers 1–2 in the leaf axils, on long pedicels (up to 15 mm):
Petals longer than the sepals; flowers distributed all over the branches ....1. *B. anagalloides*
Petals shorter than the sepals; flowers aggregated towards the apices of the branches ....2. *B. polyantha*
Flowers several to many in the leaf axils; pedicels short (up to 7 mm):
Stems erect, rarely branched, from a creeping rhizome, succulent, glabrous; valves of capsule flat and recurved at dehiscence; usually an aquatic ....3. *B. capensis*
Stems prostrate or ascending, numerous, radiating from a central base; valves of the capsule usually hollow at dehiscence; wet, sandy places:
Stamens 10; plants glandular-pubescent; petals longer than the sepals:
Sepals spathulate, midrib and margins with some long filiform processes; styles as long as the ovary ....4. *B. spathulata*
Sepals linear, unequal in length, resembling the stipules and bracts, fimbriate; style sessile or nearly so ....5. *B. glutinosa*
Stamens 5; plants pilose with whitish hairs rarely mixed with some glandular pubescence; petals shorter than the sepals ....6. *B. ammannioides*

Perennials:

Leaves small, obovate (c. 5 × 3 mm), soft; plants forming dense, woody mats; internodes short, often almost invisible; flowers solitary; a Cape endemic .... 7. *B. glomerata*

Leaves larger, linear to ovate, firm; plants with spreading or erect, usually annual branches from a persistent woody crown or branches perennial; internodes well developed; flowers several to many in each axil; summer rainfall region:

Stems repent, pilose; leaves about 1 cm long, narrowly ovate, coriaceous; flowers c. 2,5 mm long, sepals short, spine-tipped; styles short .... 8. *B. pentherana*

Stems erect or ascending, sparsely strigose or scabrid; leaves coriaceous or soft, 1,5–3 cm long; flowers c. 3–6 mm long, sepals acuminate; styles long:

Leaves linear, usually glabrous, hard and shiny when dry with a swollen margin and prominent veins, the 6–9 teeth widely spaced .... 9. *B. decumbens*

Leaves narrowly ovate, glandular-pubescent, soft, margin and nerves not prominent, the many teeth close together, usually double-serrate; stems annual or if perennial, forming untidy bushes .... 10. *B. salaria*

1. **Bergia anagalloides** *E. Mey. ex Fenzl* in Ann. Naturh. Mus. Wien 1: 344 (1836); Walp., Repert. 2: 786 (1843); Harv. in F.C. 1: 116 (1860); Thes. Cap. 2: 22 t. 133 (1863); Niedenzu in Pflanzenfam. 3, 6: 281 (1895), ed. 2, 21: 274 (1925); Roessler in F.S.W.A. 92: 3 (1968). Type: Cape, on the bank of the Gariep (Orange River) below 200 feet, *Drège* 2957 (W, holo.!; PRE, photo.).

B. alsinoides Holzhammer in Mitt. Bot. Staatssamml. München 1: 335 (1953). Type: S.W.Africa, 5 km E. of Weissenfels, *Dinter* 8054 (M, holo.; PRE).

Small, prostrate annual with many radiating branches, somewhat woody at the base. *Roots* spongy, with a long taproot (fide *Bryant*). *Stems* slender, smooth or with indistinct scabrid ridges. *Leaves* narrowly obovate, subpetiolate, up to 3 cm long and 6 mm broad, often smaller, serrulate, nerves distinct below, soft in texture; stipules minute. *Flowers* solitary, axillary, evenly distributed along the stems, on firm pedicels up to 15 mm long; bracts minute or O. *Sepals* narrowly ovate, acute, 2,5 mm long, hollow, green with the membranous margins often pink. *Petals* longer than sepals, 4 mm, oblong, obtuse, bright pink. *Ovary* ovoid with short arcuate styles. *Capsule* ovoid, 3 mm long; seed oblong, 0,4 mm, tesselate, yellow to golden brown. FIG. 6: 2.

Recorded from South West Africa, the north-western Cape and south-western Orange Free State, in damp places, usually calciphilous.

S.W.A.—2316 (Nauchas): 5 km E. of Weissenfels, *Dinter 8054*.

CAPE.—2624 (Vryburg): Armoedsvlakte, *Mogg 7946*. 2723 (Kuruman): Bathlaros near Kuruman, *Silk 64*. 2818 (Warmbad): Ramans Drift, Orange River, *Schlechter 11453*. 2824 (Kimberley): Holpan, *Acocks 2463*. 2920 (Boom Rivier): Seweperd, S.E. of Pofadder, *Acocks 15076*. 2922 (Prieska): on sandy banks of Orange River, *Bryant 783*. 2923 (Douglas): farm Mazelsfontein, junction of Orange and Vaal Rivers, *Anderson 610*. 3024 (De Aar): near Potfontein, *Acocks 2413*.

O.F.S.—2925 (Jagersfontein): Fauresmith, Waagkop, *Henrici 3910*.

2. **Bergia polyantha** *Sond.* in Linnaea 23: 16 (1850); Harv. in F.C. 1: 116 (1860); Oliv. in F.T.A. 1: 153 (1868); Niedenzu in Pflanzenfam. 3, 6: 281 (1895); ed. 2, 21: 273 (1925); Burtt Davy in Fl. Transv. 1: 147 (1926); Exell & Mendonca in C.F.A. 1, 1: 118 (1937); Wild in F.Z. 1: 375 (1961); Roessler in F.S.W.A. 92: 4 (1968). Type: Orange Free State, Rhinosterkopf by the Vaal and Sand Rivers, *Burke & Zeyher* 540 (K, holo.).

B. integrifolia Dinter ex Holzhammer in Mitt. Bot. Staatssamml. München 1: 335 (1953); Roessler in F.S.W.A. 92: 4 (1968). Type: S.W.Africa, Grootfontein, Grosshuis, in a swamp, *Dinter* 7381 (M, holo.; K; PRE).

Glabrous annuals with many radiating decumbent branches up to 20 cm long. *Roots* numerous, swollen. *Stems* terete, smooth. *Leaves* narrowly obovate, up to 3 cm long and 1 cm broad, margin entire or faintly serrulate, soft; stipules small. *Flowers* 1–2 in the leaf axils, on thin pedicels up to 2 cm long, aggregated apically where the internodes and pedicels are shorter. *Sepals* narrowly ovate-acuminate, 2 mm long, green with white to pink margins. *Petals* shorter than sepals, obovate, 1,5 mm long, pink or mauve. *Stamens* with thin filaments. *Ovary* ovoid, styles very short, arched over the ovary. *Capsule* globose, placenta often sickle-shaped, echinate; seeds oblong-cylindrical, 0,5 mm, tesselate, brown, shiny. FIG. 6: 3.

Recorded from South West Africa, the western Orange Free State, northern Cape, Botswana, Rhodesia, Mozambique, Zambia and Angola; in moist places.

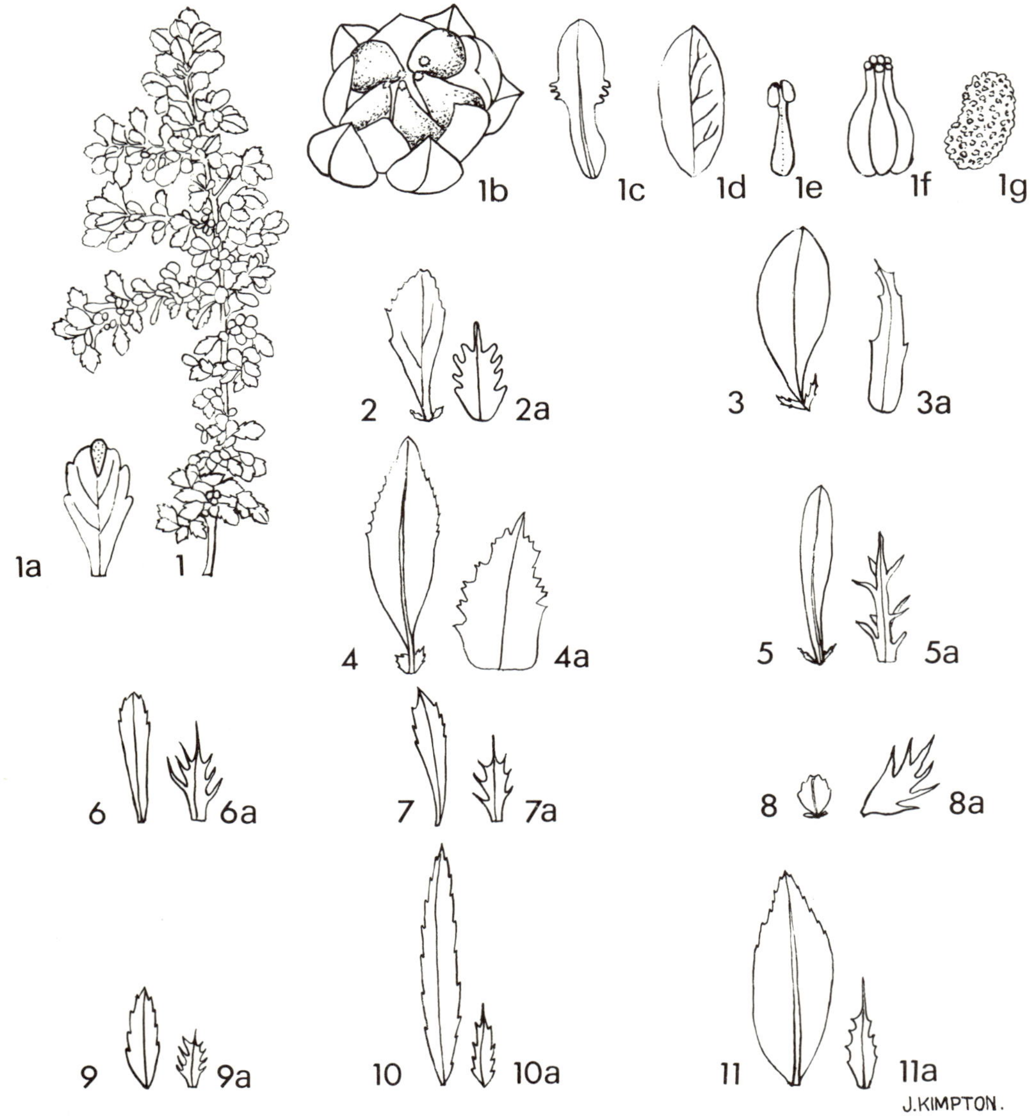

FIG. 6.—1, **Bergia glomerata,** × 1; 1a, leaf with apical gland on lower surface, × 4; 1b, flower with carpels, × 8; 1c, sepal, × 10; 1d, petal, × 10; 1e, stamen, ×10; 1f, gynoecium, × 10; 1g, seed, × 22. 2, **B. anagalloides,** leaf, × 1; 2a, stipule, × 14. 3, **B. polyantha,** leaf, × 1; 3a, stipule, × 8. 4, **B. capensis,** leaf, × 1; 4a, stipule, × 11. 5, **B. spathulata,** leaf, × 1; 5a, stipule, × 5. 6, **B. glutinosa,** leaf, × 1; 6a, stipule, × 5. 7, **B. ammannioides,** leaf, × 1; 7a, stipule, × 5. 8, **B. glomerata,** leaf, × 1; 8a, stipule, × 5. 9, **B. pentherana,** leaf, × 1; 9a, stipule, × 3. 10, **B. decumbens,** leaf, × 1; 10a, stipule, × 4. 11, **B. salaria,** leaf, × 1; 11a, stipule, × 5 (Semi-diagrammatic).

S.W.A.—1712 (Posto Velho): on the Kunene Riverbank at Otjinungua, *Giess 8871*. 1713 (Swartbooisdrift): Kunene, riverbank at Epupa Falls, *Giess & Wiss 3233;* Otjitambi, in the Hondoto River on the road to Swartbooisdrift, *Giess 3126*. 1813 (Ohopoho): Kaoko Otavi, *De Winter & Leistner 5557*. 1821 (Andara): Gam, *Story 5115*. 1918 (Grootfontein): Grosshuis, *Dinter 7381*. 2016 (Otjiwarongo): Omatjenne Experimental Farm, *Pfeiffer s.n.* 2115 (Karibib): Klein Ameib, *Dinter 7055; 7056*. 2117 (Otjosondu): Otjosondu Farm, KAR 36, Rote Berge, *Giess 3436*.

3. **Bergia capensis** *L.*, Mant. 241 (1771); Milne-Redh. in Kew Bull. 1948: 450 (1948); Subramanyam, Aquatic Angiosperms, India, Bot. Mon. No. 3: 10, f.6 (1962); Cook in Fl. Europ. 2: 295 (1968); Roessler in F.S.W.A. 92: 2 (1968); Verdcourt in F.T.E.A. Elatinaceae: 3, f.1, 1–7 (1968). Type: presumably from Asia, not from the Cape (LINN 597.1 holo.; PRE, photo.).

B. aquatica Roxb., Pl. Corom. 2: 22 (1798), nom. illegit. *B. verticillata* Willd., Sp. Pl. 2: 770 (1799); F.T.A. 1: 152 (1868), nom. illegit.

B. sessiliflora Griseb., Cat. Pl. Cub. 40 (1866); Monachino in Phytologia 5: 184 (1955). Type from Cuba.

Elatine luxurians Del., Descr. Egypt. Hist. Nat. 2: 216 (1813), nom. illegit.

Succulent, glabrous aquatic annuals with erect stems from a creeping rhizome. *Roots* numerous at each rhizome node, stout, with many long thin side roots. *Stems* up to 25 cm long, and up to 7 mm in diam., unbranched or with a few short side branches, succulent. *Leaves* reduced below, upper fairly close together, narrowly ovate, up to 4 cm long and 1 cm broad, attenuate below, shallowly serrate, thin; stipules small. *Flowers* small, in many-flowered, tight clusters, sessile or very shortly pedicellate. *Sepals* narrowly ovate, 1,5 mm long, membranous with entire margins. *Petals* narrowly ovate, slightly longer, pale lilac. *Stamens* 10. *Ovary* globose, styles short. *Capsule* globose, the valves flattened and recurved at dehiscence; seeds subcylindrical, straight or slightly bent, 0,8 mm long, tesselate, shiny yellow brown. FIG. 6: 4.

Recorded from South West Africa and the eastern Transvaal; also in tropical and subtropical Africa and Asia. An aquatic often found in semi-permanent rock pools.

S.W.A.—2115 (Karibib): Umduasbank, *Dinter 7025*. 2417 (Mariental): farm Haribes, *Leippert 4721*.

TRANSVAAL.—2531 (Komatipoort), Kruger National Park, Shabin Kop near Pretorius Kop, *Codd 6054;* Malelane byroad, 9 km from Kemps cottage, *Ihlenfeldt 2386;* Mbiamide, *Van der Schijff 1967*.

4. **Bergia spathulata** *Schinz* in Bull. Herb. Boiss. 6: 526 (1898); Niedenzu in Pflanzenfam. ed. 2, 21: 274 (1925); Exell & Mendonça in C.F.A. 1: 119 (1937); Roessler in F.S.W.A. 92: 4 (1968). Type: S.W. Africa, Olukonda, *Rautanen* 221 (Z, holo.).

B. erythroleuca Gilg in Warb., Kunene-Samb. Exped. 308 (1903). Type: Angola, on the left bank of the Kunene River near Soba Gongo, *Baum* 110 (HBG†).

Small prostrate or ascending, glandular-pubescent, aromatic annuals with numerous radiating branches, up to 40 cm tall. *Stems* thin, wiry, dark red. *Leaves* narrowly obovate, attenuate below, up to 2 cm long and 4 mm wide, minutely serrate, thin; stipules small, fimbriate. *Flowers* in dense rounded compound cymes present in every axil, the brown sepals and white petals giving the clusters a mottled appearance; bracts resembling stipules, up to 4 mm long; pedicels up to 2 mm. *Sepals* narrowly spathulate, 3 mm long, margin and midrib fimbriate, green, apex red or brown, apiculate. *Petals* narrowly obovate-acuminate, about 4 mm long, white or violet. *Stamens* 10. *Ovary* ovoid, 5-furrowed, papillate, red, the styles about as long as the ovary, erect. *Capsule* ovoid, red, papillate; seed 0,75 mm long, tesselate. FIG. 6: 5.

Known only from Angola and northern South West Africa, in moist places after rains.

S.W.A.—1713 (Swartbooisdrif): 48 km N. of Ohopoho on the road to Ruacana, *Giess & Leippert 7547*. 1817 (Tsintsabis): farm Tsintsabis, *Dinter 7668*.

Rautanen informed Schinz that the local natives sowed the seeds of this plant mixed with those of their watermelons hoping that the sweet honey-like fragrance would enter their melons.

5. **Bergia glutinosa** *Dinter & Schulze-Menz* in Notizbl. Bot. Gart. Berl. 15: 453 (1941); Roessler in F.S.W.A. 92: 3 (1968). Type: S. W. Africa, Grootfontein, Wackies, *Schoenfelder* sub herb. Dinter 7223a (B, holo.; M; PRE).

Small annuals with numerous radiating branches, scabrid, glandular-pubescent. *Stems* thin, wiry. *Leaves* narrowly ovate, up to 4 cm long and 12 mm broad, shortly petiolate, serrate, thin; stipules linear, 1,5 mm long, membranous, fimbriate. *Flowers* in several-flowered cymes, shortly pedicellate. *Sepals* linear, nearly twice as long as the petals, resembling the bracts, unequal in length, the outer longest; up to 4 mm long, green, fimbriate, margin membranous below. *Petals* linear, about 2 mm long, pink. *Stamens* 10. *Ovary* globose with the styles sessile or nearly so. *Capsule* globose; seeds 0,8 mm long, smooth or with indistinct markings. FIG. 6: 6.

Known only from two gatherings in northern South West Africa; in moist places.

S.W.A.—1918 (Grootfontein): Wackies, *Schoenfelder sub herb. Dinter 7223a;* Naruchas (Nurugas, GR 923/924), *Dinter 7223.*

Wild, in Flora Zambesiaca 1: 376 (1961), places a collection by *Drummond & Seagrief* 5186 from Botswana, Nata River, under this apparently annual species but I prefer to regard this plant as belonging to the perennial *B. salaria* Brem. The sessile flowers with the long, linear, fimbriate sepals and the globose capsule with short styles distinguish *B. glutinosa* from *B. salaria,* which has pedicellate flowers with ovate-acuminate subulate sepals only slightly longer than the petals and an ovoid capsule topped with long styles.

A collection (*Mitchell* 922) from Lake Kariba appears to be nearest *B. glutinosa* but is apparently a perennial and somewhat larger in all respects. This will need further investigation.

6. **Bergia ammannioides** *Heyne ex Roth,* Nov. Pl. Sp. 219 (1821); DC., Prodr. 1: 390 (1824); Oliv. in F.T.A. 1: 152 (1868), as *ammanoides*; Niedenzu in Pflanzenfam. 3, 6: 281 (1895); Engl., Pflanzenw. Afr. 3, 2: 524, t.237, fig. K–R (1921); Niedenzu in Pflanzenfam. ed. 2, 21: 274 (1925); Keay in F.W.T.A. ed. 2, 1: 128 (1954); Wild in F.Z. 1: 375, t.72, e (1961), pro parte; Roessler in F.S.W.A. 92: 3 (1968); Verdcourt in F.T.E.A. Elatinaceae: 3, fig. 1, 8 (1968). Type: India, *Heyne* (K, holo.).

Small prostrate annual with numerous thin radiating branches, minutely and sparsely pilose, especially in the axils, giving it a whitish appearance. *Stems* thin with long internodes. *Leaves* narrowly ovate to obovate, attenuate into a short petiole, variable in size, 15–25 mm long and 3–5 mm wide, margin serrate, coriaceous, glabrous or sparsely pilose; stipules linear, acuminate, about 2 mm long, ciliate, membranous. *Flowers* small, c. 2 mm, in dense globose, reddish cymes; pedicels up to 3 mm, pilose. *Sepals* linear-acuminate, 2 mm long, membranous with a broad green or red, pilose midrib, margin smooth. *Petals* narrowly ovate, 1,5 mm long, white or pink. *Stamens* 5 (in South African specimens seen), short, representing the outer circle. *Ovary* globose, styles short. *Capsule* with the valves often blackish red at the apex; seeds 0,5–0,75 mm long, tesselate, yellow. FIG. 6: 7.

Recorded from the warmer drier regions of Africa, Asia and Australia, where it is found in damp places. In Southern Africa recorded from South West Africa; also in Rhodesia.

S.W.A.—1918 (Grootfontein): Abenab-Gunib, *Dinter 7380;* Grootfontein, *Dinter 7627;* Ameib near Grootfontein, *Schoenfelder S635.* 2016 (Otjiwarongo): Okateitei, OTJ 157, *Dinter 7844.*

Note: The glabrous specimens cited by Wild in Flora Zambesiaca 1: 375 (1961) belong to *B. capensis* L.

7. **Bergia glomerata** *L.f.,* Suppl. 243 (1782); Thunb., Prodr. 82 (1794); Fl. Cap. ed. Schult. 395 (1823); DC., Prodr. 1: 390 (1824); Harv. in F.C. 1: 116 (1860); Niedenzu in Pflanzenfam. 3, 6: 282 (1895); ed. 2, 21: 274 (1925); Adamson in Fl. Cape Penins. 588 (1950). Type: Cape, near Swartkops River, Uitenhage, *Thunberg* (S, holo.; PRE, part of type-collection).

Perennial, woody prostrate glabrous plants forming mats often up to 1 m in diam., usually turning brown when dry. *Stems* thick, woody, with numerous short and long side branches, the internodes often telescoped together, densely leafy. *Leaves* small, obovate, 5 mm long, 3 mm broad with a few teeth towards the rounded apex or edentate, glabrous, smooth; on the lower side with a small elongate cushion-shaped gland apically; stipules large, ovate, pectinate. *Flowers* sessile, small, c. 3 mm, few to several in axillary glomerules. *Sepals* oblong to lanceolate-acute, about 2,5 mm, firm. *Petals* oblanceolate, 3 mm long, white, folded below with the upper half turned inwards. *Stamens* with the outer filaments somewhat wider and longer. *Ovary* globose, red, papillate, styles erect, terete. *Capsule* 5-partite, ovoid; seeds allantoid, 0,75 mm long, tesselate, black or brown. FIG. 6: 1.

A Cape endemic found around the coast or somewhat further inland, from Vanrhynsdorp in the west to the Albany district in the east, in temporary pools or damp places.

CAPE.—3118 (Vanrhynsdorp): farm Liebendal, 11 km N. of Vredendal, *Hall 3918.* 3119 (Calvinia): Lokenburg, *Acocks 17577; 17467.* 3318 (Cape Town): 27 km from Malmesbury to Hopefield, *Marsh 646;* N. of Tygerberg, *Compton 20061.* 3319 (Worcester): Worcester, *Van Breda 1056.* 3322 (Oudtshoorn): Kruispad, *Compton 21772.* 3325 (Port Elizabeth): at the mouths of the Koega and Swartkops Rivers, *Zeyher 1956;* near the Swartkops and Gamtoos Rivers, Uitenhage, *Zeyher 1768;* Swartkops River and Koegaskop, *Zeyher 513;* Korsten commonage, *Long 1141.* 3326 (Grahamstown): "Strowan" farm near Grahamstown, *Britten 6575.* 3420 (Bredasdorp): De Hoop Nature Reserve, *Van der Merwe 1980;* Bontebok Park, *Maguire 820.* 3421 (Riversdale): Riversdale coast, *Marloth 50; 8064.*

Of interest are the vestiges of the pectinate stipules found adnate to the bases of the bracts and outer sepals. From this it can be deduced that the thin transparent or white, often ciliate margins of the sepals are of stipular origin and now fused to the sepal. The elongate swollen dorsal apical gland has been observed in this species only; its function is unknown.

8. **Bergia pentherana** *Keissl.* in Ann. Naturh. Hofm. Wien 15: 58 (1900)*; Niedenzu in Pflanzenfam. ed. 2, 21: 274 (1925). Type: Rhodesia, Wankie Game Reserve, Bulungweti pans ("Land der 1000 Vleys, Bolongeti or Bellumbeti"), *Penther* 2317 (W, holo.!; PRE, photo.).

Bergia prostrata Schinz in Mém. Herb. Boiss. 20: 22 (15 Oct. 1900); Engl., Pflanzenw. Afr. 3, 2: 524 (1921); Wild in F.Z. 1: 376 (1961); Roessler in F.S.W.A. 92: 4 (1968). Type: S.W.Africa, Orumbo on the Nossob River, *Dinter* 1335 (Z, holo.). *B. palliderosea* Gilg in Warb., Kunene-Samb. Exped. 308 (1903); Exell & Mendonça, C.F.A. 1: 118 (1937). Type: Angola, on the left bank of the Kunene River above Humbe, *Baum* 98 (HBG †).

Prostrate perennial with long radiating annual branches, often rooting at the nodes, from a perennial crown; side branches many, short, prostrate or erect. *Stems* terete, pubescent. *Leaves* small, uniform in shape, but varying somewhat in size, narrowly ovate, 5–10 mm long and up to 3 mm broad, coriaceous, margin raised, laxly serrulate, nerves prominent below; stipules linear, serrulate. *Flowers* small, about 2,5 mm long, in 4–7-flowered axillary glomerules, pedicels about 2 mm, hairy. *Sepals* narrowly oblong, 2 mm long, acute, mucronate, firm. *Petals* ovate, 2,5 mm long, white or pink. *Stamens* 10, purple, with the outer filaments somewhat expanded and slightly longer than inner. *Ovary* ovoid, purple, styles fairly short with bulbous stigmas. *Capsule* 5-partite, globose with the valves hollow; seeds oblong-cylindrical, 0,75 mm tesselate, brown, usually 1–few developed in each carpel. FIG. 6: 9.

Recorded from Angola, northern South West Africa, Rhodesia, northern Cape, south western Transvaal and western Orange Free State, i.e. around the borders of the Kalahari; in sand, often in limestone areas; common around Kimberley and surrounding districts.

S.W.A.—1715 (Ondangua): 17 km E. of Oshikango, *De Winter & Giess 7047.* 1718 (Kuring-kuru): Omuramba Mpungu, Tsintsabis–Kuring-kuru road, *De Winter 3875.* 1719 (Runtu): Tamsoe, S.E. of Runtu, *Maguire 1633.* 1917 (Tsumeb): Otavi, *Dinter 5235.* 1918 (Grootfontein): Okavango, 16 km W. of Xeidang, *Giess 10051.* 2117 (Otjasondu): Quickborn, *Bradfield 449.* 2218 (Gobabis): farm Hester, *Merxmüller & Giess 1216;* Okasewa, *Dinter 2750.*

TRANSVAAL.—2526 (Zeerust): Panfontein Game Reserve, *Louw 2044.* 2725 (Bloemhof): 20 km S. of Bloemhof, *Acocks 12583.*

O.F.S.—2726 (Odendaalsrus): Odendaalsrus, *Goossens 1239.*

CAPE.—2624 (Vryburg): Armoedsvlakte, *Mogg 8111.* 2625 (Delareyville): Setlagodi, *Burtt Davy sub PRE 11032.* 2723 (Kuruman): Esperanza, near Kuruman, *Esterhuysen 2244.* 2824 (Kimberley): Barkley West, Pienaarsfontein, *Acocks 154.*

9. **Bergia decumbens** *Planch. ex Harv.*, Thes. Cap. 1: 15, t.24 (1859); F.C. 1: 116 (1860); Oliv. in F.T.A. 1: 153 (1868); Niedenzu in Pflanzenfam. 3, 6: 281 (1896); ed. 2, 21: 274 (1925); Engl., Pflanzenw. Afr. 3, 2: 524 (1921); Marloth, Fl. S. Afr. 2: 184, f.123 (1925); Burtt Davy, Fl. Transv. 1: 147 (1926); Wild in F.Z. 1: 377 (1960). Type: Transvaal, Magaliesberg, *Burke & Zeyher s.n.* (K, holo.; PRE; SAM).

Small undershrubs 20–50 cm tall with many annual (or rarely perennial) erect and procumbent branches arising from a perennial woody crown, red when young, soon becoming lignified, glabrous or hispidulous; side branches short or repressed to form leafy fascicles. *Leaves* linear to broadly linear, 1,5–2,5 cm long, 2–4 mm broad, uniform in shape, margin appearing swollen in the dried state with about 4–9 (–12) widely spaced teeth, the veins fairly broad, consisting of elongated cells contrasting sharply with the rest of the lamina where the cells appear small and squared to rounded, leathery in texture, shiny; stipules small, linear, serrulate. *Flowers* small, c. 4 mm long, in 2–8-flowered dichasia, pedicels pilose, c. 2 mm long. *Sepals* ovate-acuminate, 4–5 mm long, mucronate, margins laxly fimbriate. *Petals* slightly longer than sepals, narrowly obovate, white or pink. *Stamens* 10, filaments narrowly ovate-acuminate. *Ovary* ovoid-attenuate with 5 long, clavate, arcuate styles. *Capsule* narrowly ovoid, the valves usually flattened and spreading at dehiscence; seeds 0,5 mm, black, tesselate. FIG. 6: 10.

Widespread and fairly common in grassland in drier parts of the Transvaal, eastern Botswana, Rhodesia and Mozambique.

TRANSVAAL.—2329 (Pietersburg): near Pietersburg, *Bolus 10900.* 2330 (Tzaneen): Tzaneen, *Rogers 12545.* 2428 (Nylstroom): Schaaphok, c. 13 km N. of P.O. Rietbokspruit, *Story 1637; Codd 2341;* above Warmbaths, *Smuts & Gillett 3321; Bolus s.n.* 2429 (Zebediela): Pyramid Estate near Potgietersrus, *Galpin 8990;* Lunsklip, *Maguire 1396.* 2526 (Zeerust): Zeerust, *Leendertz sub TRV 11295.* 2527 (Rustenburg): Rustenburg, *Leendertz sub TRV 9743.* 2528 (Pretoria): Rietvlei Veld Reserve, *Acocks 11281.* 2530 (Lydenburg): Crocodile River, *Schlechter 3912.* 2626 (Klerksdorp): Lichtenburg, White's quarry,

* The exact date of publication could not be established. Since however the species appeared in the first of four parts of Ann. Naturh. Hofm. Wien 15 (1900), it may be assumed that it was published before October, the month in which Schinz's *Bergia prostrata* appeared.

Morris & Engelbrecht 1030. 2627 (Potchefstroom): Boskop, near Potchefstroom, *Louw 1626.* 2628 (Johannesburg): 32 km S. of Heidelberg, *Acocks 21024.*

The species possibly hybridises with *B. salaria,* which would account for the glandular pubescence found on plants collected in the northern Transvaal, northern Natal and southern Mozambique, e.g. *Galpin* M68, M289 from Mosdene near Naboomspruit in the Transvaal, *Strey & Moll* 3789 from near Maputa in northern Natal and *Drummond* 7724 from Mozambique, etc. These specimens appear to be intermediate between these two taxa. This will need further investigation.

10. **Bergia salaria** *Brem.* in Ann. Transv. Mus. 15: 248 (1933). Type: Transvaal, Soutpansberg, Soutpan, *Bremekamp & Schweickerdt* 227 (PRE, holo.).

B. mossambicensis Wild in Bol. Soc. Brot. Sér. 2, 31: 93 (1957); in F.Z. 1: 375, t.72 f.C (1961). Type: Mozambique, Sul do Save, Guijà district, Limpopo River, *Gazaland Expedition* sub TRV 15816 (SRGH, holo.; PRE).

Small glandular-scabrid undershrubs with erect or decumbent annual stems from a woody crown; if perennial, the stems elongate, produce abbreviated side branches and adopt a more straggling habit. *Leaves* on the primary branches ovate, 2–4 cm long, 1–2 cm broad, acute, teeth many, close together, often double-serrulate, fairly soft in texture, veins narrow, with elongated cells, deciduous; on perennial stems the later leaves produced on the abbreviated secondary branchlets are smaller; stipules linear-acuminate, about 5 mm long. *Flowers* in 2–7-flowered cymes; pedicels 1–7 mm long. *Sepals* ovate, subulate, 5–6 mm long, the broad midrib green, glandular-pubescent, the membranous margins red and white, denticulate. *Petals* shorter than sepals, broadly ovate-acute, 4–5 mm long, white, pink or pale mauve. *Stamens* 10 with the outer filaments somewhat longer and expanded below. *Ovary* 5-lobed, ovoid, attenuate into the styles which are about as long as the ovary. *Capsule* ovoid, the valves hollow at dehiscence, thin; seeds oblong-cylindrical, 0,5 mm, tesselate, dark brown, shiny. FIG. 6: 11.

Found in warmer parts of the Transvaal, Rhodesia and Mozambique, on sandy riverbanks or beside pans, possibly distributed by floods.

TRANSVAAL.—2229 (Waterpoort): Dongola, farm Nekel 620, *Codd 4095.* 2230 (Messina): Messina, *Rogers 19303.* 2231 (Pafuri): Machai pan, Nwambiya, *Brynard & Pienaar 4264; Van der Schijff 5688.* 2328 (Baltimore): Magalakwin, *Smuts sub PRE 1980.*

As these plants, which inhabit the warmer parts of the Transvaal and neighbouring countries, often perennate, their appearance in time becomes more lax and straggly; the primary leaves are dropped and on the short side branches smaller leaves take their place.

The type material of *B. mossambicensis* Wild, from the banks of the Limpopo River, appears to consist of young plants. It matches *Codd* 4095 from Dongola near Messina and links up with older specimens which match *B. salaria* from the Soutpan, not far from Dongola. *B. salaria* is also recorded from Rhodesia, e.g. *Drummond* 5762 from near Chiturupadzi store 40 km N.N.W. of the Bubye–Limpopo confluence, *Phipps* 2906 from the Nuanetsi district, 48 km S. of Chipanda pools, etc. Typical plants of *B. decumbens* and *B. salaria* can be distinguished easily but there are intermediates in areas where both species occur which seem to suggest that they hybridise (cf also note under *B. decumbens*).

5231 2. ELATINE

Elatine *L.,* Sp. Pl. 1: 367 (1753); Gen. Pl. ed. 5: 172 (1754); Benth. & Hook.f., Gen. Pl. 1: 162 (1862); Niedenzu in Pflanzenfam. 3, 6: 282 (1895); ed. 2, 21: 276 (1925); Hutch., Fam. Flow. Pl. ed. 2, 1: 428, f.271 (1959); Wild in F.Z. 1: 377 (1961). Type species: *E. hydropiper L.*

Small herbaceous, glabrous, aquatic or semi-aquatic annuals. *Leaves* decussate or verticillate, margin entire or serrulate; stipules small, deciduous. *Flowers* solitary in the leaf-axils, sessile or nearly so, ebracteate, very small, often cleistogamous. *Sepals* 2–4, connate below, midrib indistinct, margin not membranous. *Petals* 2–4, free, larger than sepals, white or pink. *Stamens* as many as the petals or diplostemonous. *Ovary* globose, 3–4-locular, styles 3–4, minute, situated on the depressed apex of the ovary. *Capsule* globose, membranous, the dissepiments remaining attached to central axis as wings; seeds numerous, straight or bent to horseshoe-shaped, tesselate.

Cosmopolitan with about 12 species occurring in tropical, subtropical and temperate regions as aquatics or in moist situations. One or possibly 2 species recorded from Southern Africa.

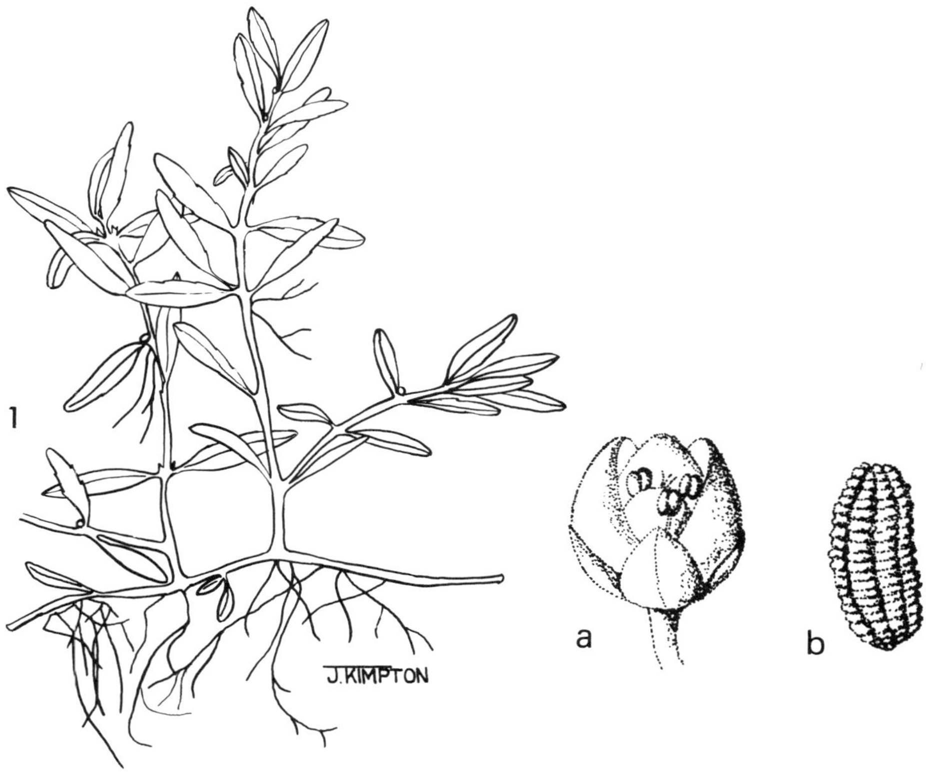

FIG. 7.—1, **Elatine triandra,** × 1; a, flower, × 50; b, seed, × 80 *(Ward* 2574).

1. **Elatine triandra** *Schkuhr,* Bot. Handb. ed. 2, 1: 345, t.1096 (1808); Niedenzu in Pflanzenfam. 3, 6: 383 f. 131 K–L (1895); ed. 2, 21: 276, 120 K–L (1925); Wild in F.Z. 1: 378 (1961); Subramanyam, Aquatic Angiosperms India, Bot. Mon. No. 3: 10 (1962). Type from Germany.

A small, soft creeping annual herb, rooting at the nodes. *Leaves* decussate, shortly petiolate, linear-oblong to narrowly ovate, up to 15 mm long and 3 mm broad, apex emarginate, narrowed at the base into a short petiole; stipules membranous, deciduous. *Flowers* sessile or with pedicels up to 1 mm long. *Sepals* 3 (rarely 2), narrowly ovate, 0,5 mm long. *Petals* 3, broadly ovate, 1–1,5 mm long, white or pink. *Stamens* 3, shorter than petals. *Ovary* 3-locular, globose, styles very short. *Capsule* membranous; seed slightly curved, oblong-terete, 0,5 mm, tesselate, light brown, shiny. FIG. 7.

Recorded from northern South West Africa and Zululand. Widespread but usually rather rare in Europe, Africa and America.

S.W.A.—1918 (Grootfontein): Grootfontein, *Volk 1977.*

NATAL.—2832 (Mtubatuba): Hluhluwe Game Reserve, *Ward 2574; 2575.*

Wild, in F.Z. 1: 378 (1961), records a possible second species (*Elatine? ambigua* Wight) from the Caprivi Strip, viz. *Stephens* 56 (BM), from pools close to the Linyanti River, Kazungula and Kasane. It differs from *E. triandra* in that it develops longer petioles (up to 2 mm long) and longer pedicels (up to 5 mm long), otherwise the two taxa are very similar.

FRANKENIACEAE*

by A. A. OBERMEYER

Perennial woody shrubs or shrublets, often prostrate, rarely annuals, richly endowed with salt secreting glands. *Leaves* decussate or in verticils of 4, often in axillary fascicles through suppression of the branchlets. *Flowers* persistent, solitary or in some species aggregated apically in terminal dichasia with the subtending leaves sometimes reduced to bracts, sessile, regular, bisexual (polygamous and sometimes dioecious in *Niederleinia*). *Calyx* tubular, induplicate, 4–6-lobed. *Petals* 4–6, free, imbricate, with a ligule attached to the claw on the inside (rarely absent), pink, mauve, red, white or yellow. *Stamens* 3–24 (mostly 6), free or filaments connivent and expanded below, filiform above, anthers bilocular, versatile, extrorse; staminodes present in *Niederleinia*. *Ovary* superior, 1-chambered, with 1–4 (usually 3) parietal or basal placentas bearing 1–numerous ovules attached to long funicles; style filiform, with as many branches as placentas, rarely capitate. *Capsule* enclosed in the calyx, dehiscing longitudinally; seeds small, ovoid-fusiform, smooth or rough, albuminous.

A cosmopolitan family consisting of 5 genera (Hutchinson, Gen. Fl. Pl., 1967); *Frankenia,* the only genus represented in Southern Africa, being the largest and most widespread; the genera *Hypericopsis, Beatsonia* and *Niederleinia* monotypic, and *Anthobryum* with 5 species, all recorded from restricted, widely separated regions; all 4 genera closely related to *Frankenia.* Halophytes, usually gregarious, found along the coast and further inland, rarely in non-saline surroundings.

Literature of general interest:

Marloth, R. Zur Bedeutung der salzabscheidenden Drüsen der Tamariscinen. Ber. Dt. Bot. Ges. 5: 319–324 (1887). (Cf. p. 323 for reference to *Frankenia*).

Bray, W. L. The geographical distribution of Frankeniaceae in connection with their systematic relationships. Bot. Jahrb. 24: 395–417 (1898).

Gundersen. The Frankeniaceae as a link in the classification of Dicotyledons. Torreya 27: 65–71 (1927).

Summerhayes, V. S. A revision of the Australian species of *Frankenia*. J. Linn. Soc. (Bot.) 48: 337–387 (1930).

*The name was preserved for the family.

5233 FRANKENIA

Frankenia *L.*, Sp. Pl.: 331 (1753); Gen. Pl. ed. 5: 154 (1754); DC., Prodr. 1: 349 (1824); Harv. in F.C. 1: 114 (1860); Oliv. in F.T.A. 1: 136 (1868); Niedenzu in Pflanzenfam. 3, 6: 286 (1895); ed. 2, 21: 279 (1925); Adamson in Fl. Cape Penins. 588 (1950); Phill., Gen. ed. 2: 508 (1951); Hutch., Gen. Flow. Pl. 2: 321 (1967); Roessler & Schreiber in F.S.W.A. 91 (1968). Type species: *F. laevis* L.

Nothria Berg., Descr. Pl. Cap. 171 (1767).

Low spreading woody shrublets, rarely annuals, often hairy and salt encrusted. *Leaves* decussate below, in verticils of 4 above, with leaf clusters (from suppressed branchlets) and flowers in their axils, sessile, subsessile or shortly petiolate, connected by an amplexicaul, usually ciliate, stipular ridge or sheath; lamina small, linear to ovate, flat or with the margins rolled back, often tightly so and thus terete or allantoid in shape, glabrous or pubescent, punctate. *Flowers* in forks of branches (determinate) sessile, small, persistent, scattered all over or aggregated in terminal dichasia (leaves subtending flowers in dichasia not differing from ordinary leaves in South African species). *Calyx* tubular, induplicate, 4–5-fid with prominent ribs, glabrous, puberulous and occasionally with long scattered setae. *Petals* 4–5 (5 in South African species), pink, mauvy pink or white, with an adnate scale or ligule attached to the inside of the claw. *Stamens* 6 (rarely 4–5), filaments connivent and expanded below, filiform above, anthers bilocular, bipartite or nearly so, versatile. *Ovary* with 2–4 (3 in South African species) parietal placentas bearing one to many ovules; style terete, with 3 stigmatic lobes. *Capsule* ovoid to narrowly ovoid, seeds small, ellipsoid-fusiform, papillate or smooth, maturing very early.

About 75 species (Hutchinson, 1967) recorded from the Mediterranean, along the Atlantic seaboard, western and southern Africa to western and southern Australia (the latter with about 45 species) to Asia and the Americas. Three species are found in Southern Africa, two endemic, the third almost cosmopolitan and weed-like in behaviour. Halophytes, found along the coast and further inland.

The genus was named in honour of Prof. Franken or Frankenius, a professor at Uppsala, who died in 1661.

Leaves ovate to lanceolate, flat or revolute, petiolate with the petiole 1 mm long; internodes usually long; flowers very numerous, solitary or bunched at the nodes, about 5 mm long; seeds small, 0,5 mm long, smooth, yellow; annual or perennial; widespread .... 1. *F. pulverulenta*

Leaves allantoid or flat and linear, subsessile; internodes usually short; flowers in apical dichasia or few, solitary and scattered, 9–11 mm long; seeds 1 mm long, rough, brown; woody perennials:

- Plants prostrate with short erect branchlets ending in many-flowered dichasia; stamens as long as petals, style longer, exserted; south-western Cape to eastern Cape (as far as East London) .... 2. *F. repens*
- Plants forming bushes up to 1 m tall; flowers few, scattered towards the tips of the branches; stamens shorter than petals; style just overtopping stamens; Namib .... 3. *F. pomonensis*

1. **Frankenia pulverulenta** *L.*, Sp. Pl. 1: 332 (1753); Harv. in F.C. 1: 115 (1860); Oliv. in F.T.A. 1: 136 (1868); Niedenzu in Pflanzenfam. 3, 6: 287 (1895); ed. 2, 21: 279 (1925); Summerhayes in J. Linn. Soc. (Bot.) 48: 386 (1930); Adamson in Fl. Cape Penins. 589 (1950). Type from Crete (LINN 457.6, holo.).

F. nodiflora Lam., Encycl. 2: 396 (1786); Ill. t.262, f.4 (1792); DC., Prodr. 1: 349 (1824); Harv. in F.C. 1: 115 (1860). Type: Cape of Good Hope (P, holo.). *F. densa* Pohnert in Mitt. Bot. Staatssamml. München 1: 447 (1954). Type: S.W. Africa, Grootfontein Süd, *Dinter* 8059 (M, holo.; PRE).

Copiously flowering annuals or rarely perennials with spreading, often reddish branches; internodes usually long and bare, glabrous or puberulous with short curly hairs, branchlets often contracted, forming leafy fascicles at the nodes. *Leaves* with a short petiole (up to 1,5 mm in the lower leaves), which is fused below to a stipular amplexicaul, ciliate (occasionally entire) sheath; lamina ovate, obovate, lanceolate or oblanceolate, 3–5 (–7) mm long, flat or with the margin recurved to tightly revolute and then compressed-terete; upper surface glabrous, punctate, lower glabrous or puberulous. *Flow-*

ers small, c. 5 mm long, very numerous, scattered and solitary in the forks of the branches (probably determinate) and flanked by side-branches, which are either well developed or repressed. *Calyx* tubular, 3,5–4,5 mm long, ribs prominent, lobes very short and acute. *Petals* cuneate, 5 mm long, apex obtuse, erose, ligule longitudinally adnate to the claw, acute at the apex with free margins, pink, mauve or white. *Stamens* just exserted from the corolla-throat, didymous, the longest 2,5 mm long, filaments connivent, linear-lanceolate below, filiform in upper half; anthers small, somewhat bipartite, medifixed, semicircular. *Ovary* with many ovules (c. 45 ovules in South African plants), style with 3 terete, short stigmatic branches, just exserted above the stamens. *Capsule* with ovoid-fusiform, papillate, reticulate seeds about 0,5 mm long, maturing early (possibly in a matter of days).

A widespread species, probably dispersed by man and beast, occurring in Europe, Africa, Asia, Australia. Recorded from S.W. Africa and along the Cape coast as far east as Port Elizabeth; further inland in the drier areas, in saline surroundings such as salt pans and banks of brackish streams. It flowers all the year round. Collections indicate that the plants may behave as annuals or occasionally persist as perennials when they become more woody and larger.

S.W.A.—2416 (Maltahöhe): Grootfontein-Süd, possibly farm Lisbon, *Dinter 8059*. 2615 (Lüderitz): Farm Sinclair Mine, *Giess 2298; 2303*. 2617 (Bethanie): Bethanien, *Dinter 8270*. 2618 (Keetmanshoop): River N. of Station Water, *Merxmüller & Giess 3606*. 2619 (Aroab): 38 km W.N.W. of Aroab, *Acocks 18072*. 2717 (Chamaites): Great Fish River, Aiais, *Oertendahl 635; Thorne sub SAM 52643*. 2818 (Warmbad): Bank of Ham River, Warmbad, *Galpin s.n.* 2819 (Ariamsvlei): Ariamsvlei, Walsersbrunn, *Oertendahl 652*.

O.F.S.—2726 (Odendaalsrus): 30 km W. of Odendaalsrus, *Acocks 14010*. 2926 (Bloemfontein): Haagenstad, *Potts sub BLFU 6615*.

CAPE.—2818 (Warmbad): Henkries, Orange River, *M. Schlechter 22*. 2824 (Kimberley): Rooipan, c. 48 km W.S.W. of Kimberley, *Leistner 1974*. 2917 (Springbok): Nababeep, *Hardy 1682*. 2918 (Gamoep): Brakrivier, *Pearson 4868*. 3118 (Vanrhynsdorp): 19 km S. of Bitterfontein, *Taylor 5505*. 3119 (Calvinia): Zwart Doorn River W. of Brandkop, *Compton 18894;* 33 km E.N.E. of Loeriesfontein, *Acocks 13202*. 3124 (Victoria West): Hutchinson, *Acocks 9589*. 3218 (Clanwilliam): Piketberg, *Compton 15042*. 3219 (Wuppertal): Beukesfontein, *Pearson 5005*. 3224 (Graaff Reinet): near Kendrew P.O., *Acocks 20616*. 3318 (Cape Town): About Salt River, *Burchell 513;* Green Point, *Ecklon & Zeyher* (Enum. 238); Robben Island, *Walgate 613*. 3320 (Montagu): Witteberge near Matjiesfontein, *Marloth 11442*. 3322 (Oudtshoorn): Oudtshoorn, *Theron 1038*. 3325 (Port Elizabeth): Swartkops, *Theron 1141*. 3421 (Riversdale): The Fisheries, *Acocks 21566*.

2. **Frankenia repens** *(Berg.) Fourc.* in Mem. Bot. Surv. S. Afr. 20: 43 (1941). Type: Cape, *Auge* (SBT, holo.; PRE, photo.).

Nothria repens Berg., Descr. Pl. Cap. 171 (1767). Type as above.

F. nothria Thunb., Prodr. 58 (1794); Fl. Cap. ed. Schult. 295 (1823). Type as for *F. repens*. *F. krebsii* Cham. & Schlechtd. in Linnaea 1: 36 (1826). Type: Cape of Good Hope, *Krebs* (B†). *F. capitata* and varieties, sensu Harv. in F.C. 1: 114 (1860). *F. hirsuta* sensu L., quoad coll. *Tulbagh* No. 195; cf. Jackson, Suppl. Proc. Linn. Soc. Session 130: 11 (1917–18); sensu Marloth, Fl. S. Afr. 2: 185, f.124 (1925); sensu Adamson in Fl. Cape Penins. 589 (1950); non L. (1753).

Prostrate woody shrublets, the spreading branches with numerous short upright branchlets usually ending in flowering dichasia, glabrous or roughly curly puberulous. *Leaves* subsessile, allantoid, 4–7 mm long, rarely flat and linear, dorsally smooth, punctate, ventrally with white papillae or minutely puberulent and punctate. *Flowers* in dense to few-flowered dichasia, congested near the apices of the branchlets. *Calyx* 7 mm long, often with long white coarse hairs on the ribs and minutely puberulent in the grooves. *Petals* obovate, c. 11 mm long, erose, pink to mauvy pink. *Stamens* with filaments 11 mm long, somewhat expanded in lower half, filiform above. *Ovary* cylindrical, with the style and 3 stigmatic branches exserted for 2–3 mm beyond stamens; ovules 21–35. *Capsule* with papillate, brown, broadly fusiform seeds c. 1,3 mm long. FIG. 8.

Endemic and fairly frequent along the coasts of the Cape from Namaqualand to Port Elizabeth, on sea-facing rock ledges, salt pans, saline flats and lagoons. Further inland it is recorded from brackish streambanks, edges of saltpans and calcareous sandveld. "Lightly browsed by cattle and sheep" fide P.A.B. van Breda.

A glabrous form is met with from Riversdale to Port Elizabeth, accompanied however by the more common pubescent form. It flowers during the summer months, from December onwards.

CAPE.—2818 (Warmbad): near Henkries, *Phillips 1636*. 2916 (Port Nolloth): between Port Nolloth and Holgat River, *Pillans 5193*. 2917 (Springbok): Steinkopf, *M. Schlechter 40*. 3017 (Hondeklip Bay): Hondeklip Bay, Spitfire Rock, *Pillans 17941*. 3118 (Vanrhynsdorp): Olifants River mouth, *Acocks 24164*. 3218 (Clanwilliam): Clanwilliam, *Leipoldt 4403;* Elands Bay, *Leach & Carp 11358*. 3317 (Saldanha): Danger Bay, *Tölken 426;* 9 km W. of Moorreesburg, *Acocks 20663*. 3318 (Cape Town): Langebaan, Saldanha Bay, *Taylor 3774;* Van Putten's vlei, *Compton 21878*. 3319 (Worcester): Goudini, *Van Breda 21*. 3325 (Port Elizabeth): at the mouths of the Coega and Swartkops Rivers, *Zeyher 1955;* near the Swartkops River, *Zeyher 1060*. 3418 (Simonstown): Noordhoek, *Wasserfall 679;* Cape of Good Hope Nature Reserve, Batsata Cave, *Taylor 6727*. 3419 (Caledon): Bot River, *Marloth 2786;*

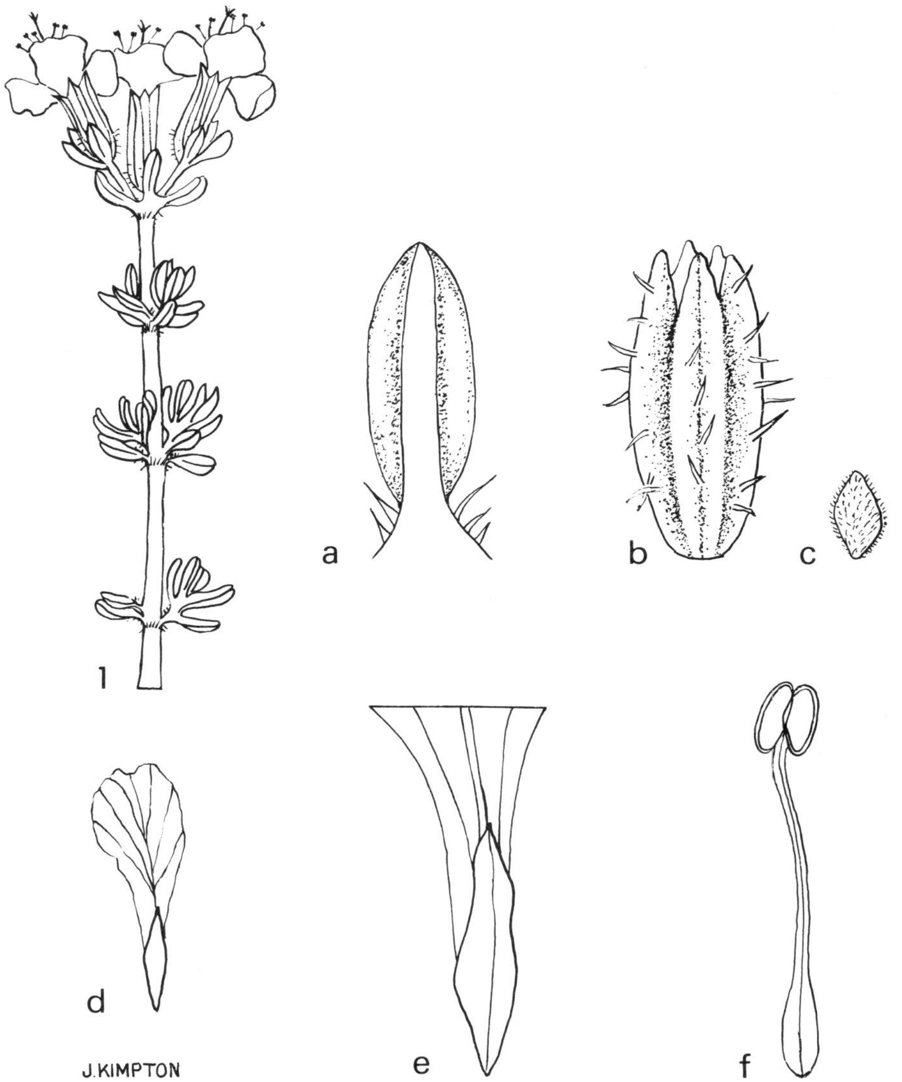

FIG. 8.—1, **Frankenia repens,** × 2½; a, leaf (dorsal view) × 7; b, calyx, × 7; c, seed, × 7; d, petal, × 3½; e, claw of petal with ligule, × 7; f, stamen, × 7 (*Taylor* 6727).

Danger Point, *Bond 768*. 3421 (Riversdale): Still Bay, *Muir 20*. 3422 (Mosselbay): Mosselbay, The Point, *Britten 645;* Great Brak River, *Schlechter 5753*.

Common names: Soutrankie; Seaheath.

This Cape species is closely related to *F. hirsuta* L. and Linnaeus placed a Cape specimen (No. 195) from a collection forwarded by the Governor Rijk Tulbagh under this Mediterranean species and reduced *Nothria repens* Berg. to synonymy. The flowers of the Cape species are, however, bigger and the stamens and style, with its 3 stigmatic lobes, well exserted. It is closely related to *F. pomonensis,* endemic in the Namib, but the latter forms small bushes and bears fewer and smaller flowers.

In Flora Capensis the species is named *F. capitata* Webb & Berth., a species described from the Canary Islands in Phyt. Canar. 1: 131 (1837). No mention is made in the original description of a ligule on the petals (which is common to most species), neither is it shown on the drawing (t. 16). The style moreover is depicted as shorter than the stamens.

3. **Frankenia pomonensis** *Pohnert* in Mitt. Bot. Staatssamml. München 1: 446 (1954); Roessler & Schreiber in F.S.W.A. 91: 1 (1968). Type: S.W. Africa, Namib, Pomona, *Dinter* 6424 (M, holo.; PRE).

Small perennial, woody, greyish brown shrublets, much branched and bushy, up to 1 m high, glabrous, densely leafy. *Leaves* subsessile, allantoid, c. 3(–6) mm long, upper surface glabrous, lower minutely and densely puberulous or papillate. *Flowers* sparse, towards the ends of the branches. *Calyx* tubular, 5–6 mm long, glabrous. *Petals* obovate, c. 9 mm long, crinkly, margin obtuse, erose, white to deep lilac or dark pink. *Stamens* just exserted from the corolla throat, anthers small, bipartite, semicircular. *Ovary* narrowly ovoid, with about 39 ovules, style with 3 short fairly thick stigmas a little longer than the stamens. *Capsule* with ovoid-fusiform, greyish brown, rough seeds 1 mm long.

Recorded only from the Namib in S.W. Africa, on brackish flats, dolomite outcrops facing the sea and bordering lagoons. Flowering during summer.

S.W.A.—2615 (Lüderitz): Lüderitzbay, *Marloth 4810b; Giess & van Vuuren 676;* Elizabeth Bay, *Merxmüller & Giess 2353*. 2715 (Bogenfels): Pomona, *Dinter 6424;* Bogenfels, *Merxmüller & Giess 2351; 6219; Dinter 4037*.

TAMARICACEAE

by A. A. OBERMEYER

Shrubs, trees or subherbaceous, often halophytic. *Leaves* exstipulate, sessile, alternate, usually scale-like, punctate with deep-seated salt secreting glands. *Flowers* usually very small, terminal, solitary, spicate, racemose or panicled, regular, hypogynous, bisexual, rarely unisexual and dioecious. *Sepals* usually 5–4, free or rarely connate below, imbricate. *Petals* 5–4, free or connate below, imbricate. *Stamens* numerous or 5, free or rarely connate, inserted on a glandular disk; anthers versatile, bilocular with longitudinal slits. *Gynoecium* one-chambered or imperfectly septate with 3–5 basal placentas; styles 3–5, free, or stigmas 3, sessile; ovules numerous—2, erect, anatropous. *Capsule* splitting into 3–5 valves, coriaceous; seeds erect with an apical hairy tuft and/or central plumose awn; embryo straight, elongate, endosperm present or 0.

A small family consisting of 4 genera and about 75 species, the genus *Tamarix* alone accounting for about 50 of these. Distributed mostly in the northern hemisphere and usually halophytic.

5239 TAMARIX

Tamarix *L.*, Sp. Pl. 1: 270 (1753); Gen. Pl. ed. 5: 131, No. 337 (1754); DC., Prodr. 3: 95 (1828); Harv. in F.C. 1: 119 (1860); Oliv. in F.T.A. 1: 151 (1868); Niedenzu in Pflanzenfam. 3,6: 293 (1895), ed. 2, 21: 285 (1925); Phill., Gen. ed. 2: 508 (1951); Hutch., Gen. Flow. Pl. 2: 323 (1967); Baum, Monogr. *Tamarix,* thesis, Hebrew University (1967). Podlech in F.S.W.A. 90: 1 (1968). Type species: lectotype, *T. gallica* L. Chromosome number: 2n = 24.

Shrubs or trees with the leaves reduced to scales, sessile, alternate, occasionally decurrent and/or auriculate, amplexicaul or vaginate, glabrous, papillose or hairy, punctate with sunken, salt secreting glands. *Inflorescence* a simple or compound raceme on old and/or on young wood; bracts more or less resembling the leaf, 1 per flower. *Flowers* in soft, erect or drooping catkin-like racemes, shortly pedicelled, very small and very numerous, bisexual or rarely unisexual and dioecious. *Sepals* 4–5, free or connate at the base, glabrous or hairy. *Petals* 4–5, free, persistent or deciduous, longer than the sepals, white, pink or red. *Disk* saucer- to cup-shaped, often fleshy, dark coloured with the margin variously lobed. *Stamens* 4–10, fused basally with the disk; filaments free, thin; anthers apiculate or obtuse; in female flowers changed to staminodes. *Ovary* flask-shaped, one-chambered with 3(4–5) basal placentas bearing several to many ovules, styles 3, short, incurved. *Capsule* usually 3-valved, pyramidal, narrowed to the apex; seeds several to many, narrowly ovoid with an apical tuft of hairs and a plumose awn, endosperm 0.

Species about 50, mainly distributed around the Mediterranean basin, westwards to England and the Canary Isles, eastwards as far as China; with one isolated species in south-western and southern Africa. Two species commonly cultivated in Southern Africa, *T. gallica* L., the French Tamarisk, and *T. ramosissima* Ledeb.

Tamarix usneoides *E. Mey. ex Bunge*, Tent. Gen. Tamaric. 74 (1852); Baum, Monogr. *Tamarix*, (thesis, Hebrew University, Jerusalem) 82, fig. 32 (1967), ined.; Podlech in F.S.W.A. 90: 1 (1968). Syntypes: Cape, Koup area between Blauwekrans and Bitterwater on the Gamka River, *Drège;* Zwartruggens area, Zondags River, on the bank and in the stream, *Drège b* (PRE!; K); Little Namaqualand, in the river near Natvoet and at the mouth of the Orange River, *Drège;* T'Kousie River and near Rietfontein in Namaqualand, *Ecklon & Zeyher*. Syntypes at LE; iso-syntypes in many herbaria.

T. usneoides E. Mey. ex Drège, Zwei Pfl. Doc. 61, 63, 92, 94, 225 (1838), nom. nud.; E. Mey. ex Niedenzu in Pflanzenfam. ed. 2, 21: 288 (1925), in clavis. *T. articulata* sensu Harv., F.C. 1: 119 (1860); sensu Oliv. in F.T.A. 1: 151 (1868), pro parte quoad specim. *Welwitsch* s.n. from Angola, non Vahl. *T. austro-africana* Schinz in Bull. Herb. Boiss. 2: 183 (1894). Syntypes: Cape, *Drège, Ecklon* 2130, *Zeyher* 725, *Rehmann;* S.W. Africa, *Wandres* 19, *Schenk* 253, 383, *Schinz* 406, *Pohle, Lüderitz* 150, 32, *Stapff, Höpfner* 15, *Fleck* 743 (all at Z, presumably). *T. angolensis* Niedenzu in Engl., Pflanzenw. Afr. 3, 2: 531 (1921), in obs.; Pflanzenfam. ed. 2, 21: 288 (1925), in clavis.; Baum, l.c. 75, fig. 28 (1967). Type: Angola, *Welwitsch* 1086 (sphalm. 86) (BM, lecto.; K). *T. engleri* Arendt, Beitr. *Tamarix* (thesis, Berlin University) 51 (1926), ined.; Baum, l.c. in synonymy.

A gregarious, copiously branched shrub or tree up to 5 m high, with a deep taproot and spreading superficial roots from which adventive buds may give rise to new plants; bark brownish grey with rough transverse scars; upper branchlets congested, soft, pendulous, giving the tree a plumose appearance. *Leaves* reduced to sharp pointed, amplexicaul, vaginate scales, completely covering the young stems, pale glaucous-green. *Inflorescences* massed apically, the upper parts of the branchlets turning into simple many-flowered racemes; leaves changing into semi-amplexicaul, auriculate bracts with long patent apices, the flowers solitary in their axils. *Flowers* very small, shortly pedicellate, greyish or creamy white, unisexual and dioecious or rarely bisexual. *Sepals* 5, free, ovate, 1 mm, spreading. *Petals* 5, free, oblong, 2 mm, obtuse, erectopatent. *Stamens* in male or bisexual flowers on long, thin filaments exceeding the petals, basally attached to a 5-lobed dark red, fleshy disc; anthers obtuse, pink; in the female flowers the flattened staminodes vary in shape, from nearly as long as the ovary to minute, the apex spathulate (if vestigial anthers are present) or apiculate. *Ovary* flask-shaped, 1,5 mm long, with 3 thick, short incurved stigmas; ovules several; in the male flowers the ovary reduced in size. *Capsule* pyramidal-attenuate, about 6 mm long; seeds c. 3 mm, grain-like with a plumose apical awn and a tuft of hairs as long as the seed. FIG. 9.

Recorded from the S.E. and N.W. Cape, South West Africa and Angola, frequenting dry sandy areas with subterranean, often saline water, such as salt pans and riverbanks; often the dominant tree of the community.

S.W.A.—1712 (Posto Velho): Kunene River at Otjinungua, *Giess 8908; Story 5817.* 1713 (Swartbooisdrift): Ososouu River at Otjiwero, *De Winter & Leistner 5358.* 1914 (Kamanjab): Huab River, farm Alettasrust about 40 km S. of Kamanjab on road to Fransfontein, *De Winter 3114.* 2016 (Otjiwarongo): farm Onguati, *Schwerdtfeger 1/168.* 2114 (Uis): Brandberg, *Liebenberg 5009;* Hochbrandberg, Numas-Binsenwiesen, *Wiss 1459.* 2115 (Karibib): Erongo Mts. near Usakos, *Schlieben 10321.*

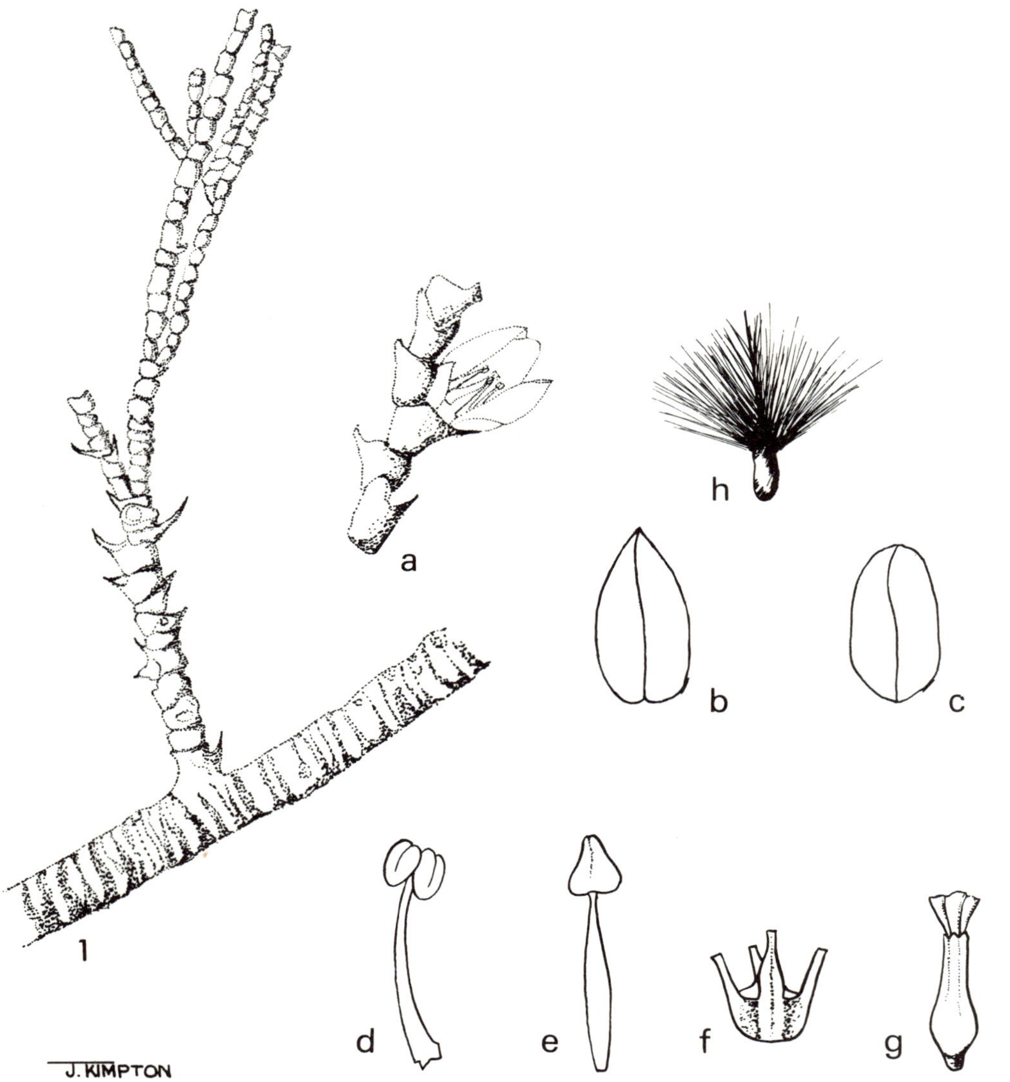

FIG. 9.—1, **Tamarix usneoides,** branches, × 2½; a, branchlet with leaves and flower (one sepal removed), × 8; b, sepal, × 25; c, petal, × 12; d, stamen, × 24; e, staminode, × 36; f, disc with bases of filaments, × 12; g, ovary, × 12; h, seed, × 9 *(Hardy & Ihlenfeldt* 1429).

2214 (Swakopmund): Kuiseb River at Gobabeb, *Hardy & Ihlenfeldt 1429* (☿). 2216 (Otjimbingwe): Okapuka camp on farm Otjisewa, Klein Windhoek River, *Wiss & Kinges 718.* 2314 (Sandwich Harbour): 10 km S. of Walvis Bay in Kuiseb River, *De Winter 3179.* 2417 (Mariental): near Mariental, *Basson 221.* 2517 (Gibeon): 40 km S.S.E. of Stampriet, *Leistner 1815* ☿. 2617 (Bethanie): Leeu River near Seeheim, *D'Ewes sub NBG 84305;* along tributary of Fish River, Bethanie, *Gerstner 6287.* 2816 (Oranjemund): Oranjemund, *Merxmüller & Giess 2286;* Arrisdrift, *Schenk 253.* 2818 (Warmbad): Ramansdrift, *Kruger 6.* 2819 (Ariamsvlei): Klein Karas, Groendoorn, *Oertendahl 363;* Ariamsvlei, Walsersbrunn, *Oertendahl 323.*

CAPE.—2816 (Oranjemund): Richtersveld, Sendlingsdrift, Orange River banks, *Werdermann & Oberdieck 592.* 2817 (Vioolsdrif): Kuboos Mission, *Marloth 12331.* 2819 (Ariamsvlei): Velloorsdrift, *Van Son sub TRV 31820.* 2820 (Kakamas): near Augrabies Falls, *Van Zinderen Bakker 1368;* 1,5 km N. of Kakamas near Orange River, *Leistner & Joynt 2820; Botha 2979.* 2821 (Upington): between Upington and Keimoes, *Theron 828* (☿). 3018 (Kamiesberg): near Garies, *Rodin 1374.* 3019 (Loeriesfontein): farm Brackfontein near Loeriesfontein, *Schlieben & Van Breda 9851.* 3119 (Calvinia): Zwartdoorn River W. of Brandkop, *Compton 18895.* 3222 (Oudtshoorn): Dwyka River, 16 km S.W. of Prince Albert Road, *Theron 1316;* Beaufort West, *Marloth 2149.* 3324 (Steytlerville): Grootrivierpoort, *Scharf 1364; 1365; 1366; 1369.*

Common names: Tamarix; Abikwa tree.

Tamarix usneoides is a gregarious, sand-loving halophytic tree or shrub dependent on an ample supply of ground-water. In South Africa it is common along the banks of the Orange River from near Upington to its mouth and also occurs along its tributaries and around saline depressions in Namaqualand and further south-east. When pushed over by floods, the prostrate trunks send up straight young stems, while broken off branches may also take root where they are deposited further down stream. Zeyher collected it in the south-eastern Cape in 1826 in the Grootrivierpoort (3324 Steytlerville). The tree from which Zeyher collected his material could very likely be the solitary tree next to the old road at the drift, now a farm road (*cf.* notes by Scharf at PRE). It is still common along the rivers of that region.

In South West Africa the species is quite common in the low-lying western desert areas, bordering sandy riverbeds and brackish shallow pans. It stretches further north to southern Angola. Although it is usually unisexual and dioecious, an occasional plant in a colony will revert to the (ancestral?) bisexual form, when fertile stamens are produced. Niedenzu separated these as *T. angolensis.*

In and around Swakopmund the species hybridises with the introduced garden species *T. ramosissima* Ledeb., which elsewhere has also been found growing wild. It can be distinguished from the endemic species by its green colour and pink flowers; the leaves moreover are not amplexicaul.

T. usneoides is closely related to *T. aphylla* (L.) Karst. from north Africa and Arabia, which is also occasionally cultivated in South West Africa, but this species has smaller flowers which are always bisexual, the anthers apiculate, and the branches appear to be thinner and longer, resembling *Casuarina.*

The Cape specimens are frequently festooned with red galls and the flowers are often attacked by insects which damage the ovaries causing them to become sterile and globose in shape.

CANELLACEAE

by L. E. CODD

Glabrous, often aromatic trees. *Leaves* alternate, simple, entire, gland-dotted, exstipulate, penninerved. *Inflorescence* axillary or terminal, cymose or flowers solitary. *Flowers* hermaphrodite, regular. *Sepals* 3, imbricate, persistent. *Petals* 5–12, free or united into a tube at the base, rather thick, in 1 or 2 whorls. *Stamens* 6–12, hypogynous; filaments united in a tube; anthers adnate to the outer surface of the tube with longitudinal extrorse dehiscence. *Ovary* superior, 1-locular; ovules 2–many on 2–6 parietal placentas. *Style* short, thick; stigmas 2–5. *Fruit* a berry. *Seeds* 2–many, shining; endosperm oily and fleshy.

A family of 5 genera and about 11 species, natives of tropical north and south America, east tropical Africa and Madagascar. The African genus *Warburgia* extends to Southern Africa. Several species are known to have aromatic or pungent bark which is used medicinally.

The relationships of the Canellaceae are probably with the woody Ranales (Magnoliales of Hutchinson), near to the Myristicaceae or Winteraceae. The structures described as 3 sepals above are, by some authors (Verdcourt, Hutchinson), regarded as 3 bracts, and the 2 rows of petals as calyx and corolla, respectively. On the other hand Wilson supports the interpretation given here.

5256 WARBURGIA

Warburgia *Engl.*, Pflanzenw. Ost-Afr. C: 276 (1895); Burtt Davy, Fl. Transv. 1: 218 (1926); Phill., Gen. ed. 2: 509 (1951); Verdcourt in F.T.E.A. Canellaceae: 1 (1956); Melchior & Schultze-Motel in Pflanzenfam. ed. 2, 17a II: 223 (1959); Hutchinson, Gen. Flow. Pl. 1: 64 (1964); Wilson in Am. J. Bot. 53: 336 (1966); nom. conserv. (Taxon 17: 328, 1968). Type species: *W. stuhlmannii* Engl.

Chibaca Bertol.f. in Mem. Accad. Sci. Istit. Bologna 4: 545 (1853). Type species: *C. salutaris* Bertol.f.

Evergreen trees. *Leaves* dark green, shiny above. *Inflorescence* axillary, flowers solitary or in reduced 3 or 4-flowered cymes. *Sepals* 3, suborbicular. *Petals* 10, in 2 rows, the outer larger and thicker than the inner, pellucid gland-dotted. *Stamens* 10; staminal tube equalling or subequalling the petals; anthers sessile on the upper part of the tube, about as long as the staminal tube, linear, with a fleshy apical appendage. *Ovary* ellipsoid; ovules 10–30 in 1 or 2 rows on 5 placentas. *Style* enclosed in the staminal tube or just protruding, truncate at the apex, bearing 5 oval stigmatic patches round the sides. *Fruit* a globose or ovoid berry; pericarp coriaceous.

An African genus of 3 species, 1 of which extends into Southern Africa.

Warburgia salutaris (*Bertol.f.) Chiov.* in Nuov. Giorn. Bot. Ital. n.ser. 44: 680 (1937); Mendes in Bol. Soc. Brot. Sér. 2, 43: 338 (1969); Ross, Fl. Natal 247 (1972); Palmer & Pitman, Trees S. Afr. 3: 1523 (1972). Type: Mozambique, *Fornasini* s.n. (BOLO†).

Chibaca salutaris Bertol.f. in Mem. Accad. Sci. Istit. Bologna 4: 545, t.23 (1853).

Warburgia breyeri Pott in Ann. Transv. Mus. 6: 60 (1918); Gerstner in Bantu Studies 12: 220 (1938); Doke & Vilakazi, Zulu-English Dictionary 20 (1948). Type: Transvaal, Letaba District, Makhutswi (''Macoutsie'') River, *Breyer* sub TRV 17573 (PRE, holo.!).

Tree up to 10 m tall; bark rough and mottled, red on inner side; branches terete, striate, lenticellate. *Leaves* coriaceous, shiny above, paler below, densely pellucid-dotted, shortly petiolate; petiole 2–5 mm long, channelled above; blade oblong to oblong-lanceolate or elliptic, 4,5–11 cm long, 1,5–2,5 cm broad, veins obscure above, inconspicuous below; apex acute; base cuneate; margin slightly involute. *Flowers* axillary, solitary or in abbreviated 3-flowered cymes; peduncle 2–3 mm long; bracts 0,5 mm long, deciduous, leaving conspicuous scars; pedicel up to 1,5 mm long. *Sepals* 3, suborbicular, 2 mm long, 3 mm broad, minutely ciliate. *Petals* in 2 whorls; outer 5 subcoriaceous, obovate, concave, 4–5 mm long, 3 mm broad; inner petals 5, yellow, thinner-textured, spathulate, 4 mm long, 1,5 mm broad. *Staminal-tube* 3–4 mm long, apex protruding slightly beyond the stamens; stamens 10, 1,5 mm long. *Ovary* oblong-ovoid, 3 mm long, 1 mm diam.; stigma subsessile, truncate, obscurely 5-lobed; ovules about 15–20. *Fruits* ovoid to subglobose, 2–3 cm long, 1,5–2,5 cm in diam.; pericarp coriaceous, wrinkled; seeds several, flattened. FIG. 10.

A rare constituent of coastal forest in northern Zululand and of submontane forest in eastern and northern Transvaal.

TRANSVAAL.—2229 (Waterpoort): Soutpansberg, *Van Warmelo s.n.* 2328 (Baltimore): Blouberg, *Codd 8733; Strey & Schlieben 8611.* 2330 (Tzaneen): near Tzaneen, *McCallum sub PRE 15483.* 2430 (Pilgrims Rest): Macoutsie River, *Breyer sub TRV 17573;* Mariepskop, *Van der Schijff 5196; 5503; 5950; Govt. Forester s.n.;* Blyde River Poort, *Van der Schijff 4307.*

NATAL.—2731 (Louwsburg): Magut, *Gerstner 3416.* 2732 (Ubombo): Lake Sibayi area, *Tinley 197;* False Bay, *Gerstner 6974; Ward 3004.* 2832 (Mtubatuba): Corridor, Umfolozi Game Reserve, *Ward 3067.*

Known as isiBaha (Zulu), Shibaha (Tsonga) and Molaka (Sotho), the bark, especially the root bark, is pungent and bitter and is widely used as a medicine against colds and fever. Gerstner reports that this has led to regular cutting down of trees which may have an adverse effect on their regeneration. The leaves are also bitter and this helps in identifying the species in the field.

As yet no flowering or fruiting material has been seen from Zululand but the specimens from this area agree well in vegetative characters with those occurring in the Transvaal. However, fertile material is important for a correct assessment to be made.

Verdcourt, l.c., united the Transvaal *W. breyeri* Pott with the tropical African species, *W. ugandensis* Sprague, but the slight differences in texture, size and shape of the leaves suggest that this view may not be correct. The fruits of Transvaal specimens (possibly immature) also appear to be smaller and more wrinkled than those from Tropical Africa.

The type of *W. salutaris* is from Mozambique and, according to Verdcourt in Regnum Vegetabile 40: 27 (1965), is no longer extant. Good material from Mozambique is thus also required in order to establish its affinity. At present it seems to be related rather to the southern than the tropical populations.

FIG. 10.—1, **Warburgia salutaris,** flowering twig, × 1 *(Strey & Schlieben* 8611); a, fruit, × 1 *(Van Warmelo* s.n.); b, flower, × 4; c, staminal column, × 7; d, ovary, × 7; e, sepal, × 4; f, outer petal, × 4; g, inner petal, × 4 *(Strey & Schlieben* 8611).

VIOLACEAE

by A. A. OBERMEYER

Herbaceous perennials, rarely annuals, shrubs or trees. *Leaves* alternate, rarely opposite, simple, margin entire, dentate or incised; stipules 2, foliaceous, small or minute, persistent or caducous. *Flowers* solitary, axillary, or in compound, often reduced, inflorescences, bisexual or rarely polygamous, occasionally also with cleistogamous flowers, actinomorphic or usually zygomorphic. *Sepals* 5, usually free, persistent, quincuncial, sometimes produced beyond point of insertion. *Petals* 5, free, deciduous or persistent, imbricate or contorted, hypogynous; in zygomorphic flowers the anterior petal sometimes spurred. *Stamens* 5 (–15), usually hypogynous; anthers connivent in a ring around the ovary, bilocular, introrse, opening by longitudinal slits, thecae and connective often produced into dorsal and apical appendages; filaments 0 or short, simple or variously appendaged; the 2 anterior often forming caudate glandular spurs. *Ovary* superior, sessile, 1-locular with 3 (–5) parietal placentas bearing 1–∞ ovules; styles completely united, straight, oblique or sigmoid, often somewhat swollen in upper half, stigmas usually entire, apical. *Capsule* loculicidal often with contractile valves; rarely a berry or a nut. *Seed* with a short funicle, sometimes winged, tomentose or smooth, sometimes hygroscopic becoming glutinous when wetted, occasionally with a gibbous aril, with abundant endosperm.

Genera 20; species about 800. Mostly pantropical with the large cosmopolitan genus *Viola* predominantly temperate.

Flowers actinomorphic; trees or shrubs ....1. **Rinorea**

Flowers zygomorphic, the anterior petal longer or shorter, often spurred; herbs or undershrubs:

  Pedicels articulate; stipules entire, small, linear-acuminate; anterior petal much longer than the other 4; with or without a spur ....2. **Hybanthus**

  Pedicels inarticulate; stipules small, with 2 basal teeth or foliaceous, laciniate; anterior petal shorter than the other 4, spurred ....3. **Viola**

5262 1. RINOREA

Rinorea *Aubl.*, Hist. Pl. Guian. Franc. 1: 235, t. 93 (1775); Melchior in Planzenfam. ed. 2, 21: 349 (1925); Phill., Gen. ed. 2: 510 (1951); Robson in F.Z. 1: 246 (1960); Hutch., Gen. Flow. Pl. 2: 330 (1967). Type species: *R. guianensis* Aubl.

Alsodeia Thouars, Hist. Vég. Rec. 55 (1806); Oliv. in F.T.A. 1: 106 (1868).

Trees or shrubs. *Leaves* alternate, rarely opposite or whorled, petiolate, simple, serrate, rarely entire, with reticulate, arched, closed veins; stipules small, covering very young, pointed leaf-buds, caducous. *Inflorescence* simple or compound, axillary and/or terminal; pedicels articulated; bracts small, caducous. *Flowers* actinomorphic, small. *Sepals* 5, quincuncial, persistent, short and hard. *Petals* equal or subequal, soft, erect or reflexed, white, greenish white, yellow or red. *Stamens* 5, the filaments free or connate below, with or without appendages; anthers with the connective variously appendaged and occasionally also bearing thecal appendages. *Ovary* with 3 parietal placentas, style straight, stigma terminal, small, ovules 1–several on each placenta. *Capsule* 3-valved, the inner tissues woody, tearing apart from the outer skin, rarely semi-succulent or a berry. *Seed* 1–3 per valve, compressed, hard, shiny.

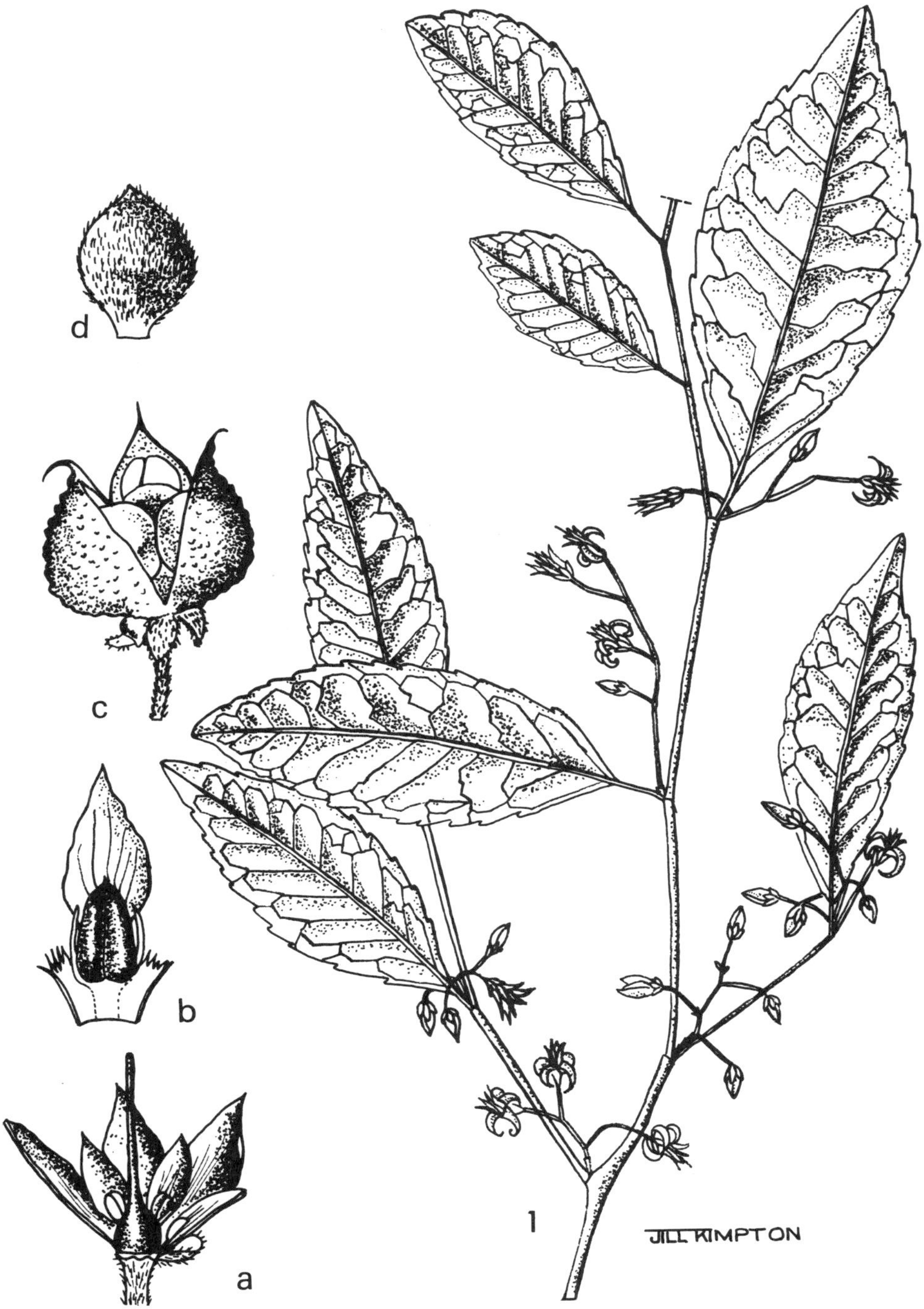

FIG. 11.—1, **Rinorea angustifolia,** flowering twig, × 1; a, longitudinal section of flower, × 4; b, stamen, × 7; c, capsule and seed, × 2; d, bud, × 4 *(Guy* 129).

Species about 250, tropics and subtropics; 2 species in Southern Africa.

Leaves 4–8 cm long, soft, dark green, closely dentate, on slender, shortly hairy dark branchlets; leaf scars not conspicuous; petals spreading to recurved .... 1. *R. angustifolius*
Leaves 8–14 (–22) cm long, leathery, glaucous, coarsely spinoso-dentate, on stout pale green glabrous branches; transverse leaf scars prominent; petals erect to spreading .... 2. *R. ilicifolius*

1. **Rinorea angustifolia** *(Thouars) Baill.* in Bull. Soc. Linn., Paris 1: 582 (1886); Kuntze, Rev. Gen. Pl. 1: 42 (1891); Melchior in Pflanzenfam. 2, 21: 352 (1925); Perr. in Mém. Inst. Sci. Madag., sér B, 2, 1949: 328 (1950); in Humbert, Fl. Madag. Fam. 139, Violacées: 36 (1955); Tennant in Kew Bull. 16: 412 (1963). Type: Madagascar, without precise locality, *Du Petit Thouars* s.n. (P, holo.).

Alsodeia angustifolia Thouars, Hist. Vég. Rec. ed. 2: 57, t. 18, fig. 1 (1806); DC., Prodr. 1: 313 (1824). Type as above. *A. ardisiiflora* Welw. ex Oliv. in F.T.A. 1: 108 (1868). Type: Angola, Cuanza Norte, Pungo Andongo, Mata de Pungo, *Welwitsch* 885 (LISU, lecto; BM; K). *A. natalensis* (Engl.) Bak.f. in J. Linn. Soc. (Bot.) 40: 22 (1911). Syntypes: Natal, Inanda, *Wood* 1001 (B†; BM; GRA; K; NH; SAM); Pondoland, Intsubana, *Bachmann* 1005 (B†).

Rinorea natalensis Engl. in Bot. Jahrb. 33: 135 (1902); Brandt in Bot. Jahrb. 50, Suppl.: 411 (1914); Bews, Fl. Natal & Zululand 140 (1921); Melchior in Pflanzenfam. ed. 2, 21: 350 (1925); Henkel, Woody Pl. Natal 84 (1934). Type as above. *R. ardisiiflora* (Welw. ex Oliv.) Kuntze, Rev. Gen. Pl. 1: 42 (1891); Sim, For. Fl. Cape Col. 125, t. 15, 1 (1907); For. Fl. P.E. Afr. 11 (1909); Bews, Fl. Natal & Zululand 140 (1921). Type as above.

Evergreen trees or shrubs up to 10 m high with a smooth, mottled grey bark, the juvenile parts and inflorescences puberulous. *Branches* slender, terete, brown. *Leaves* alternate, widely spaced, elliptic, 5–9 cm long, 2–4 cm broad, base cuneate, apex acuminate, margin serrate, the teeth gland-tipped, soft, dark glossy green, paler below; petiole short, stipules small, caducous. *Inflorescences* axillary, short, subcorymbose, several–1-flowered; peduncle and pedicels slender, bracts minute. *Flowers* sweetly scented, small; buds conical, reddish. *Sepals* ovate, 2 mm, persistent. *Petals* oblong-elliptic, c. 7 mm long, upper half reflexed, glabrous or velvety towards the base, white. *Stamens* with the filaments fused below and with a dorsal appendage forming a short fringed or irregularly lobed cup, or appendage absent (in specimens seen from Zululand and Transvaal); anthers with an apical, erect, exserted, ovate-acuminate, petaloid, reddish-purple connective-appendage, thecae with an apical, small lingulate appendage or this reduced to a mucro or 0. *Ovary* globose, glabrous (in S. Afr. specimens examined), style erect, terete, exserted, c. 4 mm long; stigma minute, apical. *Capsule* ellipsoid, c. 1,3 cm long, obtusely 3-lobed, hard, muricate, glandular-punctate; seed 1 (–2) per valve, compressed-ovoid, 5 mm, smooth, pale yellow. FIG. 12.

Common as an understorey tree (rarely up to 10 m tall) or shrub in evergreen coastal forests of Pondoland, Natal, Zululand and north eastern Transvaal; tropical Africa and Madagascar. Flowering August to October; occasionally earlier or later.

TRANSVAAL.—2230 (Messina): Soutpansberg, Entabeni Forest Reserve, *Obermeyer sub TRV 30275.* 2330 (Tzaneen): Westfalia Estate, *Scheepers 766;* Woodbush Forest Reserve, De Hoek, *Botha sub PRF 7411;* Magoebaskloof, *Gerstner 5862.* 2531 (Komatipoort): road from Barberton to Piggs Peak, *West 3013.*

NATAL.—2731 (Louwsburg): Ngome Forest Reserve, *Gerstner 5113; Sim 2918;* Sokosoko Forest, *Gerstner 4911.* 2732 (Ubombo): Gwalaweni Forest, *Moll 4451.* 2830 (Dundee): Qudeni Forest Reserve, *Strey 9317; Wood 7717.* 2831 (Nkandla): Nkandla Forest, *Acocks 11799; Gerstner 3583;* Ngoye Forest, *Nicholson 434.* 2832 (Mtubatuba): Hluhluwe Game Reserve, *Ward 1426.* 2930 (Pietermaritzburg): Karkloof, *Wylie sub Wood 10046;* Hilton Road, *Wood 10081.* 2931 (Stanger): Durban, *Oliver 6.* 3030 (Port Shepstone): Umtamvuna River, *Strey 5834;* Paddock, *Strey 5997.*

CAPE.—3129 (Port St Johns): Egossa, *Sim 2419; 2439;* Mkambati Leper Institute, *Marais 1192;* Port St Johns, *Hafström & Acocks 1932.* 3328 (Butterworth): The Haven, *Gordon-Gray 1340.*

In collections from the Gwalaweni Forest and the Transvaal the stamens have short free filaments; the dorsal appendage which fuses the filaments into a cup with a free ciliate margin which is found in typical material, is apparently suppressed.

2. **Rinorea ilicifolia** *(Welw. ex Oliv.) Kuntze,* Rev. Gen. Pl. 1: 42 (1891); Brandt in Bot. Jahrb. 50, Suppl.: 412 (1914); Keay in F.W.T.A. ed 2, 1: 101 (1954); Robson in F.Z. 1: 250 (1960); Tennant in Kew Bull. 16: 409 (1963). Type: Angola, Cuanza Norte, Pungo Andongo, Barrancos de Catete, *Welwitsch* 889 (LISU, lecto; BM; PRE).

Alsodeia ilicifolia Welw. ex Oliv. in F.T.A. 1: 108 (1868). Type as above.

Small understorey evergreen tree or shrub up to 5 m high. *Branches* terete, stout, pale green with dark transverse scars of the caducous stipules. *Leaves* alternate, coriaceous, glaucous, paler below, elliptic, usually about 8–16

FIG. 12.—1, **Rinorea ilicifolia,** flowering twig, × 1; a, petal, × 7; b, stamen, × 7; c, sepal, × 9; d, capsule and seeds, × ¾; e, flower, × 5 *(Tinley* 474).

cm long and 4–8 cm broad, margin serrate to spinoso-dentate; petiole 1–2 cm, stout. *Flowers* very small, arranged in compound, terminal or axillary, elongated panicles which are shorter than the leaves, bearing small clusters of brown velvet buds, bracts deciduous, small; pedicels short. *Sepals* broadly ovate, 2–3 mm long, obtuse, finely ribbed fan-wise with a membranous margin, persistent. *Petals* erect, narrowly elliptic, about 5 mm long, yellow, deciduous. *Stamens* on very short filaments, fused at the base; apical connective-appendage ovate-acuminate, petaloid; thecal appendage small to minute. *Ovary* globose, glabrous, style erect, terete, exserted, stigma apical, minute. *Capsule* globose, obtusely 3-lobed, about 1–2 cm in diam., hard, brown, verrucose; seeds angled, about 6 mm long, yellowish. FIG. 12.

Recorded from tropical and subtropical Africa as far south as Zululand, in damp evergreen coastal forests, where it may be the dominant understorey shrub. Flowering August to October.

NATAL.—2732 (Ubombo): Gwalaweni Forest, *Moll & Nel 5536; Tinley 474; Dutton & Tinley 21.* 2832 (Mtubatuba): Hluhluwe Game Reserve, *Ward 2313.*

5271 2. HYBANTHUS

Hybanthus *Jacq.*, Enum. Syst. Pl. Ins. Carib. 2 (1760), nom. conserv.; Melchior in Pflanzenfam. ed. 2, 21: 357 (1925); Phill., Gen. ed. 2: 510 (1951); Robson in F.Z. 1: 254 (1960); Roessler in F.S.W.A. 87: 2 (1969). Type species: *H. havanensis* Jacq.

Ionidium Vent., Jard. Malm. 1: sub t.27 (1803); Sond. in F.C. 1: 74 (1860).

Annuals, perennials, shrublets or shrubs. *Leaves* alternate and/or opposite, elliptic, ovate or linear, sessile or petiolate; stipules small, subulate or ovate, usually persistent, margin serrate to subentire, the teeth gland-tipped. *Flowers* solitary, axillary, rarely massed in terminal inflorescences, pendulous, persistent; pedicels articulated, the 2 bracteoles opposite or subopposite at or below articulation. *Sepals* free, more or less equal, not produced below. *Petals* unequal, the anterior (in S. Afr. species) much larger than other 4, with or without a basal pouch, clawed in the middle, expanded above into a spathulate or broad lamina; the other 4 somewhat longer than the sepals. *Stamens* 5, free or connate at the base, similar or the 2 (4) anterior each with a recurved caudate nectariferous spur projected into the pouch of the anterior petal; anthers connivent with the connective produced above, petaloid. *Ovary* with 3 parietal placentas, 3–many ovules; style terete, short or long, straight or curved. *Capsule* loculicidal with 3 elastic valves; seed broadly fusiform or ovoid-globose, apex discoid, raphe distinct or indistinct, aril usually prominent; testa striate, punctate or smooth, hygroscopic, swelling to form a glutinous covering, when wetted.

Species about 100, distributed in tropical and subtropical regions, predominantly in South and Central America. Four species in southern Africa, 2 of these endemic.

Common Names: Lady's Slipper, Pink Lady's Slipper.

Anterior petal more than 3 mm long; flowers often showy, blue, pink or violet, rarely white; seed white or pale yellow, fusiform with a truncate, discoid apex, or globose:

- Anterior petal with a basal pouch protruding from between the anterior sepals; filaments of 2 anterior stamens villous, bearing a short recurved caudate appendage which is inserted in the pouch; capsule glabrous or scabridulous; seed fusiform:
  - Perennial herbs or suffrutices (flowering in their first year), glabrous to pubescent, very variable in growth-form and leaf-shape; leaves and flowers not congested; capsule smooth, glabrous; widespread . . 1. *H. enneaspermus*
  - Annual, small, erect, scabridulous herbs with long linear leaves and shortly pedicelled flowers congested on the short stems; capsule scabridulous; Kalahari region . . . . 2. *H. densifolius*
- Anterior petal without a basal pouch; filaments glabrous, entire, without an appendage; capsule glabrous or pubescent; seed globose, pitted . . . . 3. *H. capensis*

Anterior petal 3 mm long; flowers very small, white; ripe seed black, smooth, shiny, globose to ovoid, 2 mm in diam., with a white protruding aril . . . . 4. *H. parviflorus*

1. **Hybanthus enneaspermus** *(L.) F. Muell.*, Fragm. 10: 81 (1876); Eyles in Trans. R. Soc. S. Afr. 5: 422 (1916), as *H. enneaspermum* (Vent.) Torre & Harms; Engl. in Bot. Jahrb. 55: 398 (1919): Robson in Bol. Soc. Brot. Sér. 2, 32: 164 (1958); in F.Z. 1: 254 (1960); Roessler in F.S.W.A. 87: 2 (1969). Tennant in Kew Bull. 16: 430 (1963). Type: Ceylon, *Hermann* 317 (BM).

Viola enneasperma L., Sp. Pl. 2: 937 (1753). Type as above.

Ionidium enneaspermum (L.) Vent., Jard. Malm. 1, sub t.27 (1803); Oliv. in F.T.A. 1: 105 (1868). Type as above. *I. thymifolium* Presl, Bot. Bemerk. 11 (1844), nom. illegit., non Presl (1835); Sond. in F.C. 1: 74 (1860). Type: eastern Cape—Natal, "Omsamwubo, Omsamcaba, Omtendo, towards Port Natal, in grassy places", *Drège* (S; TCD). *I. caffrum* Sond. in Linnaea 23: 13 (1850); in F.C. 1: 74 (1860). Type: Port Natal, *Gueinzius* 94 (S, holo; K; PRE, photo.). *I. hirtum* Klotzsch in Peters, Reise Mossamb. Bot. 1: 148 (1861). Type: Mozambique, Rios de Sena, *Peters* (B†). *I. enneaspermum* var. *hirtum* (Klotzsch) Oliv. in F.T.A. 1: 106 (1868), (sphalm. *hirta*).

Hybanthus enneaspermus var. *serratus* Engl. in Bot. Jahrb. 55: 398 (1919); Burtt Davy, Fl. Transv. 1: 130 (1926). Syntypes: Transvaal, Pienaars River, *Schlechter* 4218 (B†; GRA; K; PRE); Mozambique, Ungulubi, *Schlechter* 12140 (B†; BM; K; PRE). —var. *caffer* (Sond.) Robson in Bol. Soc. Brot. Sér. 2, 32: 169 (1958); in F.Z. 1: 257 (1960). *H. caffer* (Sond.) Engl., l.c. 400 (1919). —var. *angustifolius* Engl., l.c. 400 (1919); Hulme, Wild Flow. Natal t.1, fig. 12 (1954). Type: No specimens cited. *H. hirtus* (Klotzsch) Engl., l.c. 399 (1919). —var. *glabrescens* Engl., l.c. 399 (1919); Burtt Davy, Fl. Transv. 1: 131 (1926). Syntypes: numerous collections from tropical Africa and incl. *Schlechter* 11729 (B†; GRA) from Komatipoort. *H. thymifolius* (Presl) Engl., l.c. 400 (1919).

Perennial (flowering in first year) with the stems often deciduous, trailing or erect, clustered from a woody crown, 20–50 cm high, with few side branches. *Stems* terete or ridged, glabrous, scabrid or pubescent. *Leaves* alternate, very variable, linear to ovate, 1–9 cm long, 1–2,5 cm wide, apex acute to obtuse, base attenuate into a petiole or petiole 0, soft, glabrous, puberulous to sericeous, margin serrate to more or less entire, the teeth gland-tipped (if teeth repressed, glands still persist); stipules small, subulate, 0,5–2 mm, gland-tipped. *Flowers* on long to short, thin pedicels; size of flower variable. *Sepals* free, subequal, linear-acuminate, to ovate-acuminate, 3–6 mm long, membranous, keeled, somewhat pubescent. *Petals* pink to bluish or violet; anterior much exceeding other 2 pairs in length, 5–20 mm long with a narrow claw about as long as the quadrate, emarginate lamina, at the base with a broad saccate bulge protruding from between the lower sepals; lateral petals asymmetrical, triangular, curved upwards, about half as long as anterior petal; posterior petals smaller and narrower. *Stamens* short, the filaments (and sometimes connective) of the 2 anterior tomentose, producing short caudate hairy appendages which descend into the pouch of the anterior petal; anthers connivent, connective produced into an ovate petaloid apical appendage. *Ovary* ovoid, style terete, stigma capitate. *Capsule* globose to ovoid, about 6–10 mm long, smooth with c. 3–4 seeds ripening on each placenta; seed fusiform to ovoid, apex and base discoid, c. 1,5 mm long, longitudinally ribbed, glabrous shiny, yellow to white, testa hygroscopic. FIG. 13: 1.

Widely distributed in Africa, Asia, Australia; widespread in South West Africa, Transvaal and Natal, often invading disturbed areas such as road verges, ploughed lands, riverbanks, forest margins, etc.

S.W.A.—1918 (Grootfontein): Aukas-Kreyfontein, *Dinter 811;* near Grootfontein, *Schoenfelder S 385.*

TRANSVAAL.—2229 (Waterpoort): Waterpoort, *Bremekamp & Schweickerdt 249.* 2231 (Pafuri): Mazanji, 35 km N.E. of Punda Milia, *Codd & Dyer 4618.* 2329 (Pietersburg): near Mara, about 29 km from Louis Trichardt on Mara–Vivo road, *Meeuse 10194.* 2428 (Nylstroom): Mosdene, *Galpin M576.* 2430 (Pilgrims Rest): Shiluvane, *Junod sub TRV 10205.* 2431 (Acornhoek): Skukuza Camp, *Acocks 16743.* 2528 (Pretoria): Codrington, *Thode A392.* 2529 (Witbank): 3 km N.E. of Middelburg, *Meeuse 10263.*

NATAL.—2632 (Bela Vista): Ndumu, edge of Pongola flood plain, *Pooley 111.* 2832 (Mtubatuba): Hluhluwe Game Reserve, *Ward 2403.* 2930 (Pietermaritzburg): Pinetown district, Everton, *Ward 5242;* Isipingo, *Wood 12474.* 2931 (Stanger): near Durban, *Wood 1105.* 3029 (Kokstad): Port Shepstone, Umtamvuna Bridge, *Strey 7106.*

CAPE.—3029 (Kokstad): 9 km from Umtamvuna Pont on Bizana Road, *Acocks 10930.* 3129 (Port St. Johns): Lombazi River Flats, *Mogg 13563;* Ntsubane, *Strey 8562.* 3130 (Port Edward): 8 km inland from Umtamvuna River, *Codd 10700.*

Common name: Pink Lady's Slipper.

Apparently the species is extremely variable and adaptable. The varieties *caffer* and *serratus* described for southern Africa and found elsewhere are here regarded as ecotypes. The flower, fruit and seed characters are stable throughout the range of the species but the leaf shows great variation, being narrow or broad, small or large, glabrous or pubescent, entire or serrate.

2. **Hybanthus densifolius** *Engl.* in Bot. Jahrb. 55: 398 (1919); Melchior in Pflanzenfam. ed. 2, 21: 359 (1925); Robson in F.Z. 1: 258 (1960): Roessler in F.S.W.A. 87: 1 (1969). Type: S.W. Africa, Otjihua, *Dinter* 430 (B†; K; SAM!; GRA!).

Small, compact scabridulous annuals up to 15 cm tall, the side branches mostly ab-

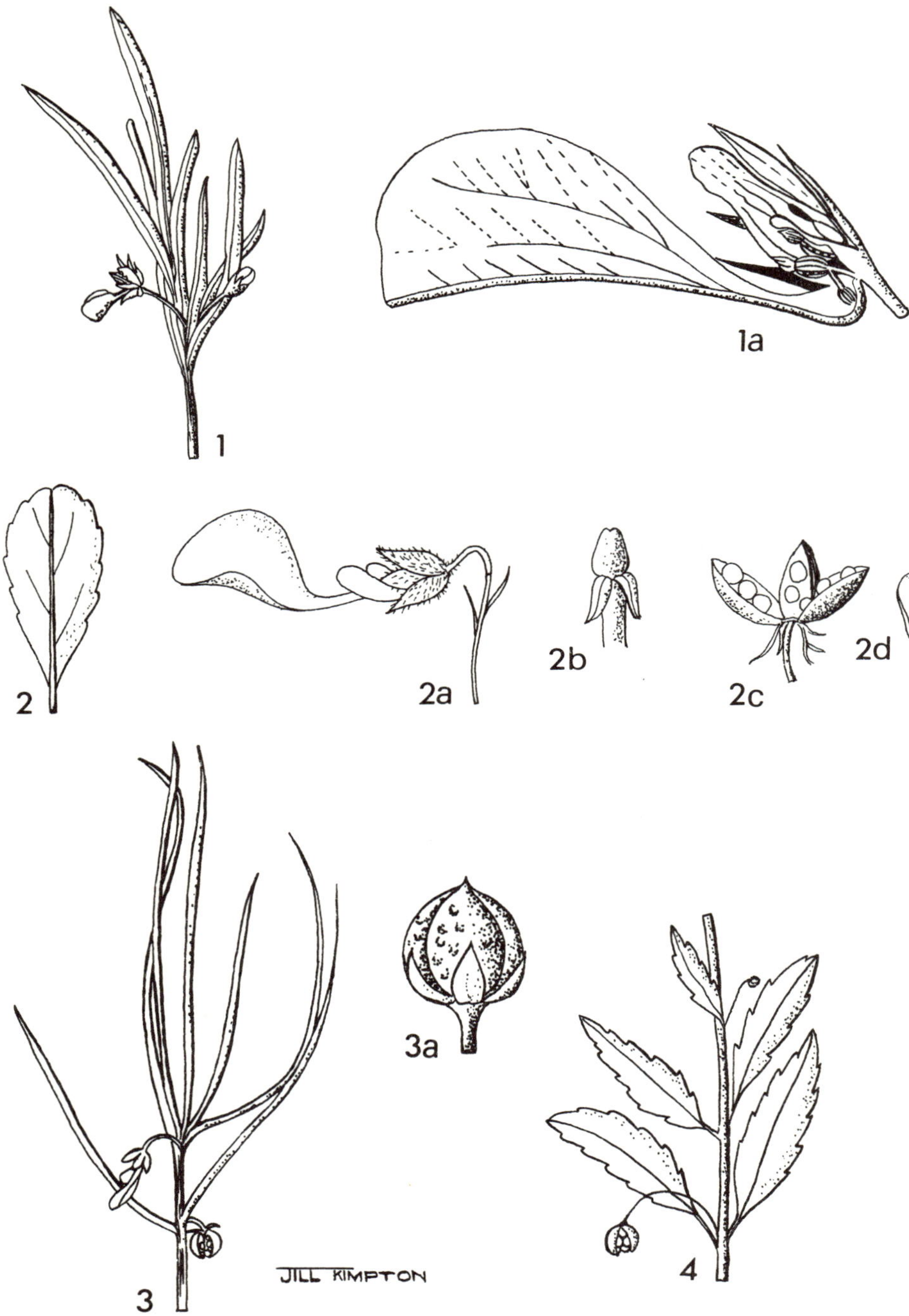

FIG. 13.—1, **Hybanthus enneaspermus,** × ½; 1a, longitudinal section of flower, × 3 *(Van der Schijff* 1852). 2, **H. capensis,** leaf, × 1; 2a, flower, × 4; 2b, stamen, × 3 *(Killick & Marais* 2013). 2c, capsule, × 1; 2d, seed, × 2 *(Flanagan* 2692). 3, **H. densifolius,** × 1; 3a, capsule, × 3 *(Dinter* 6870). 4, **H. parviflorus,** × 1 *(Hilliard* 3147).

breviated. *Stems* with short ridged internodes. *Leaves* crowded, alternate, narrowly linear, 3–8 cm long, 1–1,5 mm wide, apex acuminate, base attenuate into a thin petiole, margins inrolled with glands evenly spaced at intervals, forming shallow teeth. *Flowers* numerous, massed along the main stem in the leaf axils of the stem and abbreviated branches, on short pedicels, declinate, small. *Sepals* linear-lanceolate, 3 mm, ciliate. *Petals* whitish, pink or purplish; anterior petal forming a basal pouch, clawed and with a spathulate lamina c. 7 mm long. *Stamens* with the 2 anterior tomentose, bearing a recurved spur. *Ovary* globose, scabridulous, style short, stigma obtriangular, pointing downwards. *Capsule* globose, c. 7 mm in diam., scabridulous; seeds stoutly fusiform, apex discoid, striate, yellowish. FIG. 13: 3.

A small annual confined to the Kalahari region; South West Africa, Botswana, north-western Transvaal and northern Cape in sandy areas. A stable species, possibly self-fertile.

S.W.A.—1918 (Grootfontein): W. of Grootfontein, *Schoenfelder S316*. 2115 (Karibib): Karibib, Omaheke Insel, *Dinter 6973*. 2116 (Okahandja): Okahandja, *Bradfield 324*. 2217 (Windhoek): Lichtenstein, *Dinter 4354*. 2316 (Nauchas): Namib border towards Kuiseb, *Strey 2598*. 2317 (Rehoboth): Kalkrand on road to Rehoboth, *De Winter 3535*. 2417 (Mariental): 10 km N. of Mariental, *Acocks 18126*.

TRANSVAAL.—2229 (Waterpoort): Langjan Nature Reserve, *Zwanziger 742*.

CAPE—2624 (Vryburg): Armoedsvlakte, *Henrici 36*. 2822 (Glen Lyon): 6 km W. of Padkloof Pass, *Acocks 2062*.

3. **Hybanthus capensis** *(Thunb.) Engl.* in Bot. Jahrb. 55: 400 (1919); Robson in Bol. Soc. Brot. Sér. 2, 32: 165 (1958), in clavis; Batten & Bokelmann, Wild Flow. E. Cape Prov. 106, t.86: 6 (1966), as *H. capensis* (Roem. & Schult.) Engl. Type: Cape, Uitenhage district, Galgebosch, *Thunberg* (LINN 1052.20).

Viola capensis Thunb., Prodr. 40 (1794); Fl. Cap. ed. Schult. 186 (1923). Type as above.

Ionidium capense (Thunb.) Roem. & Schult., Syst. Veg. 5: 393 (1819); Sond. in F.C. 1: 74 (1868). Type as above. *I. natalense* Harv. in F.C. 2: 585 (1862). Type: Port Natal, *Sanderson* 415 (K, holo.; PRE!, fragment of holotype). *I. thorncroftii* N.E. Br. in Kew Bull. 1921: 289 (1921). Type: Transvaal, Barberton, Reimers Creek, *Thorncroft* 1086 (K, holo.!; PRE, photo.).

Hybanthus natalensis (Harv.) Burtt Davy, Fl. Transv. 1: 131 (1926); Robson in Bol. Soc. Brot. Sér. 2, 32: 165 (1958), in clavis. *H. thorncroftii* (N.E. Br.) Burtt Davy in Kew Bull. 1924: 225 (1924); Fl. Transv. 1: 131 (1926); Robson in Bol. Soc. Brot. Sér. 2, 32: 165 (1958), in clavis. *H. sp.* Hulme, Wild Flow. Natal t.41, f. 1 (1954).

Small hispid suffrutices from a perennial rootstock. *Stems* terete, ribbed when young. *Leaves* alternate, ovate to obovate (1,5) 3(–7) cm long, (0,5) 1(–3) cm wide, apex obtuse, base attenuate into a petiole, lower surface lighter coloured, margin shallowly crenate; stipules ovate-acuminate, membranous. *Flowers* declinate, on long thin pedicels, the sub-opposite bracts usually arising some distance below articulation. *Sepals* narrowly ovate-acuminate, fimbriate, the 2 lower slightly convex at the base. *Petals* white with lilac stripes, violet or mauve and a cream throat, the anterior 1,5 cm long, glabrous or tomentose near the base, without a pouch, claw short, lamina broadly spathulate; the other 4 somewhat longer than the sepals. *Stamens* short, with glabrous filaments, the connective-appendage oblong, obtuse to emarginate, orange. *Ovary* glabrous or pubescent, globose, style as long as ovary. *Capsule* pubescent or glabrous; seed obovoid with a small discoid apex and a distinct raphe and basal aril, smooth or faintly punctate. FIG. 13: 2.

Recorded from the Uitenhage district in the eastern Cape, northwards along the coast to Natal and the north-eastern Transvaal; in grassland (often ravaged by fires) or damp sheltered places, in valleys or margins of subtropical forests; rare.

The collections from the Uitenhage district possess pubescent capsules. These plants appear to be depauperate, never exceeding 10 cm in height. The leaves are small and pubescent. It would seem to be an adaptation to more harsh surroundings. Several collections found in the more lush vegetation of the Transkei and further north are up to 40 cm tall and much bigger in all respects.

TRANSVAAL.—2430 (Pilgrims Rest): The Downs, *Rogers 21878*. 2531 (Komatipoort): Barberton, *Williams sub TRV 10389*.

NATAL.—2731 (Louwsburg): Hlomohlomo Road near Ngotshe, *Acocks 13051;* Vryheid district, Ngwibi Mt., *Hilliard & Burtt 5907*. 2831 (Nkandla): Eshowe, *Gerstner 2387*. 2930 (Pietermaritzburg): Worlds View, Pietermaritzburg, *Fischer 909;* Inanda, *Wood 421*. 2931 (Stanger): Avoca, *Schlechter 3006*. 3030 (Port Shepstone): Campbellton, Dumisa, *Rudatis 1744*.

CAPE.—3029 (Kokstad): Clydesdale, *Tyson 1062*. 3127 (Lady Frere): Engcobo Mt., *Flanagan 2692*. 3128 (Umtata): Hill slopes at Baziya, *Baur 290*. 3228 (Butterworth): near Bashee River mouth, *Acocks 12271;* Kentani, *Pegler 157;* The Haven, *Gordon Gray 1135*. 3325 (Port Elizabeth): Uitenhage, *Prior s.n.;* hill slopes at Algoa Bay, *Zeyher 1919*. 3326 (Grahamstown): Howiesons Poort, *Dyer 1725*. 3327 (Peddie): East London, *Bokelmann 3;* Cambridge, *Wormald 41*.

4. **Hybanthus parviflorus** *(L.f.) Baill.*, Bot. Med. 2: 481 (1884); Hilliard in Notes R. Bot. Gard. Edinb. 30: 128 (1970). Type: Central America, *Mutis* (LINN 1052.25).

Viola parviflora L.f., Suppl. 396 (1782). Type as above.

Perennial with prostrate woody branches. *Stems* with vertical lines of pubescence alternating at the nodes. *Leaves* opposite or alternate, narrowly elliptic, 1–3 cm long, 5–9 mm broad, apex acute, base attenuate into a short to long petiole, margin serrate; stipules subulate. *Flowers* axillary, declinate, very small, c. 2–3 mm long, on thin pedicels up to 1 cm long in fruit. *Sepals* 2 mm long, curved upwards, pubescent. *Petals* white, short, the anterior 3 mm, with a broadly obovate, emarginate lamina. *Stamens* sessile, with an ovate, dark apical appendage. *Ovary* globose, with about 9 ovules; style short, oblique, stigma obtriangular. *Capsule* ovoid, 3 mm in diam.; seed globose-ovoid, c. 2 mm in diam., black when mature, smooth, shiny, with a white spongy aril, hygroscopic. FIG. 13: 4.

A native of South and Central America, apparently recently introduced. Recorded from several localities in Natal, usually in disturbed places.

NATAL.—2829 (Harrismith): Oliviershoek Pass, Highland Sourveld, *Acocks 23823*. 2930 (Pietermaritzburg): Pietermaritzburg, *Hilliard 3147; Barker 5128;* Isipingo Flats, *Ward 6634*.

5274 3. VIOLA

Viola *L.*, Sp. Pl. 2: 933 (1753); Gen. Pl. ed. 5: 402 (1754): Sond. in F.C. 1: 73 (1860); Becker in Pflanzenfam. ed. 2, 21: 363 (1925); Burtt Davy, Fl. Transv. 1: 130 (1926); Phill., Gen. ed. 2: 510 (1951); Robson in F.Z. 1: 258 (1960); Hutch., Gen. Fl. Pl. 2: 334 (1967). Type species: *Viola odorata* L.

Herbs, rarely shrublets, erect or trailing and rooting at the nodes. *Leaves* alternate, simple, petiolate, entire or usually serrate; stipules persistent, minute to foliaceous. *Flowers* usually single, axillary, declinate (cleistogamous flowers absent in African species); pedicels long, thin, inarticulate, bibracteate. *Sepals* 5, subequal, produced at the base below the point of insertion. *Petals* 5, unequal, obovate, the anterior spurred or saccate. *Stamens* with very short free filaments, anthers usually coherent in a ring around ovary, the 2 lower each bearing a recurved caudate nectariferous appendage protruded into the spur of the anterior petal; the anther connectives produced apically into a petaloid appendage. *Ovary* unilocular with 3 parietal placentas, multi-ovulate; style variously shaped, stigma terminal. *Capsule* 3-valved; seeds globose, pointed above and below, with a spongy aril.

A very large cosmopolitan genus with probably about 300 species (*Hutchinson*, 1967), mainly north temperate; *V. odorata* L. (Violet) and *V. tricolor* L. (Pansy or Heartsease) are popular garden plants and forms of the latter have been recorded as adventives in Southern Africa. Two closely related species occur in tropical Africa, one of these extending south to the Transvaal (Woodbush); one species with two varieties is endemic in the Cape Winter-rainfall Region.

Herbaceous with creeping stems rooting at the nodes; leaves cordiform, soft, crenate; northern Transvaal .... 1. *V. abyssinica*
Suffruticose with woody erect stems; leaves narrowly linear, somewhat succulent, hard; southern Cape .... 2. *V. decumbens*

1. **Viola abyssinica** *Steud. ex Oliv.* in F.T.A. 1: 105 (1868); Becker in Pflanzenfam. ed. 2, 21: 364, t.159, fig. 34 (1925); Burtt Davy in Fl. Transv. 1: 130 (1926); Keay in F.W.T.A. ed. 2, 1: 107, fig. 30 (1954); Robson in F.Z. 1: 258, t.41 (1960). Type: Ethiopia, *Schimper* (K; PRE, photo.).

Creeping or sometimes scrambling herbs from a perennial base, the long stems mostly unbranched, rooting at the nodes. *Stems* thin, narrowly winged. *Leaves* alternate, cordate, 1–2 cm long and broad, crenate, hispid, lower side usually with scattered resinous bodies; petiole slender, up to 2 cm long, stipules foliaceous, laciniate, setulose. *Flowers* axillary, solitary, declinate, persistent; pedicels thin, up to 5 cm long, bibracteate, inarticulate. *Sepals* more or less equal, narrowly ovate-acuminate, c. 5 mm long, with rudimentary basal bulges produced below the point of insertion. *Petals* with the 2 upper pairs whitish, obovate, 5–10 mm long; the anterior shorter, pale to dark blue or violet, veined, with a cylindrical large blunt spur adjoining the curved pedicel-apex. *Stamens* connivent, the apical connective appendage ovate, orange. *Ovary* globose, glabrous;

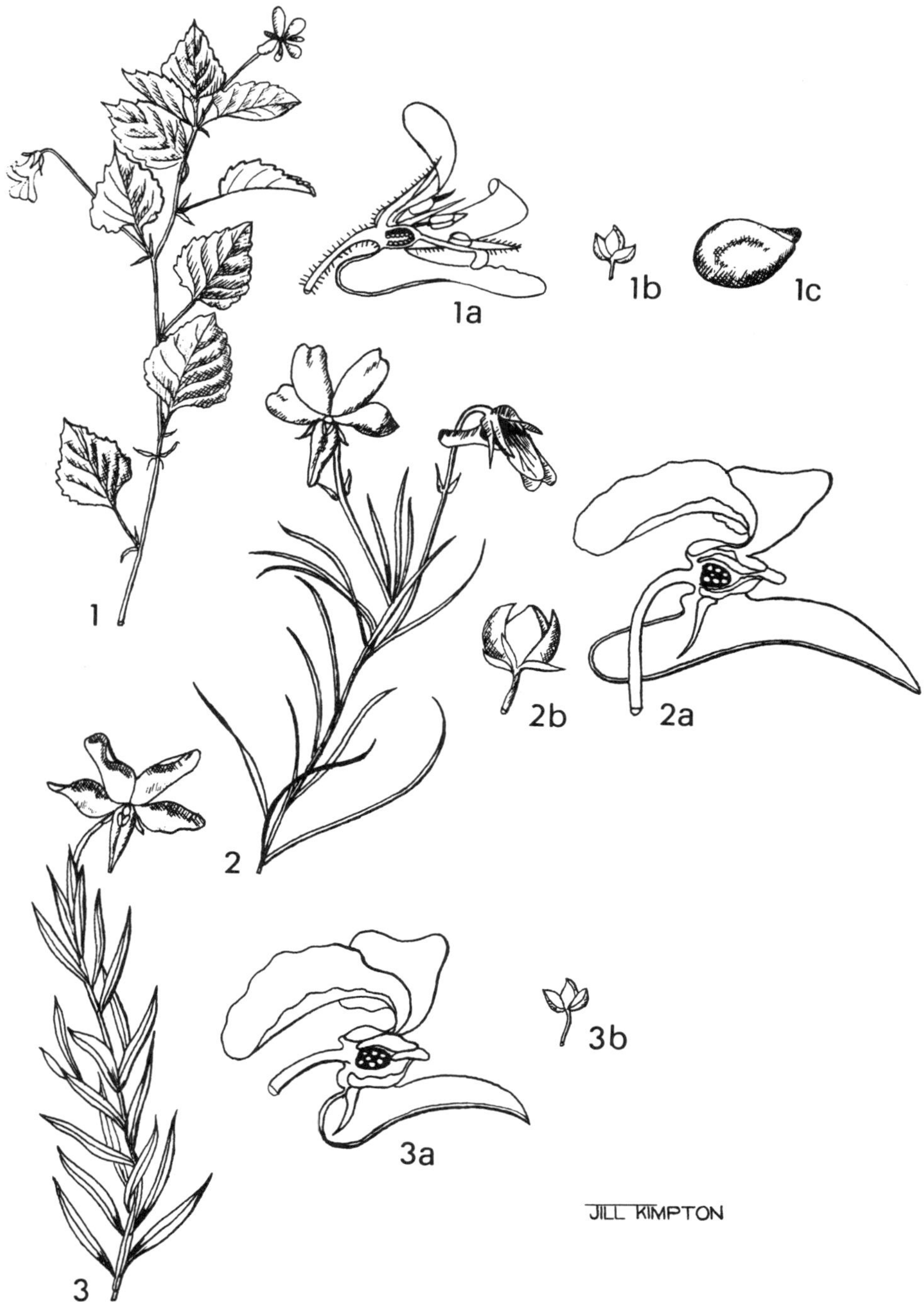

FIG. 14.—1, **Viola abyssinica,** × 1 (*Bos* 1147); 1a, longitudinal section of flower, × 3; 1b, capsule, × 1; 1c, seed, × 7. 2, **V. decumbens** var. **decumbens,** × 1 (*Acocks* 21093); 2a, longitudinal section of flower, × 2; 2b, capsule, × 1. 3, **V. decumbens** var. **scrotiformis,** × 1 (*Thompson* 192); 3a, longitudinal section of flower, × 2; 3b, capsule, × 1.

style narrowly funnel-shaped, oblique. *Capsule* ovoid, c. 5 mm long, glabrous, pale; seed ovoid, 1–2 mm, several to many, with a distinct spongy aril. FIG. 14: 1.

Widespread in tropical Africa. Recorded from Soutpansberg and Woodbush forests, locally common in clearings and along paths.

TRANSVAAL.—2230 (Messina): Soutpansberg, Entabeni Forest Reserve, *Obermeyer sub TRV 30317*. 2329 (Pietersburg): Woodbush Forest Reserve, *Hilliard 4753;* Houtbosberg, *Schlechter 4395*. 2330 (Tzaneen): Duiwelskloof, Westfalia Estate, *Bos 1147; Scheepers 843*. 2430 (Pilgrims Rest): The Downs, *Junod 4260*.

2. **Viola decumbens** *L.f.,* Suppl. 397 (1782). Type: Locality and collector unknown (LINN 1052.19).

Small perennial shrublet up to about 25 cm high with erect or spreading-erect, woody, minutely papillate branched stems. *Leaves* alternate, linear, 15–25 mm long, 0,5–2 mm broad, apex acute, somewhat succulent, minutely papillate; stipules adnate to the sessile base of lamina, large and resembling the leaf or reduced, with a small tooth on either side near the base. *Flowers* axillary, solitary, faintly scented, overtopping the leaves, on thin, usually long pedicels: bracts 2, subopposite in upper half or towards the apex, swollen below with 2 minute lateral teeth. *Sepals* narrowly ovate-acuminate, 3–5 mm long, produced below point of insertion. *Petals* purple to violet, veined, the upper 2 pairs subsimilar, narrowly oblong, 5–10 mm long; anterior shorter with a basal tubular blunt spur 2–5 mm long. *Stamens* with connivent sessile anthers, the apical connective appendage ovate, firm, orange; the 2 anterior stamens each with a long caudate nectariferous appendage projecting into the spur. *Ovary* globose with numerous ovules, style somewhat sigmoid. *Capsule* ovoid, 5–9 mm long; seed ovoid, 3 mm, yellow, minutely punctate.

Restricted to the southern Cape, from Worcester to the Cape Peninsula and eastwards as far as Riviersonderend Mountains; a montane species in fynbos, flowering July–December.

The type of *V. decumbens* L.f. in the Linnaean Herbarium gives no information on locality or collector but it may be a duplicate of a Thunberg specimen which was collected in the Hottentots Holland Mountains.

Two closely related varieties with a similar distribution are recognized.

Leaves up to 50 mm long and 1 mm broad; flowers usually congested above; sepals 5–9 mm long; spur tubular, 3–5,5 mm long . . . .(a) var. *decumbens*

Leaves about 20 mm long, 2–3 mm broad; flowers few, scattered along the branches; sepals 3,5–5 mm long; spur saccate, 3 mm long and broad . . . .(b) var. *scrotiformis*

(a) var. **decumbens.**

V. decumbens L.f., Suppl. 397 (1782); Thunb., Prodr. 40 (1794); DC., Prodr. 1: 299 (1824); Harv., Thes. Cap. 1: 30, t.46 (1859); Sond. in F.C. 1: 73 (1860); Marloth, Fl. S. Afr. 2: 189, t.64 d, fig. 126, B5–6 (1925); Becker in Pflanzenfam. ed. 2, 21: 373 (1925); Adamson in Fl. Cape Penins. 590 (1950); Jessop in Flow. Pl. Afr. 38, t.1500 (1967). Type: Without locality or collector (LINN 1052.19). —var. *tenuis* Bartl. in Linnaea 7: 540 (1832). Syntypes: Cape, Klein River Mountain, *Ecklon & Zeyher* (SAM); *Zeyher* 1923 (PRE; S; TCD).

Leaves linear, 3–5 cm long, 0,5–1 mm broad. *Flowers* aggregated at the top of the branchlets, exserted beyond the leaves. *Sepals* 5–9 mm long. *Spur* tubular, blunt, up to 5,5 mm long. *Capsule* 9 mm long. FIG. 14: 2.

CAPE.—3318 (Cape Town): Table Mountain, *Schlechter 12*. 3418 (Simonstown): Steenbras, *Rogers 11025,* Steenbras Siding, *Esterhuysen 2692;* Hottentots Holland Mountains, *Zeyher 1923;* Betty's Bay, *White 5171;* Pringle Bay, *Compton 22721*. 3419 (Caledon): Palmiet River Mouth, *Acocks 21093;* Hermanus, Fern Kloof, *Anderson 8;* Paardeberg, Paviesvlei, *Taylor 3603;* Hangklip, *Compton 13502;* Awila, E. of Witbaken, *Thompson 1198*.

(b) var. **scrotiformis** *(DC.) Jessop* in Flow. Pl. Afr. 38, t.1500 (1967).

V. scrotiformis DC., Prodr. 1: 299 (1824); Sond. in F.C. 1: 73 (1860); Becker in Pflanzenfam. ed. 2, 21: 373 (1925); Marloth, Fl. S. Afr. 2: 189, t.64 c, fig. 126 A 1-4 (1925). Type: Cape Peninsula, *Masson* (BM, holo.; PRE, photo.). *V. decumbens* var. *stipulacea* Bartl. in Linnaea 7: 540 (1832). Syntypes: Caledon, Swarteberg, kloof near the warm baths, *Ecklon & Zeyher* (SAM); *Zeyher* 1922 (PRE; S; TCD).

Leaves shorter and broader than in the typical form, 2 cm long, 2 mm broad; flowers fewer, produced along the branches. *Sepals* 3–5 mm long. *Spur* saccate, 3 mm long and broad. *Capsule* 5 mm long. FIG. 14: 3.

CAPE.—3318 (Cape Town): Sneeukop near Wellington, *Marloth 660;* Stellenbosch, Dwarsberg, Jonkershoek, *Thompson 191*. 3319 (Worcester): Bainskloof, *Schlechter 1589; 9192; Compton 17492;* Great Drakenstein Mountains, E. slopes of Bullers Kop, *Esterhuysen 11873;* Robertson district, Boesmans Kloof Pass at McGregor, *Esterhuysen 5241*. 3418 (Simonstown): Worcester district, Baviaans Kloof, *Leighton 1355;* Sir Lowry's Pass, *Compton 9004;* Kogelberg Forest Reserve, near Rooiels River, *Kruger 788*. 3419 (Caledon): Elgin, *Stokoe sub SAM 63573;* Zwarteberg among stones, *Ecklon & Zeyher 1922; Galpin 3748; Schlechter 5537;* Dasberg near Stormsvlei, Riviersonderend Mountains, *Stokoe sub SAM 63577*.

FLACOURTIACEAE
(including Samydaceae)

by D. J. B. KILLICK*

Trees or shrubs, armed and unarmed. *Leaves* alternate, rarely opposite or whorled, simple, entire or toothed, teeth often glandular. *Stipules* present or 0. *Inflorescence* axillary or terminal, racemose, corymbose or paniculate, or flowers fasciculate or solitary. *Flowers* actinomorphic, bisexual or unisexual, monoecious or dioecious, occasionally polygamous. *Sepals* 3–6 or more, free or connate below into a calyx-tube, contorted or imbricate, rarely valvate, often persistent or accrescent. *Petals* as many as sepals or more or 0, usually small and inconspicuous, free, contorted or imbricate, with or without a scale on inner face. *Disc* extrastaminal or with free glands between stamens or of staminode-like scales inserted at base of petals, often adnate to the receptacle. *Stamens* 5–∞, free or rarely connate, sometimes alternating with staminodes; anthers 2-thecous, dehiscing by longitudinal slits, rarely by apical pores. *Ovary* superior or more rarely semi-inferior, 1-locular with 1 to many ovules on 2–8 parietal placentas, rarely ovules axile; styles 1–10, free or connate. *Fruit* a capsule or berry, 1 to many-seeded. *Seeds* usually small, sometimes arillate; testa woody, leathery or chartaceous; embryo straight; cotyledons usually broad; endosperm fleshy, copious, rarely thin or 0.

Genera 89 with about 1250 species, mainly tropical and subtropical.

Leaves opposite . . . . 6. **Pseudoscolopia**
Leaves alternate:
  Petals with a fleshy gland on the inner face or staminodes petaloid with fleshy glands:
    Stamens numerous; petals spirally arranged . . . . 1. **Rawsonia**
    Stamens 10–12; petals whorled . . . . 4. **Kiggelaria**
  Petals or petaloid staminodes without fleshy glands; petals sometimes absent:
    Petals present:
      Petals more than sepals:
        Branches spiny; fruit globose, indehiscent . . . . 2. **Oncoba**
        Branches without spines; fruit ovoid, beaked, dehiscing into 4–8 valves . . . . 3. **Xylotheca**
      Petals as many as sepals:
        Stamens numerous, not collected in bundles opposite petals; fruit a berry . . . . 5. **Scolopia**
        Stamens equal in number to petals or in bundles opposite petals; fruit usually a capsule:
          Leaves with 5–9 nerves from base; flowers dioecious; seeds with a red aril . . . . 9. **Trimeria**
          Leaves penninerved; flowers bisexual; seeds without an aril:
            Style simple with a capitate or minutely bilobed stigma; flowers in pedunculate cymes; stipules deltoid . . . . 7. **Gerrardina**
            Styles 2–6 or 2–6-cleft; flowers in racemes or panicles; stipules absent or large and orbicular or reniform . . . . 8. **Homalium**

*Except for *Dovyalis* by J. E. Langenegger.

Petals absent:
Leaves with linear or circular pellucid glands ....13. **Casearia**
Leaves without pellucid glands:
Stamens in bundles of 3–10 alternating with disc glands ....9. **Trimeria**
Stamens not in bundles:
Flowers bisexual; style very short; stigma peltate; leaves narrowly elliptic to oblanceolate ....10. **Aphloia**
Flowers unisexual or more rarely bisexual; styles 2–8; leaves elliptic to obovate or orbicular:
Stamens surrounded by a ring of disc glands; ovary incompletely 4–8-locular with 2 ovules per loculus one above the other ....11. **Flacourtia**
Stamens intermingled with disc glands and alternating with them; ovules 1–6 per placenta ....12. **Dovyalis**

5275 1. **RAWSONIA**

Rawsonia *Harv. & Sond.* in F.C. 1: 67 (1860); Harv., Thes. Cap. 1: 20, t.31 (1860); Sim, For. Fl. P. E. Afr. 12 (1909); Engl., Pflanzenw. Afr. 3,2: 559 (1921); Gilg in Pflanzenfam. ed. 2, 21: 394 (1925); Phill., Gen. ed. 2: 512 (1951); Wild in F.Z. 1: 262 (1960); Bamps in F.C.B. Flacourt. 4 (1968). Type species: *R. lucida* Harv. & Sond.

Evergreen shrubs or small trees. *Leaves* alternate, simple, shortly petioled, apparently stipulate (scar present); blade usually elliptic, coriaceous, margins spinulose-serrate. *Flowers* in axillary spike-like racemes, bisexual or male. *Sepals* 4–5, free, unequal, imbricate, concave. *Petals* similar to sepals but larger, with a petaloid scale opposite each petal with a gland on the inner face at the base. *Stamens* numerous, in several rows, the inner hypogynous, the outer attached to the base of the scales. *Ovary* superior, 1-locular, with a thick fleshy wall; ovules many on 4–5 parietal placentas; style almost none; stigmas 4–5. *Fruit* a globose berry. *Seeds* few, subglobose; testa leathery; embryo straight; cotyledons thin, elliptic; endosperm fleshy, copious.

A genus of probably not more than 2 species occurring in southern and tropical Africa. The genus commemorates the Hon. Rawson W. Rawson, Secretary to the Cape Government and ''ardent promoter and efficient patron of botany in South Africa''.

Rawsonia lucida *Harv. & Sond.* in F.C. 1: 67 (1860); Harv., Thes. Cap. 1: 20, t.31 (1860); Sim, For. Fl. Cape Col. 128 (1907); Gilg in Bot. Jahrb. 40: 449 (1908); Sim, For. Fl. P.E. Afr. 12 (1909); Bak.f. in J. Linn. Soc. (Bot.) 40: 23 (1911); Eyles in Trans. R. Soc. S. Afr. 5: 422 (1916); Engl., Pflanzenw. Afr. 3,2: 559 (1921); Gilg in Pflanzenfam. ed. 2,21: 394 (1925); Burtt Davy, Fl. Transv. 1: 215 (1926); Henkel, Woody Pl. Natal 83 (1934); Wild in F.Z. 1: 262 (1960); Bamps in F.C.B. Flacourt. 4 (1968). Type: Port Natal, *Sanderson* 118 (TCD, holo!; K!; PRE, photos).

Tree up to 15 m high. *Branches* glabrous. *Leaf-blade* elliptic, occasionally obovate, 10–15 cm long, 3–5,5 cm wide, apex acuminate, rarely acute, base narrowly cuneate to rounded, margins spinulose-serrate with incurved teeth, midrib prominent on both sides, lateral veins 6–9, more prominent below, curved, looping near margin, rigidly coriaceous, shiny, dark green above, paler below, glabrous; petiole 6–9 mm long, channelled, incurved. *Flowers* few; peduncles 1,2–2 cm long; pedicels 1–2 mm long. *Sepals* broadly ovate-triangular, 2–5 mm diam., concave, leathery, ciliolate along margin. *Petals* 3–5, unequal, broadly ovate-rotund, 3,5–8 mm long, 4–7 mm wide, concave, ciliolate along margin, ''deciduous''; petaloid scales elliptic-subrotund, 7 mm long, 4–5 mm wide, with somewhat fleshy tumid pubescent gland on inner face at base. *Stamens:* filaments 3–5 mm long; anthers narrowly ovate, slightly divaricate at base, 3,5 mm long. *Ovary* ellipsoid, 4–8 mm

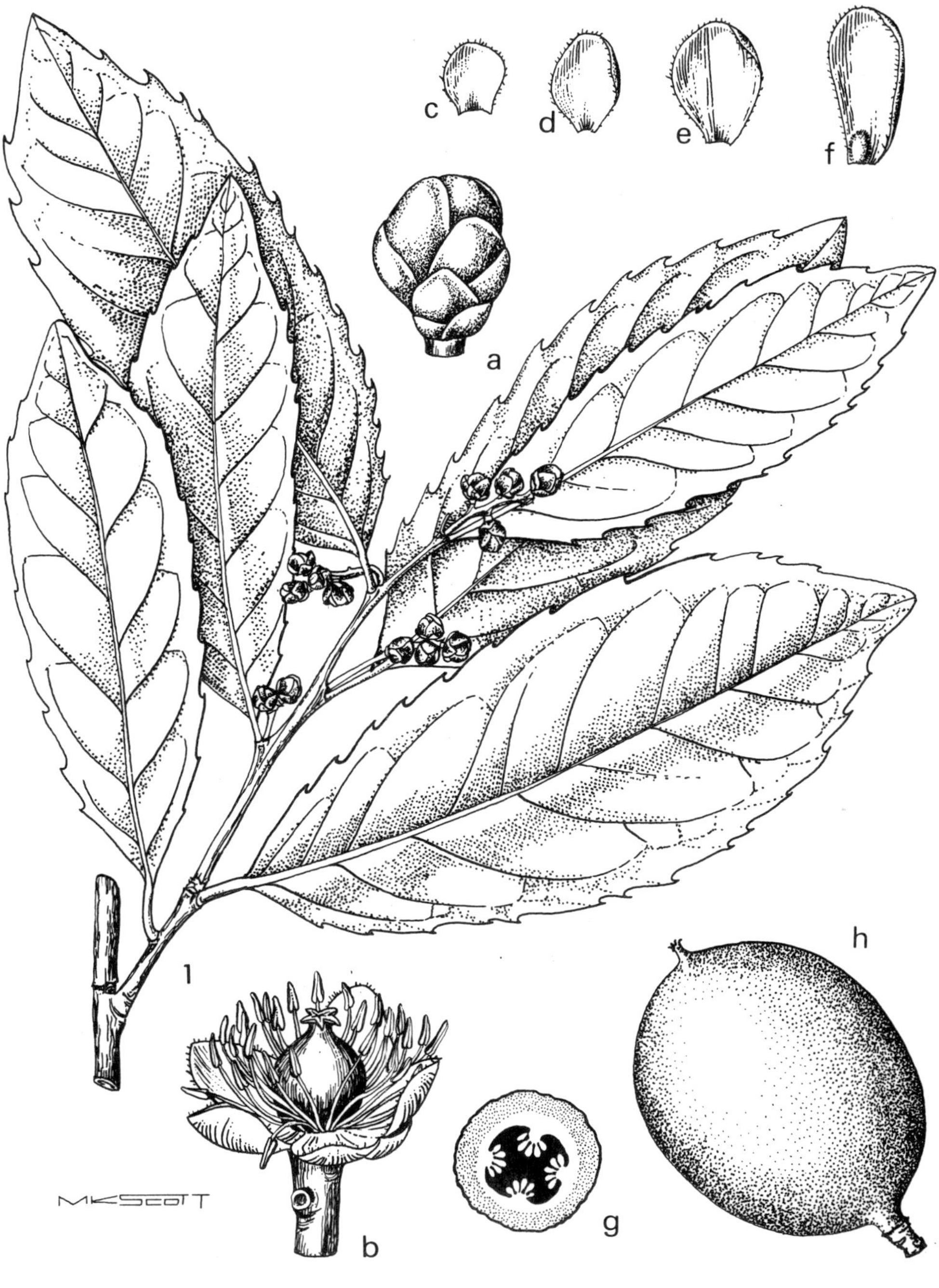

FIG. 15.—1, **Rawsonia lucida,** twig with flower buds, × 1 (*Killick & Strey* 2466); a, bud, × 6 (*PRF* 3727); b, flower, × 6 (*PRF* 3727); c, d, outer and inner sepals, × 6 (*PRF* 3727); e, petal, × 6 (*PRF* 3727); f, petaloid scale, × 6 (*PRF* 3727); g, transverse section through ovary, × 6 (*PRF* 3727); h, fruit, × 1 (*Story* 5389).

long, 4–5 mm wide, ridged; style almost none; stigmas 4–5, 1–1,5 mm long, spreading. *Fruit* yellow or green, 2,5–4 cm diam., with persistent style, "tardily dehiscent into 4–5 longitudinal sections when dry" (Wild, l.c.). *Seeds* few, subglobose, c.1 cm diam. FIG. 15.

A scrub-forest and forest tree occurring in the coastal areas of the eastern Cape and Natal, on the slopes of the Transvaal Drakensberg and in the northern Transvaal, Swaziland and tropical Africa.

TRANSVAAL.—2230 (Messina): Makonde Mission, 9 km N.E. of Sibasa, *Codd 6819*. 2330 (Tzaneen): Woodbush, *Hutchinson 2237*. 2430 (Pilgrims Rest): Blyde River, Mariepskop Forest Station, *Killick & Strey 2466;* The Downs, *Marais 131*.

SWAZILAND.—2531 (Komatipoort): Kings Forest, Piggs Peak, *Dlamini s.n.* 2631 (Mbabane): Ubombo Mountains, 2,4 km S. of Stegi, *Keith s.n.*

NATAL.—2732 (Ubombo): Gwalaweni Forest, *Moll 4452*. 2831 (Nkandla): Eshowe Forest, *Codd 1857*. 2832 (Mtubatuba): Hluhluwe Game Reserve, *Young 128*. 2930 (Pietermaritzburg): Krantzkloof Nature Reserve, *Moll 4033;* Table Mountin, *Killick 170; 687*. 2931 (Stanger): 6 km from Tongaat/Wartburg, *Moll 919*. 3030 (Port Shepstone): Elliott's Farm, Paddock, *Strey 7164*.

CAPE.—3129 (Port St. Johns): on way to Eagle's Nest, Port St. Johns, *Marais 1221;* 1,8 km N of Umtata Mouth, *Acocks 13573*. 3228 (Butterworth): Kentani, *Pegler 805* (SAM).

With its rigidly coriaceous leaves with spinulose-serrate margins and incurved petioles, *R. lucida* is a clear-cut and easily recognizable species with several synonyms in tropical Africa.

5284 2. ONCOBA

Oncoba *Forsk.,* Fl. Aegypt.-Arab. CXIII, 103 (1775); Gilg in Bot. Jahrb. 40: 455 (1908); Pflanzenfam. ed. 2, 21: 402, t.172 (1925); Phill., Gen. ed. 2: 512 (1951), pro parte; Wild in F.Z. 1, 1: 275 (1960); Bamps in F.C.B. Flacourt. 16 (1968); Wild & Vidigal in Fl. Mocamb. Flacourt. 15 (1973). Type species: *O. spinosa* Forsk.

Lundia Schumach. & Thonn., Beskr. Guin. Pl. 231 (1827).

Trees or shrubs, unarmed or with axillary spines; branches glabrous. *Leaves* alternate, simple, petiolate, stipulate; blade elliptic or ovate, margin serrate. *Flowers* bisexual, terminal or axillary, often on short sideshoots, usually large and showy. *Sepals* 3 or 4, free or united at base, imbricate. *Petals* 5–10 or more, exceeding the sepals, whorled. *Stamens* numerous, free; anthers linear. *Ovary* superior, 1-locular with 2–10 multiovulate parietal placentas; style simple. *Fruit* large, globose, with woody pericarp, smooth, indehiscent, many-seeded. *Seeds* with a leathery testa; embryo with leafy cotyledons.

Forty species according to Hutchinson (1967), but possibly not more than 4 distributed in tropical Africa, southern Africa and Arabia. The generic name is derived from *Onkob*, the Arabic name for *O. spinosa*.

Oncoba spinosa *Forsk.*, Fl. Aegypt.-Arab. CXIII, 103 (1775); Oliv. in F.T.A. 1: 115 (1868); Gibbs in J. Linn. Soc. (Bot.) 37: 429 (1906); Sim, For. Fl. P.E. Afr. 12, t.2, fig. B (1909); Bak.f. in J. Linn. Soc. (Bot.) 40: 23 (1911); R.E. Fr., Schwed. Rhod.–Kongo–Exped. 1: 155 (1914); Eyles in Trans. R. Soc. S. Afr. 5: 422 (1916); Gilg in Pflanzenfam. ed. 2,21: 402, t. 172 (1925); Burtt Davy, Fl. Transv. 1: 214, t.30 (1926); Hutch. in F.W.T.A. 1: 161 (1927); Henkel, Woody Pl. Natal 79, 91 (1934); Exell & Mendonça in C.F.A. 1, 1: 81 (1937); Verdoorn in Flow. Pl. Afr. 28: t.1111 (1950–51); Keay in F.W.T.A. ed. 2, 1: 188, t.71 (1954); Wild in F.Z. 1, 1: 275 (1960); Bamps in F.C.B. Flacourt. 16 (1968); Wild & Vidigal in Fl. Mocamb. Flacourt. 16 (1973). Type: Yemen, Hadiè and Wadi Surdus, *Forskål* Herb. 626 & 627 (C, syn.).

Lundia monacantha Schumach. & Thonn., Beskr. Guin. Pl. 231 (1827). Type: Ghana, *Thonning* 296 (missing).
Oncoba monacantha (Schumach. & Thonn.) Steud., Nom. Bot. ed. 2, part 2: 212 (1841).

Shrub or small tree up to 5 m high. *Branches* glabrous, conspicuously lenticillate with slender, sharp spines up to 5 cm long. *Leaf-blade* elliptic or ovate, 7–11 cm long, 3,5–5 cm wide, apex acuminate, base cuneate to rounded, margin serrulate or crenate-serrate, midrib and lateral veins sub-prominent above, prominent below, lateral veins 4–8 pairs, curved, glabrous, membranous to sub-coriaceous, dark green and shiny above, duller below; petiole 5–10 mm

FIG. 16.—1, **Oncoba spinosa,** flowering twig, × 1; a, fruit, × 1 (after Plate 1111 in Flowering Plants of Africa).

long, channelled above. *Flowers* solitary and terminal or lateral on short axillary shoots, scented; pedicels 1,5–2,7 cm long. *Calyx* globose in bud; sepals 4, joined at base, imbricate, elliptic, 1,5–1,8 mm long, c. 8 mm wide, strongly concave, green dorsally and white ventrally. *Petals* white, 7–13, spreading, whorled, elliptic to obovate, the outer 3–4 mm long, 1,3–1,7 cm wide, slightly narrowed to a claw, the inner narrower. *Stamens:* filaments 4–8 mm long; anthers yellow, linear, c. 2 mm long with the connective produced into a triangular or minutely lobed tip or appendage. *Ovary* broadly obovoid to subglobose, 4,5 mm long, 3,5 mm wide, longitudinally sulcate; style c. 1 cm long; stigma patelliform, lobed at margin. *Fruit* yellow, globose, c. 5 cm diam., smooth, marked with longitudinal lines, calyx and androecium persistent at base and gynoecium at apex, rind woody, hard, pulp yellow. *Seeds* numerous, ovoid-flattened, 6 mm long, 3 mm wide, testa shiny, brown. FIG. 16.

A decorative shrub or small tree found on riverbanks, in woodland and scrub forest in South West Africa, the eastern Transvaal, Swaziland, Natal, tropical Africa and Arabia.

S.W.A.—1725 (Livingstone): Mpilila Island, eastern Caprivi, *Killick & Leistner 3397.*

TRANSVAAL.—2230 (Messina): Palmaryville, *Codd & Dyer 4486;* Sibasa, *Van der Schijff 5255.* 2330 (Tzaneen): Duiwelskloof, *Scheepers 1113.* 2430 (Pilgrims Rest): Calais, *Renny 107.* 2531 (Komatipoort): Barberton, *Thorncroft H11255;* Sabi River, E. of Skukuza, *Wolhuter 4606.*

SWAZILAND.—2631 (Mbabane): Lwandle, near Manzini, *Miller S/176;* Mbuluzi Poort, Stegi, *Compton 31579.*

NATAL.—2632 (Bela Vista): Ndumu Game Reserve, *Moll 4235 A.* 2831 (Nkandla): Eshowe, *Lawn 2009* (NH). 2930 (Pietermaritzburg): near Durban, *Goldie sub Herb. Austro-Africanum 1702* (K, Z).

Known as the Snuffbox Tree, Kafferklapper and um Thongwane (Z). The fruits are edible, but not palatable. The seeds yield a drying oil, but the mechanical separation of the seeds from the pulp makes this an uneconomical proposition. The fruits are used to make snuff boxes, rattles for children and arm or ankle bands for dancers. The roots and leaves are said to be used medicinally by some tribes. The tree flowers between October and November.

5284a 3. XYLOTHECA

Xylotheca *Hochst.* in Flora 26: 69 (1843); Gilg in Bot. Jahrb. 40: 455 (1908); Pflanzenfam. ed. 2, 21: 402 (1925); Wild in F.Z. 1,1: 272 (1960); Wild & Vidigal in Fl. Mocamb. Flacourt. 11 (1973). Type species: *X. kraussiana* Hochst.

Shrubs or trees, unarmed. *Leaves* alternate, simple, petiolate, stipulate; blade elliptic-obovate to broadly elliptic-obovate. *Flowers* solitary, male or bisexual, cymose or sub-umbellate in the upper leaf axils or terminal on the branchlets, large, sweet-scented. *Calyx* of usually 3 sepals, concave, free or almost free, imbricate, glabrous or pubescent. *Petals* 7–14, free or narrowed to base, imbricate. *Stamens* numerous; filaments free; anthers linear, dehiscing longitudinally from above. *Ovary* rudimentary in male flowers, 1-locular, multiovulate; ovules pendulous from c. 7 parietal placentas; style terminal; stigmas equalling number of placentas, short, spreading. *Fruit* a woody capsule, splitting into c. 8 longitudinal valves; style persists as hard woody point. *Seeds* numerous, sometimes with a resinous aril.

A tropical and southern African genus of about 10 species some of which are doubtfully distinct. The generic name refers to the woody pericarp of the fruit.

Xylotheca kraussiana *Hochst.* in Flora 26: 69 (1843); Gilg in Bot. Jahrb. 40: 455 (1908); Pflanzenfam. ed. 2, 21: 402 (1925); Wild in F.Z. 1, 1: 274 (1960); Wild & Vidigal in Fl. Mocamb. Flacourt. 12 (1973). Type: Natal, near Umlaas River, *Krauss* s.n. (352 in Herb. Kew, iso.!; PRE, photo.).

Oncoba kraussiana (Hochst.) Planch. in Hook., Lond. J. Bot. 6: 296 (1847); Harv. in F.C. 1: 66 (1860); Sim, For. Fl. P.E. Afr. 12 (1909). *O. tettensis* sensu Hook.f. ex Harv. in F.C. 2: 584 (1862) quoad specim. *Forbes* (K!), excl. specim. *Kirk* (K!); Schinz in Mem. Herb. Boiss. 10: 52 (1900). *O. macrophylla* sensu Schinz l.c.

FIG. 17.—1, **Xylotheca kraussiana,** flowering twig, × 1; a, pistil, × 3; b, transverse section of ovary, × 5; c, dehiscing fruit, × 1; d, arillate seed, × 1; e, fully dehisced fruit, × ½ (after Plate 1535 in Flowering Plants of Africa).

Xylotheca lasiopetala Gilg in Bot. Jahrb. 40: 457 (1908); Pflanzenfam. ed. 2, 21: 402 (1925). Syntypes: Mozambique, Delagoa Bay, *Monteiro* 12 (K!); *Junod* (BR!; Z); *Schlechter* 11578 (COI; K!; L!); *Quintas* 68 (COI). *X. kotzei* Phill. in Kew Bull. 1922: 193 (1922). Type: Natal, Port Durnford Plantation, *Kotze* sub PRE 1479 (PRE, holo.!; K; sub PRF 3632). *X. kraussiana* Hochst. var. *glabrifolia* Wild in Bol. Soc. Brot. Sér. 2, 32: 53 (1958); in F.Z. 1, 1: 274 (1960); Killick in Flow. Pl. Afr. 39, t.1535 (1968); Wild & Vidigal in Fl. Mocamb. Flacourt. 13 (1973). Type: Mozambique, Quissico, *Exell, Mendonça & Wild* 698 (BM, holo.!).

Tree or shrub up to 10 m tall; branchlets glabrous or pubescent. *Leaf-blade* elliptic to broadly elliptic or obovate, 4–12 cm long, 1,7–4 cm wide, apex acuminate, acute to rounded, base narrowly cuneate to rounded, margin entire, midrib prominent, veins 6–8 pairs, dividing near margin to form conspicuous submarginal vein, membranous to chartaceous, glabrous or pubescent; petiole 0,5–1 cm long, glabrous or pubescent. *Flowers* in 1–3-flowered cymes in axils of leaves or on the branches; peduncles 2–3 cm long; pedicels up to 4,5 cm long, glabrous or pubescent. *Sepals* 3, ovate to obovate, 1,5–1,8 cm long, 1 cm wide, deeply concave, apex rounded, green but margins white, pubescent outside. *Petals* white, 7–12, obovate, 2–3,5 cm long, 0,5–1,5 cm wide, narrowed at base, glabrous or sparsely woolly. *Stamens* numerous; filaments 6 mm long, with scattered hairs; anthers linear, 5 mm long. *Ovary* ellipsoid, sulcate, densely hairy; style 0,8–1,0 cm long, columnar; stigmas 4, spreading, 1 mm long. *Fruit* woody, ovoid or ellipsoid with 8 longitudinal ridges, beaked, splitting into 4–8 segments. *Seeds* numerous, reddish black, obovoid, 7 mm long, 5 mm wide, with fleshy red aril along one side. FIG. 17.

Usually a forest margin shrub or tree found along the coast of Natal from the Port Shepstone area northwards into Mozambique. Also occurring in the north-eastern Transvaal.

TRANSVAAL.—2231 (Pafuri): Nwambiya-sandveld, *Brynard & Pienaar 4254; Van der Schijff & Marais 3684;* Nyanda Bos, *Van Wyk 4577.*

NATAL.—2632 (Bela Vista): Kwamuzimba, near Ndumu Game Reserve, *Stephen 807.* 2732 (Ubombo): 16 km Pongola Bridge/Maputa, *Moll 4378.* 2831 (Nkandla): Eshowe, *Thode A1223.* 2832 (Mtubatuba): Enseleni River, *Bayer 1456.* 2930 (Pietermaritzburg): Shongweni Dam, *Morris 805.* 2931 (Stanger): Oqaqeni, Mapumulo, *Edwards 1791; 1708.* 3030 (Port Shepstone): The Valleys, *Mogg 13841;* Gibraltar, *Strey 10565.*

A small tree or shrub with elliptic to broadly elliptic or obovate leaves, 4–12 cm long, with entire margins and a submarginal vein and large, showy flowers. Wild (1.c.) described var. *glabrifolia* and distinguished it from var. *kraussiana* by its glabrous vegetative parts. Examination of a wide range of material shows that this difference does not hold and the distribution is practically the same, therefore var. *glabrifolia* is not upheld in this treatment. It is interesting to note that *X. kotzei,* which Wild sinks under *X. kraussiana* var. *kraussiana,* has a holotype with sparsely pubescent vegetative parts, while the paratypes, apparently from the same tree, are completely glabrous.

5296 4. KIGGELARIA

Kiggelaria *L.*, Sp. Pl. 2: 1037 (1753); Gen. Pl. ed 5: 459 (1754); Harv. in F.C. 1: 7 (1860); Phill., Gen. ed. 2: 513 (1951); Wild in F.Z. 1: 265 (1960). Type species: *K. africana* L.

Unarmed shrub or small tree. *Leaves* alternate, simple, petiolate, usually elliptic or ovoid, variously toothed, sometimes discolorous; stipules 0. *Plants* dioecious. *Male flowers* in axillary cymes, on longish pedicels. *Female flowers* solitary on long pedicels. *Sepals* 5-parted, segments narrowly elliptic, valvate, deciduous, puberulous. *Petals* 5, often very similar to sepals, elliptic, imbricate, each with a fleshy gland adnate at the base to the petals and free above, puberulous. *Stamens* 8–10, free; filaments short; anthers 2-thecous, opening by terminal pores. *Ovary* sessile, ellipsoid, 1–2–5 parietal placentas, densely pubescent; style short and thick, divided into 5 branches. *Fruit* a globose, woody, several-seeded capsule; valves 2–5, pubescent, sometimes tuberculate. *Seeds* subglobose; testa somewhat woody; embryo straight; cotyledons flat, suborbicular; endosperm fleshy.

1 species (doubtfully more) occurring in Southern Africa and tropical Africa. The genus was named in honour of Francis Kiggelaer, a Dutch botanist.

FIG. 18.—1, **Kiggelaria africana,** twig with male flowers, × 1, *(Thode* A2603); a, male flower, × 3 *(Thode* A2603); b, female flower, × 3 *(Fourcade* 1591); c, fruit, × 1 *(Mogg* 2319); d, dehiscing fruit, × 1 *(Mogg* 2319).

Kiggelaria africana *L.*, Sp. Pl. 2: 1037 (1753); Pappe, Silv. Cap. 4 (1854); Harv. in F.C. 1: 71 (1860); Sim, For. Fl. Cape Col. 128, t. iv (1907); Bak.f. in J. Linn. Soc. (Bot.) 40: 24 (1911); Eyles in Trans. R. Soc. S. Afr. 5: 422 (1916); Engl., Pflanzenw. Afr. 3,2: 571 (1921); Gilg in Pflanzenfam. ed. 2, 21: 413, t.179 (fig. f–h) (1925); Burtt Davy, Fl. Transv. 1: 217 (1926); Henkel, Woody Pl. Natal 32, 73 (1934); Brenan, Checklist Tang. Terr. 233 (1949); Milne-Redh. in Mem. N.Y. Bot. Gdn. 8,3: 219 (1953); Wild in F.Z. 1: 265 (1960). Type: "Aethiopia" (LINN 1191.3, holo.!)

K. ferruginea Eckl. & Zeyh., Enum. 15 (1834–1835); Burtt Davy in Kew Bull. 1921: 337 (1921). Type: Namaqualand, Kamiesberg and near mouth of Orange River, *Ecklon & Zeyher* 118 (K, iso.!; PRE, photo.; B!; L!; P!; SAM!). *K. integrifolia* sensu Eckl. & Zeyh. l.c., non Jacq. (1788). *K. dregeana* Turcz., Animadv. 33 (1855). Type: Cape, "Zuurberg en by Bontjiesrivier", *Drège 6722a* (K, iso!; BM!; L!; P!; PRE!; S!; TCD!).—var. *acuta* Harv. in F.C. 1: 71 (1860). Type: without locality, *Drège 6722a* (TCD, holo!; PRE, photo.).—var. *obtusa* Harv., l.c. Type: Cape, "sylvis in Sitzikamma (George), Olifantshoek (Uitenhage) et Kafferland", *Ecklon & Zeyher* 117 (TCD, holo.!; PRE, photo.; K!). *K. grandifolia* Warb. in Engl., Pflanzenw. Ost–Afr. C: 278 (1895); Engl., Pflanzenw. Afr. 3: 571 (1921). Gilg in Bot. Jahrb. 40: 468 (1902); Pflanzenfam. ed. 2,4: 413 (1925); Type: Malawi, *Buchanan* 1469 (B, holo.†; BM!; K!; PRE, photo.). *K. africana* L. var. *obtusa* (Harv.) Burtt Davy in Kew Bull. 1921: 336 (1921); Fl. Transv. 1: 217 (1926).

Shrub or tree up to 17 m high, semi-deciduous; bark pale grey, smooth; young branches stellately tomentulose, often rusty. *Leaf-blade* elliptic or ovoid, 4–11 cm long, 1,5–4 cm wide, apex acuminate, acute or obtuse, apiculate, base cuneate to round, margin entire, serrulate, serrate or duplicate-serrate, midrib prominent below, lateral veins 5–9, prominent below, looping near margin, chartaceous to coriaceous, glabrous or white or rusty stellate, with tomentulose hairy pockets (acarodomatia) in axils of veins below, sometimes discolorous; petiole 0,7–2,5 cm long, channelled, glabrous or tomentulose. *Male flowers* yellowish-green, in 3–many-flowered axillary cymes; peduncle 0,3–2 cm long, tomentulose; pedicels slender, 0,5–1,5 cm long, tomentulose, minutely bracteate at base. *Sepals* narrowly elliptic, 3–4 mm long, 1,5 mm wide, apex acute, slightly keeled, stellate-pubescent. *Petals* elliptic, 4,25 mm long, 2,25 mm wide, apex obtuse to rounded, slightly keeled dorsally, stellate, puberulous; basal scale oblong, 2 mm long, 1,5 mm wide, toothed, fleshy. *Stamens* 10; filaments up to 1,7 mm long; anthers elliptic, 1,5 mm long, stellately puberulous. *Female flowers* yellowish-green, solitary in upper axils; pedicels slender, 1,7–2,5 cm long, tomentulose. *Sepals* narrowly elliptic, 5,5 mm long, 1,5 mm wide, stellately puberulous. *Petals* elliptic 5,5–6 mm long, 2mm wide, stellately puberulous; scales ovoid 2–2,5 mm long, 1–1,25 mm wide. *Ovary* ellipsoid, 2,5–3 mm long, 1,5–2 mm diam., densely pubescent; style short, divided into (4) 5 branches, c. 1,5 mm long, divergent, canaliculate, glabrous in upper part, minutely lobed. *Fruit* up to 1,5 cm diam., splitting from apex into 5 valves, yellowish-green, densely tomentulose, tuberculate. *Seeds* bright orange-red, 4–6 mm diam. FIG. 18.

Occurring in all four provinces of the Republic in forest, usually fairly open forest, kloofs, along streams, in scrub and even in inhospitable places like the mountains of the Karoo. Also found in South West Africa, Rhodesia, Lesotho, Swaziland, Malawi, Mozambique and tropical east Africa.

S.W.A.—2317 (Rehoboth): Rehoboth, *Strey 540.*

TRANSVAAL.—2328 (Baltimore): Blauwberg, *Pole Evans 893.* 2329 (Pietersburg): near Haenertsburg, *Legat 133.* 2330 (Tzaneen): Modjadji's Reserve, *Krige 142.* 2429 (Zebediela): Schoonoord, *Van Warmelo 25.* 2430 (Pilgrims Rest): The Downs, *Rogers 21936.* 2527 (Rustenburg): Jacksonstuin, *Van Vuuren 262.* 2528 (Pretoria): Groenkloof, *Mogg 14190.* 2530 (Lydenburg): 6 km S.E. of Sewefontein, *Codd 8101.* 2531 (Komatipoort): Almeida, *Nel 252.* 2626 (Klerksdorp): Goedgedacht, Ventersdorp, *Sutton 698.* 2627 (Potchefstroom): Jack Scott Nature Reserve, *Wells 2444.* 2628 (Johannesburg): Suikerbosrand, *Prosser 1312.* 2629 (Bethal): Ermelo, *Leendertz sub TRV 7784.* 2630 (Carolina): 2 km E. of Forthill, *Schlieben & De Winter 7980.* 2730 (Vryheid): Oshoek, *Devenish 176.*

O.F.S.—2826 (Brandfort): Willem Pretorius Game Reserve, *Leistner 3017.* 2827 (Senekal): Doornkop, *Goossens 768;* Braamhoek, *Galpin 13938.* 2828 (Bethlehem): 2,4 km S.E. of Bethlehem, *Scheepers 1432.* 2829 (Harrismith): Platberg, *Puttrill s.n.* 2926 (Bloemfontein): ThabaNchu, *Roberts 1823A.* 3027 (Lady Grey): Zastron, *Heydorn 18.*

SWAZILAND.—2631 (Mbabane): Hlatikulu Forest, *Compton 28162;* Forbes Reef Bush, *Burtt Davy 2799* (BM).

NATAL.—2730 (Vryheid): Tweekloof and Altemooi, *Thode A180.* 2731 (Louwsburg): Ngome Forest, *Gerstner 4491.* 2829 (Harrismith): Cathedral Peak, *Killick 986.* 2830 (Dundee): Mfongosi, *Edwards 1290.* 2831 (Nkandla): Nkandla Forest, *Gerstner 3596* (NH). 2832 (Mtubatuba): Hluhluwe Game Reserve, *Ward 2064.* 2929 (Underberg): Bushmen's Nek, *Killick & Vahrmeijer 3977.* 2930 (Pietermaritzburg): Table Mountain, *Killick 426.* 3030 (Port Shepstone): Ifafa, *Rudatis 390.*

LESOTHO.—2828 (Bethlehem): Leribe, *Dieterlen 180.* 2927 (Maseru): near Cannibals Cave, Lekhatsi, *Jacot Guillarmod 843;* Mamathes, *Jacot Guillarmod 522.* 3027 (Lady Grey): Thaba Tsueu, *Page sub BOL 16790.*

CAPE.—3018 (Kamiesberg): Kamiesberg, *Taylor 5538;* "Kamiesberge and near mouth of Orange River", *Ecklon 118* (B, K, L, P). 3026 (Aliwal North): near Burghersdorp,

Flanagan 1550. 3125 (Steynsburg): Bo-Rietpoort, *Van der Walt s.n.* 3126 (Queenstown): mountain tops, Queenstown, *Galpin 1564*. 3127 (Lady Frere): Cala, *Ferreira 2*. 3128 (Umtata): Pot River Berg, Maclear, *Galpin 6577*. 3129 (Port St. Johns): Egossa Forest, *Strey 8864;* between Morley and Umtata River, *Drège 6722 b* (K). 3218 (Clanwilliam): on banks of Olifants River, *Galpin 11114*. 3219 (Wupperthal): Klein Koupoort Nek, *Taylor 7481*. 3222 (Beaufort West): Nieuwveld Mountains, *Marloth 8304*. 3225 (Somerset East): near Somerset East, *Bolus 326* (K). 3227 (Stutterheim): Komga, *Flanagan 303;* Happy Valley, *Comins 1766*. 3228 (Butterworth): Kentani, *Pegler 738*. 3318 (Cape Town): Table Mountain, *Flanagan 2464*. 3319 (Worcester): Klein Drakenstein Mountains near Salem, *Galpin 10594*. 3321 (Ladismith): between Calitzdorp and Cango Caves, *Hutchinson 1145*. 3322 (Oudtshoorn): 5 km E. of George, *Hutchinson 1261*. 3323 (Knysna): Groot River, *Fourcade 1459* (K). 3324 (Steytlerville): Klein River, Hankey, *Long 1350*. 3325 (Port Elizabeth): Enon, Uitenhage, *Thode A1091;* "Zuurbergen by Bontjiesrivier", *Drège 6722a*. 3326 (Grahamstown): Olifantshoek, Alexandria, *Johnson 1119*. 3418 (Simonstown): Somerset West, *Parker 3734*.

An extremely variable species as regards leaf shape, size, texture and degree of pubescence. In forest the species may have large, chartaceous leaves with little pubescence, while a few hundred metres away in a boulder group in the open, the species may have small, coriaceous leaves which are densely pubescent. *Kiggelaria flavo-velutina* Sleumer, described from East Africa, probably represents the extreme in pubescence of this species: it is doubtfully distinct from *K. africana*.

5304 5. SCOLOPIA

Scolopia *Schreb.* in L., Gen. Pl. ed. 8, 1: 335 (1789), nom. conserv.; Sim, For. Fl. Cape Col. 125 (1907); Gilg in Pflanzenfam. ed. 2, 21: 418 (1925); Phill., Gen. ed. 2: 513 (1951); Wild in F.Z. 1: 276 (1960); Bamps in F.C.B. Flacourt. 39 (1968); Sleumer in Blumea 20: 26 (1972); Wild & Vidigal, Fl. Mocamb. Flacourt. 17 (1973). Type species: *S. pusilla* (Gaertn.) Willd.

Phoberos Lour., Fl. Cochinch. 317 (1790).
Eriudaphus Nees in Eckl. & Zeyh., Enum. 271 (1835).
Adenogyrus Klotzsch in App. Ind. Sem. Hort. Berol. 1 (1854).
Rhamnicastrum Kuntze, Rev. Gen. 1: 45 (1891).

Armed or unarmed shrubs or trees. *Leaves* alternate, simple, petiolate or subsessile; blade elliptic, ovate or subrotund, entire or toothed; stipules minute, caducous. *Flowers* bi-sexual, in axillary racemes, spikes, fascicles or flowers solitary in axils. *Receptacles* usually funnel-shaped. *Sepals* 4–6. *Petals* 4–6, but not always present, alternating with sepals. *Disc* fleshy, with marginal glands. *Stamens* many, borne on surface of disc, exceeding petals in length at anthesis; anthers 2-thecous, often apiculate. *Ovary* superior, 1-locular, with few ovules on 2–5 parietal placentas; style simple, longish; stigma subcapitate or 2–5-lobed. *Fruit* a fleshy berry with several seeds and remains of sepals, petals and stamens at base and style at apex. *Seeds* suborbicular; testa leathery; embryo straight; cotyledons flat, elliptic-oblong; endosperm fleshy or somewhat horny.

About 37 species, natives of Africa, Madagascar, Comores, Mascarenes, Malaysia and Australia; 5 species in Southern Africa. The generic name is derived from the Greek for "thorn".

Flowers racemose:
- Leaves usually narrowly to broadly obovate, sometimes elliptic or ovate with narrowly cuneate base, apex acute, margins entire, repand or distantly crenate-serrate with teeth in upper half, upper surface frequently with flaky, waxy covering .... 1. *S. zeyheri*
- Leaves elliptic or ovate with cuneate base, apex acuminate, margins closely and regularly callose-serrate or serrulate throughout whole length of blade, upper surface without flaky, waxy covering .... 2. *S. mundii*

Flowers solitary or fasciculate:
- Leaves 6–11 mm long; stamens 50 or more .... 3. *S. stolzii*
- Leaves 2–4 cm long; stamens 10–25:
  - Leaves with 7–13 pairs of veins, apex acute to blunt; flowers pedicellate, 2 or 3 per axil .... 4. *S. flanaganii*
  - Leaves with 5 or 6 pairs of veins, apex acuminate; flowers sessile, solitary .... 5. *S. oreophila*

1. **Scolopia zeyheri** *(Nees) Harv.* in F.C. 2: 584 (1862), in text; Szyszyl., Polypet. Thalam. Rehm. 111 (1887); Warb. in Pflanzenfam. 3,6a: 29 (1893); Dur. & Schinz, Consp. Fl. Afr. 1: 220 (1898); Sim, For. Fl. Cape Col. 126, t.2 (1907); Gilg in Bot. Jahrb. 40 : 481 (1908); Pflanzenfam. ed. 2,21: 418 (1925); Henkel, Woody Pl. Natal 73, 90 (1934); Phill. in Bothalia 1: 84 (1922); Wild in F.Z. 1: 276 (1960); l.c. 565 (1961); Bamps in F.C.B. Flacourt. 40 (1968); Exell in F.Z. 3: 141 (1970); Sleumer in Blumea 20: 60 (1972); Wild & Vidigal in Fl. Mocamb. Flacourt. 17 (1973). Type: Cape, Olifantshoek and Addo, *Ecklon & Zeyher* 1756 (B, holo.†; P!; L; SAM!).

Eriudaphus zeyheri Nees in Eckl. and Zeyh., Enum. 2: 272 (1836). *E. ecklonii* Nees, l.c. 271 (1836). Type: Cape, Kat River, *Ecklon & Zeyher* 1754 (B, holo.†; G!; K!; P!; SAM!).

Phoberos zeyheri (Nees) Arn. in Hook., J. Bot. 3: 150 (1841); Presl, Bot. Bemerk. 70 (1844); Pappe, Silv. Cap. 4 (1854); Harv. in F.C. 1: 68 (1860). *P. ecklonii* (Nees) Arn. ex Presl, Bot. Bemerk. 70 (1844); Pappe, Silv. Cap. 3 (1854); Harv. in F.C. 1: 68 (1860).

Scolopia gerrardii Harv. in F.C. 2: 584 (1862); Dur. and Schinz, Consp. Fl. Afr. 1: 220 (1898): Gilg in Bot. Jahrb. 40: 481 (1908); Engl., Pflanzenw. Afr. 3, 2: 576 (1921); Gilg in Pflanzenfam. ed. 2, 21: 418 (1925). Type: Natal, Nototi River, *Gerrard & McKen* 26 (TCD, holo.!; PRE, photo.). *S. ecklonii* (Nees) Harv. in F.C. 2: 584 (1862), in text; Szyszyl., Polypet. Thalam. Rehm. 111 (1887); Warb. in Pflanzenfam. 3, 6a: 29 (1893), as *S. ecklonii* (Arn.) Warb.; Dur. and Schinz, Consp. Fl. Afr. 1: 220 (1898); Gilg in Bot. Jahrb. 40: 481 (1908); Phill. in Bothalia 1: 86 (1922); Gilg in Pflanzenfam. ed. 2, 22: 420 (1925); Burtt Davy, Fl. Transv. 1: 215 (1926); Henkel, Woody Pl. Natal 58, 73, 80 (1934). *S. engleri* Gilg in Bot. Jahrb. 40: 481 (1908); Engl., Pflanzenw. Afr. 3, 2: 577 (1921); Gilg in Pflanzenfam. ed. 2,21: 420 (1925). Type: Transvaal, Wonderfontein, *Engler* 2882a (B†). *S. ecklonii* (Nees) Harv. var. *gerrardii* (Harv.) Phill. in Bothalia 1: 86 (1921). —var. *engleri* (Gilg) Phill., l.c. 86 (1921); Burtt Davy, Fl. Transv. 1: 215 (1926). *S. thorncroftii* Phill., l.c. 84 (1921); Burtt Davy, l.c. 215 (1926.) Type: Transvaal, Barberton, *Thorncroft* 811 (NH, holo.!; BM!).

Rhamnicastrum zeyheri (Nees) Kuntze, Rev. Gen. 1: 45 (1891). *R. ecklonii* (Nees) Kuntze, l.c. 45.

Shrub or tree up to 13 m high, unarmed or with straight branched or unbranched rather stout spines up to 15 cm long. *Bark* dark grey, fissured. *Leaf-blade* narrowly to broadly obovate, sometimes elliptic or broadly elliptic to ovate, 3,5–10 cm long, 1,2–6 cm wide, apex acute, obtuse or emarginate, base narrowly cuneate to cuneate, margin entire, repand or distantly crenate-serrate, teeth usually in upper half only, coriaceous, dull green above, paler below, often bright pink when young, upper surface frequently covered with a thin flaky, waxy covering, veins 3–5 pairs, sometimes appearing 3-nerved at base, somewhat projecting, glabrous; petioles up to 1,5 cm long. *Flowers* in axillary racemes, 2,5–5 cm long; pedicels 3–5 mm long with small triangular bracts at base. *Receptacle* broadly funnel-shaped. *Sepals* 5–6, elliptic, 1,3–1,5 mm long, 0,75–1 mm wide, acute to round, margin sparsely ciliate. *Petals* (when present) narrowly elliptic, 1,2 mm long, 0,5–0,75 mm wide, acute. *Disc* annular, with small fleshy lobes at margin, densely villous. *Stamens* numerous; filaments 2–2,5 mm long; anthers oblong, 0,75 mm long, arcuate, apiculate. *Ovary* ovoid to subglobose, 1,3 mm diam., sulcate, glabrous; ovules 4; style 2 mm long; stigma bifid, subsessile. *Fruit* globose, up to 9 mm in diam., glabrous, with persistent style; seeds 4, sub-ovoid, 4 mm long, 3 mm wide, angular, minutely reticulate.

A tree found in bush clumps, scrub and forest from the coast to the upland regions of the Transvaal, Orange Free State, Swaziland, Natal and the Cape. Also widespread in tropical Africa occurring as far north as Cameroon, Uganda and Kenya.

TRANSVAAL.—2329 (Pietersburg): Houtbosch, *Rehmann 6464* (K; BM). 2330 (Tzaneen): Westfalia Estates, Duiwelskloof, *Scheepers 1069*. 2428 (Nylstroom): Mosdene, Naboomspruit, *Galpin M691*. 2430 (Pilgrims Rest): The Downs, *Renny DB23*. 2526 (Zeerust): Wonderfontein, *Burtt Davy 7564* (K). 2527 (Rustenburg): "Rainhill" near Rustenburg, *Codd 1083*. 2529 (Witbank): Maleeuw Kop, *Acocks 23351*. 2531 (Komatipoort): Klokwene, Kruger National Park, *Van der Schijff & Marais 3784*. 2628 (Johannesburg): *Prosser 1551* (K). 2630 (Carolina): Mooihoek, *Devenish 881*.

O.F.S.—2627 (Potchefstroom): 8 km W. of Parys, *Stapleton s.n.* 2827 (Senekal): Willem Pretorius Game Reserve, *Muller 895*. 2828 (Bethlehem): Farm Suzanna, *Scheepers 1788*.

SWAZILAND.—2631 (Mbabane): near Hlatikulu, *Compton 31337; Mafuteni Hill, Manzini, *Compton 31664;* Tungulu Mountain, Mankaiana, *Miller S/266*.

NATAL.—2632 (Bela Vista): Manguzi Forest W. of Maputa, *Tinley 338*. 2729 (Volksrust): Farm Glendale, Normandien, *Biggs 51*. 2731 (Louwsburg): Ngome Forest, *Gerstner 4881*. 2732 (Ubombo): False Bay Park, *Ward 7059*. 2830 (Dundee): 19 km from Muden, *Moll 3242*. 2831 (Nkandla): Ngoye Forest, *Wells & Edwards 66*. 2832 (Mtubatuba): Hluhluwe Game Reserve, *Ward 2635*. 2929 (Underberg): Estcourt, above New Formosa, *Acocks 10676*. 2930 (Pietermaritzburg): Westville, *Wood 8597* (K). 2931 (Stanger): Groutville, *Moll 2531*. 3030 (Port Shepstone): Ifafa River Lagoon, *Moll 3587*.

CAPE.—3028 (Matatiele): Pot River Berg, *Galpin 6830*. 3029 (Kokstad): Mt. Ingeli, *Nicholson 759*. 3126 (Queenstown): Fincham's Nek, *Galpin 1822*. 3129 (Port St. Johns): Intafufu, *Cooper 3129*. 3130 (Port Edward): Nyameni Mouth, *Strey 8617*. 3225 (Somerset East): Bruintjies Hoogte, *Burchell 3069* (K). 3226 (Fort Beaufort): Kat River, *Ecklon & Zeyher 1754* (K). 3227

(Stutterheim): Keiskammahoek, *Killick 894*. 3228 (Butterworth): Kentani, *Pegler 2109*. 3322 (Oudtshoorn): Wilderness, *Compton 10710* (NBG). 3323 (Willowmore): hills S. of De Vlugt, *Fourcade 3774* (K). 3325 (Port Elizabeth): 1,6 km N. of Zuurberg Inn, *Johnson 738* (K). 3326 (Grahamstown): near Port Alfred, *Schlechter 2733* (K; BM; P; G). 3327 (Peddie): Green Point, East London, *Smith 3779*. 3422 (Mossel Bay): Mossel Bay, *Prior s.n.* 3423 (Knysna): Buffalo Bay, Knysna, *Taylor 442* (NBG).

A variable species as regards growth form and leaf-shape. This variability accounts for the extensive synonymy of the species, especially for tropical Africa. See Sleumer's monograph for full synonymy. There has been considerable inconsistency in the citation of authors for *S. zeyheri*. The correct citation is (Nees) Harv., but the following have been frequently used: (Arn.) Warb., (Arn.) Harv., (Nees) Szyszyl. *S. zeyheri* is known by several common names, the best-known probably being Thorn Pear. The wood is very hard and was once used in wagon-making chiefly for axles, felloes and spokes. The extreme hardness of the wood also rendered it useful for the manufacture of teeth for mill wheels.

2. **Scolopia mundii** *(Eckl. & Zeyh.) Warb.* in Pflanzenfam. 3,6a: 29 (1893), as (Arn.) Warb.; Dur. & Schinz, Consp. Fl. Afr. 1: 220 (1898); Sim, For. Fl. Cape Col. 26, t.1. (1907); Gilg in Bot. Jahrb. 40: 481 (1908); Engl., Pflanzenw. Afr. 3,2: 576 (1921); Phill. in Bothalia 1: 85 (1922); Gilg in Pflanzenfam. ed. 2,21: 418 (1925); Burtt Davy, Fl. Transv. 1: 215 (1926); Henkel, Woody Pl. Natal 73, 77 (1934); Sleumer in Blumea 20: 59 (1972); Killick in Bothalia 11: 280 (1974), Type: Cape, "Sitsikamma", *Mund* s.n. (B, holo.†; K!; S!).

Eriudaphus mundii Eckl. & Zeyh., Enum. 2: 272 (1836); Steud., Nom. Bot. ed. 2: 589 (1840); Walp., Repert. 1: 548 (1842). *E. serratus* Harv., Gen. Pl. 1: 417 (1838); Dur. & Schinz, Consp. Fl. Afr. 1: 220 (1898). Types: Cape, Van Stadensberg, apparently syntypes are *Zeyher* 788 (K!; BM!; TCD); *Zeyher* 3785 (LD!; P!; S!; SAM!; TCD; Z!).

Phoberos mundii (Eckl. & Zeyh.) Presl, Bot. Bemerk. 70 (1844); Pappe, Silv. Cap. 3 (1854); Harv. in F.C. 1: 68 (1860).

Adenogyrus krebsii Klotzsch, App. Ind. Sem. Hort. Berol. 1 (1854); Walp., Ann. 4: 227 (1857). Type: "In promontorio bonae spei", *Krebs* s.n. (B†).

Rhamnicastrum mundii (Eckl. & Zeyh.) Kuntze, Rev. Gen. 1: 45 (1891).

Shrub or tree up to 16 m high (27 m *fide* Sim, *l.c.*), unarmed or with simple, slender spines up to 4 cm long. *Bark* thin, grey or light brown, smooth when young, thick, grey, brown or yellowish, rough, fissured when old. *Branches* sometimes with axillary galls resembling fruits. *Leaf-blade* elliptic, ovate or rarely subrotund, acuminate at apex, cuneate to rounded at base, (2,5) 4–7 cm long, 2–4 cm wide, margin closely and regularly callose serrate or serrulate, sometimes with spinose teeth, midrib channelled above, prominent below, veins 4–6, ½-angled, looping some distance from margin, subprominent, dark green and shiny above, paler below, coriaceous, glabrous; petiole 3–10 mm long, reddish. *Flowers* greenish white, in racemes up to 3 cm long, pedicels up to 6 mm long. *Sepals* 4 or 5, ovate to broadly ovate, apex acute or obtuse, 2–2,2 mm long, 1,5–2 mm wide. *Petals* when present, deltoid, 1,2 mm long, 1 mm wide. *Disc* annular, with small, orange, fleshy lobes, 0,75 mm long, densely villous. *Stamens* numerous, filaments 4–5,5 mm long, glabrous; anthers oblong, 0,7 mm long, arcuate, apiculate. *Ovary* ovoid, 1,5 mm long, 1,5 mm wide at base, sulcate, glabrous; ovules several; style 3,5 mm long, sulcate; stigma bilobed with each lobe again faintly bilobed. *Fruit* globose, 7–10 mm diam., glabrous, with persistent style, red; seeds 3 or 4, angular, 4 mm long, 2 mm wide. FIG. 19.

A forest tree occurring in the Transvaal, Orange Free State, Swaziland, Natal, Lesotho and the Cape as far south as the Cape Peninsula.

TRANSVAAL.—2430 (Pilgrims Rest): Mt. Sheba Forest, *Jones 67*. 2527 (Rustenburg): Tonguane Kloof, *De Winter s.n.* 2730 (Vryheid): Mooihoek, *Devenish 1024*.

O.F.S.—2828 (Bethlehem): farm of Mr Naude, *Theron 2183*. 2829 (Harrismith): Maweni Heights, *Van Zinderen Bakker Jnr. 12*.

SWAZILAND.—2631 (Mbabane): 5 km E. of Mbabane, *Miller S/215*.

NATAL.—2730 (Vryheid): Donkerhoek, *Devenish 1311*. 2828 (Bethlehem): Royal Natal National Park, *Hutchinson 4497* (K); *Galpin 9514*. 2829 (Harrismith): Cathedral Peak Forest Station, *Killick 1739*. 2830 (Dundee): Qudeni, *Davis 89* (NH). 2832 (Mtubatuba): Enseleni Nature Reserve, *Venter 594*. 2929 (Underberg): Ntabamhlope, *Pentz 312*. 2930 (Pietermaritzburg): Howick, *Moll 1070;* Dargle, *Moll 864*. 2931 (Stanger): Hawaan Forest, *Ross & Moll 2296*. 3030 (Port Shepstone): Valleys Nature Reserve, *Nicholson 922*.

LESOTHO.—2927 (Maseru): Mamathes, *Marais 1084; Jacot Guillarmod 469;* bank of Mohale River, *Jacottet sub Dieterlen 1062;* Matukeng, *Dieterlen 852*.

CAPE.—3028 (Matatiele): 5,3 km S. of Mt. Frere, *Story 942*. 3029 (Kokstad): 6,9 km N. of Emagusheni Store, *Marais 1164*. 3126 (Queenstown): Hangklip Mountain, *Roberts 2121*. 3128 (Umtata): Baziya Forest, *Marais 497*. 3226 (Fort Beaufort): E. of Abies Grove, Hogsback, *Wells 3779*. 3227 (Stutterheim): Mt. Kemp, *Killick 908*. 3228 (Butterworth): near Willowvale, *Acocks 12278*. 3318 (Cape Town): Kirstenbosch, *Esterhuysen 18637*. 3323 (Willowmore): *Mund s.n.* (K). 3325 (Port Elizabeth): Van Stadens Mountain, *Zeyher 788* (BM; K); Zuurberg Sanatorium, *Long 1251*. 3326 (Grahamstown): Grahamstown, *MacOwan 957* (BM; K). 3327 (East London): Fort Grey, Forest Reserve, *Wells 3917*. 3423 (Knysna): Knysna, *Keet 512*.

Known as the Red Pear (Rooipeer) or Mountain Safraan (Berg Safraan). The wood is hard, heavy and close grained and was formerly used in wagon making, chiefly for felloes. It is recommended by Galpin (fide Burtt Davy l.c.) as a hedge plant.

FIG. 19.—1, **Scolopia mundii,** flowering twig, (*Ranger* s.n.); a, fruiting twig, × 1 (*Johnson* 979); b, flower, × 4 (*Ranger* s.n.); c, portion of flower with sepals excised to show annular disc, × 5 (*Ranger* s.n.); d, anther, × 20 (*Ranger* s.n.).

3. **Scolopia stolzii** *Gilg* in Engl., Pflanzenw. Afr. 3,2: 577 (1921) Pflanzemfam. ed. 2,21: 420 (1951); Sleumer in Notizbl. Bot. Gart. Berl. 12: 142 (1936); Brenan, Checklist Tang. Terr. 236 (1949); Wild in F.Z. 1: 278, t.47A (1960); F. White, For. Fl. N. Rhod. 267 (1962); Bamps in F.C.B. Flacourt. 39 (1968); Sleumer in Blumea 20: 51 (1972); Wild & Vidigal in Fl. Mocamb. Flacourt. 18 (1973); Killick in Bothalia 11: 280 (1974). Type: Tanzania, Kyimbila, Kiwira River, *Stolz* 1742 (B, holo.†; BM; K; PRE!).

S. riparia Mildbr. & Sleumer in Notizbl. Bot. Gart. Berl. 11: 1077 (1934); Brenan, Checklist Tang. Terr. 235 (1949). Type: Tanzania, Tandala Kinga Mts., Iletile River, *Stolz* 2200 (B, holo.†; L; M!). *S. stolzii* Gilg var. *riparia* (Mildbr. & Sleumer) Sleumer in Blumea 20: 52 (1972).

Tree up to 14 m high, 25 cm d.b.h., unarmed, trunk brownish with irregularly shaped exfoliation sloughing to leave dark brown shallow basins; twigs white to pale grey, lenticellate. *Leaves* exstipulate, elliptic or ovate, 6–11 cm long, 3,4–5,7 cm wide, apex obliquely acuminate to obtuse, base broadly cuneate, margin remotely callose subserrate to crenulate, 5–9 pairs of nerves, basal 2 pairs acutely angled (curved ascending), remainder obtusely angled, looping before margin, immersed above, projecting below, coriaceous, glabrous, darkish green above, paler below, 5–8 mm long, recurved. *Flowers* solitary or in pairs, sessile, in upper leaf axils. *Bracts* 5, deltoid, 0,5 mm long, somewhat cartilaginous. *Sepals* 5, approximately triangular, 1,4–2 mm long, ciliate. *Petals* 5, resembling sepals, but sometimes rounded, early caducous. *Disc* annular, regularly lobed, 0,5 mm long, "salmon pink". *Stamens* numerous, sometimes over 100; filaments 7,5 mm long; anthers 0,75 mm long, slightly apiculate. *Ovary* ovoid, 3,5 mm long with longish white hairs; placentae 5–6 with numerous ovules; style short, 5 or 6 branched, branches 1,8 mm long, channelled above, glabrous. *Fruit* subglobose, 2,5 cm diam., fleshy, often crowned with persistent style; seeds 8–17, oblong, c. 5 mm long, brown. FIG. 20.

Fairly widespread in tropical Africa occurring as far north as Cameroon. In Southern Africa known only from the Sihadla River Crossing and Mtunzini in Natal, where it grows in swamp and dune forest respectively.

NATAL.—2732 (Ubombo): Sihadla River Crossing, *Killick & Vahrmeijer 4061; Moll & Nel 5608; Moll 5724; Strey 8187; Strey & Moll 3890.* 2831 (Nkandla): Mtunzini, Umlalazi Lagoon, *Venter 6013.*

Only recently discovered in Southern Africa. According to a note on *Moll & Nel* 5608 the foliage is pendent with the young leaves dark pink and highly conspicuous and the slash is pale pink. A variable species as regards leaf size, texture, margin and pubescence of ovary.

S. stolzii Gilg is treated as a *nomen nudum* by Wild & Vidigal (1973), but the present author has followed Sleumer (1972) in accepting the 3-line description as adequate for valid publication.

4. **Scolopia flanaganii** *(H. Bol.) Sim,* For. Fl. Cape Col. 127 t.3 (1907); Gilg in Bot. Jahrb. 40: 484; Engl., Pflanzenw. Afr. 3,2: 576 (1921); Phill. in Bothalia 1: 84 (1921); Gilg in Pflanzenfam. ed. 2, 21: 418 (1925); Henkel, Woody Pl. Natal 80 (1934); Sleumer in Blumea 20: 50 (1972). Type: Cape, near Komga, *Flanagan* 682 (BOL, holo.!; K!; NH!; PRE!; SAM!).

Xylosma flanaganii H. Bol. in J. Bot. Lond. 34: 17 (1896).

Tree or shrub up to 5 m high with trunk up to 17 cm d.b.h., armed with slender spines up to 3 cm long chiefly on coppice shoots and young growth. *Branchlets* glabrous or puberulous. *Leaf-blade*: adult leaves narrowly elliptic to elliptic or narrowly obovate to obovate, 2–3,5 cm long, 0,6–2,2 cm wide, apex acute to obtuse, base narrowly to broadly cuneate, margins serrulate or remotely serrulate, often entire in lower $^{1}/_{3}$, teeth callose, coriaceous, glabrous, midrib prominent on lower surface, lateral nerves 7–13, subimmersed; petiole 2–4 cm long; juvenile leaves ovate, 1,8–3,4 cm long, 1,1–2,2 cm wide, apex acute, base broadly cuneate or cordate, margins serrate, sometimes shortly spine-tipped, coppice shoot leaves broadly elliptic or circular, 2,6–3,3 cm long, 2–2,4 cm wide, apex acute or rounded, base broadly cuneate or rounded, margins serrate. *Flowers* solitary or fasciculate in leaf axils, 5-merous; pedicels 1,25–1,75 mm long, stoutish, bracteate. *Sepals* broadly ovate, 0,5–0,75 mm long, 0,75–1 mm wide, concave, somewhat keeled dorsally, scariose, ciliate on margin. *Petals* early caducuous, elliptic, ovate or circular, 0,75–1 mm long, 0,5–1 mm wide. *Disc* annular of quadrate fleshy lobes, glabrous. *Stamens* 10–20; filaments 2–2,5 mm long; anthers oblong, 0,5 mm long. *Ovary* ovoid, 1–2,5 mm long, 1–1,25 mm wide at base, pubescent or glabrous; stigma obscurely 2–4-lobed. *Fruit* subglobose, up to 7 mm long, 5 mm wide, pubescent, red. *Seeds* 1 or 2. FIG. 21.

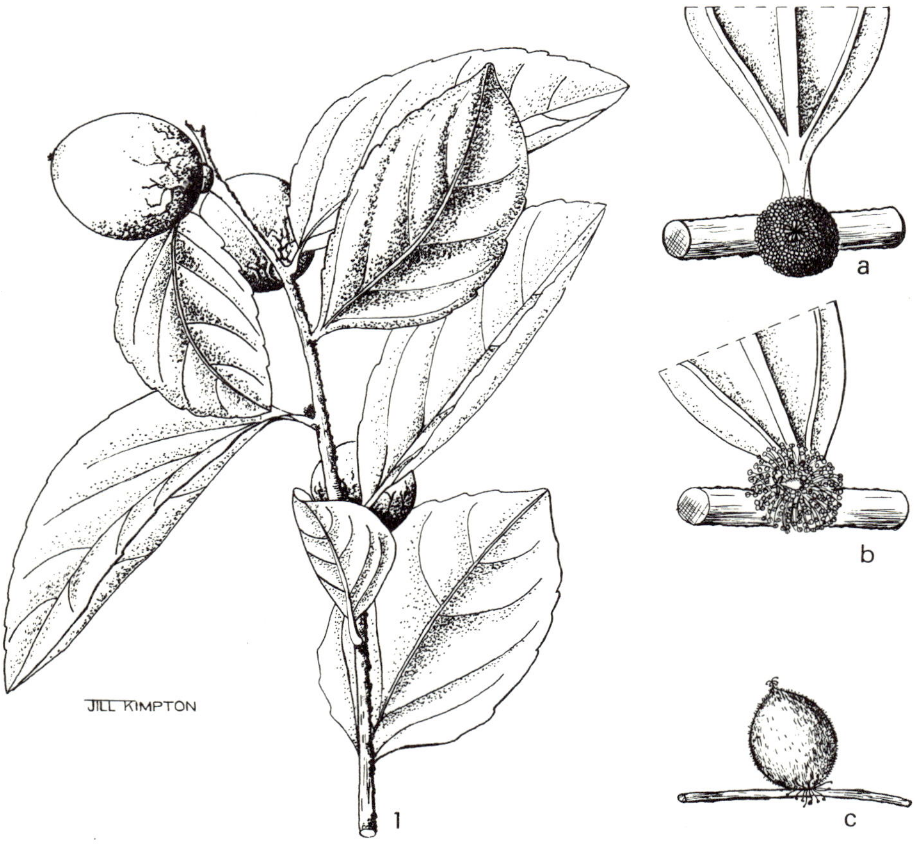

FIG. 20.—1, **Scolopia stolzii,** fruiting twig; a, flower in late bud; b, mature flower; c, young fruit with persistent styles; all × 4/5 *(Killick & Vahrmeijer* 4061).

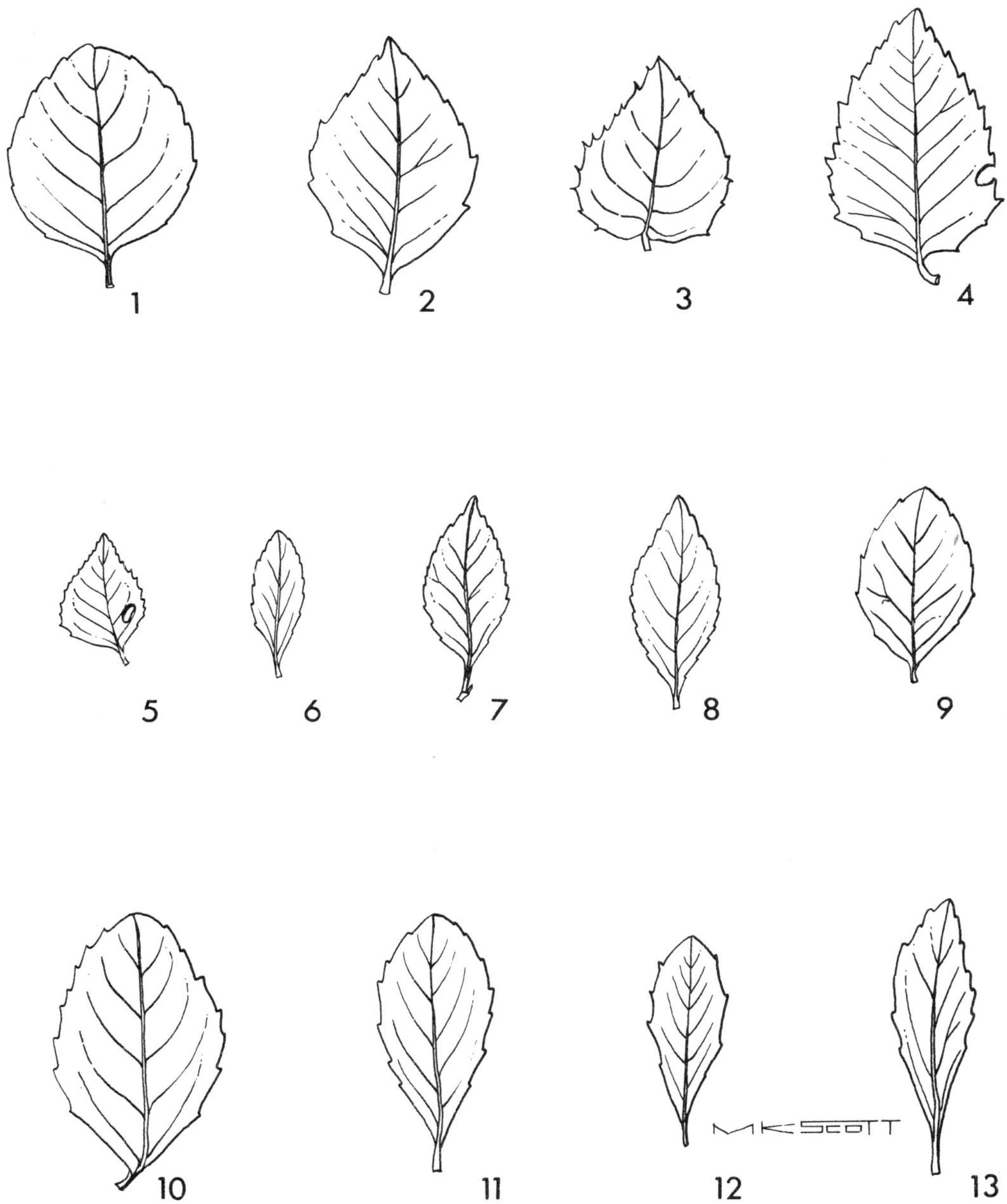

FIG. 21.—1, **Scolopia flanaganii,** series showing variation in leaf shape, all × 1. 1 and 2, coppice shoot leaves *(Killick* 4076 and 4068); 3–5, juvenile leaves *(Killick* 4056, *Edwards* 4313 and *Killick* 1957); 6–13, adult leaves *(Acocks* 10519, *Killick* 4056, 4075, 4075, 4070 and 4070, *Flanagan* 682 and *Killick* 4069).

Occurring inside and along the margin of forest and scrub from the eastern Cape to the midlands and uplands of Natal.

NATAL.—2929 (Underberg): Giants Castle Game Reserve, *Killick 4056;* Hlatikulu Forest, *Killick 1957;* 11 km from Bulwer on road to Impendhle, *Killick & Marais 2103.* 2830 (Dundee): Qudeni, *Gerstner 667:* Umhlumba Mountain, *West 1470.* 2930 (Pietermaritzburg): Keerom Farm, Greytown, *Cooper 60* (NH).

CAPE.—3128 (Umtata): All Saints Nek, *Marais 502;* 40 km S. of Ugie, *Edwards 4313.* 3227 (Stutterheim): near Komga, *Flanagan 682;* Fort Cunynghame, *Killick 4075;* Kei Road, *Killick 4069;* Mount Kemp, *Killick 902.*

S. flanaganii is an extremely variable species especially in the shape of the adult, juvenile and coppice shoot leaves. Also, the ovaries vary from completely glabrous to pubescent. A feature of the leaves is the frequent presence of small shot-holes apparently caused by a blue-green alga.

5. **Scolopia oreophila** *(Sleumer) Killick* in Bothalia 11: 515 (1975). Type: Natal, Utrecht District, Farm Retirement, *Devenish* 1319 (K, holo.!; PRE!; S!).

S. flanaganii (H. Bol.) Sim var. *oreophila* Sleumer in Blumea 20: 51 (1972).

Tree up to 13 m high, 23 cm d.b.h., spines on young plants and coppice shoots. *Branchlets* glabrous, puberulous at apex. *Leaf-blade* rhombic-elliptic (coppice shoot leaves ovate), 2–3 cm long, 1–1,7 cm wide, apex acuminate, base cuneate (wedge-shaped), margin distantly serrate-crenate, teeth callose, midrib prominent below, lateral veins 5–6 pairs, subimmersed, coriaceous, glabrous; petiole 2–4 mm long, red, thickish. *Flowers* solitary in leaf axils, sessile, bracteolate at base, (4) 5-merous. *Sepals* narrowly elliptic to broadly ovate, slightly concave, 2–2,5 mm long, 1–2 mm wide. *Petals* not seen. *Disc* annular, of orange-coloured, fleshy lobes, 0,5 mm long, glabrous. *Stamens* 15–25; filaments 2,5–3 mm long, glabrous; anthers oblong, 0,5 mm long, sometimes with short black bristles. *Ovary* ellipsoid to ovoid, 2–3 mm long, 2 mm wide, glabrous; ovules 3; style 2,5–3 mm long, glabrous; stigma bilobed with each lobe faintly bilobed. *Fruit* not seen.

Known only from the Utrecht District of Natal, where it occurs along the margin of streambank scrub and forest.

NATAL.—2730 (Vryheid): Farm Donkerhoek, *Devenish 528;* Farm Retirement, *Devenish 1319; Killick 4057;* Farm Naauwhoek, *Devenish 1141.*

Can be distinguished from *S. flanaganii* by the leaves which are somewhat rhombic in shape with an acuminate apex, 5 or 6 pairs of veins and reddish petioles and the sessile, solitary flowers.

5304a 6. PSEUDOSCOLOPIA

Pseudoscolopia *Gilg* in Bot. Jahrb. 54: 343 (1917); Phill., Gen. ed. 2: 513 (1951). Type species: *P. polyantha* Gilg.

Pseudoscolopia Phill., Gen. 416 (1926), nom. illegit. Type species: *P. fraseri* Phill.

Shrub or small tree. *Leaves* opposite, petiolate, exstipulate; blade elliptic, serrate. *Inflorescence* a lax, axillary cyme. *Flowers* bisexual. *Sepals* 4, valvate in bud, persistent. *Petals* 4, subsimilar to sepals, imbricate in bud, persistent. *Stamens* many; filaments free; anthers arcuate. *Ovary* superior, ovoid-globose, 1-locular with 2 or 3 parietal placentas and 1 ovule on each placenta, pubescent; style terete, 2 or 3-fid at apex. *Fruit* a 2 or 3-valved capsule, ovoid, with persistent style. *Seed* ellipsoid, covered with stellate hairs; embryo straight; cotyledons flat, somewhat obovate; endosperm sparse.

A monotypic genus recorded from Natal and the Cape. The genus resembles and is closely related to *Scolopia,* hence the name *Pseudoscolopia.*

Pseudoscolopia polyantha *Gilg* in Bot. Jahrb. 54: 343 (1917); Phill. & Hutch. in Hook. Icon. Pl. t.3119 (1933). Syntypes: Cape, Enkweni and Egossa Forest, *Bachmann* 1712 and 1713 (B†); *Beyrich* 116 and 125 (B†); Mkambati River, *Marais* 969 (PRE, neo.!).

P. fraseri Phill., Gen. 416 (1926). Type: not indicated, probably Cape, Transkei, Ntsubane Forest, *Fraser* sub PRE 1417 (PRE, holo.!).

Shrub or tree up to 5 m high. *Branchlets* angled, glabrous. *Leaf-blade* narrowly elliptic to elliptic, 4–6,5 cm long, 2–3 cm wide, apex acuminate or acute, base cuneate, margin serrate with cartilaginous teeth or nearly entire, midrib prominent below, lateral veins 5 or 6, not very conspicuous, coriaceous, glabrous; petiole 3–8 mm long. *Inflorescence* 4–6-flowered, 2–3 cm long. *Flowers* bisexual;

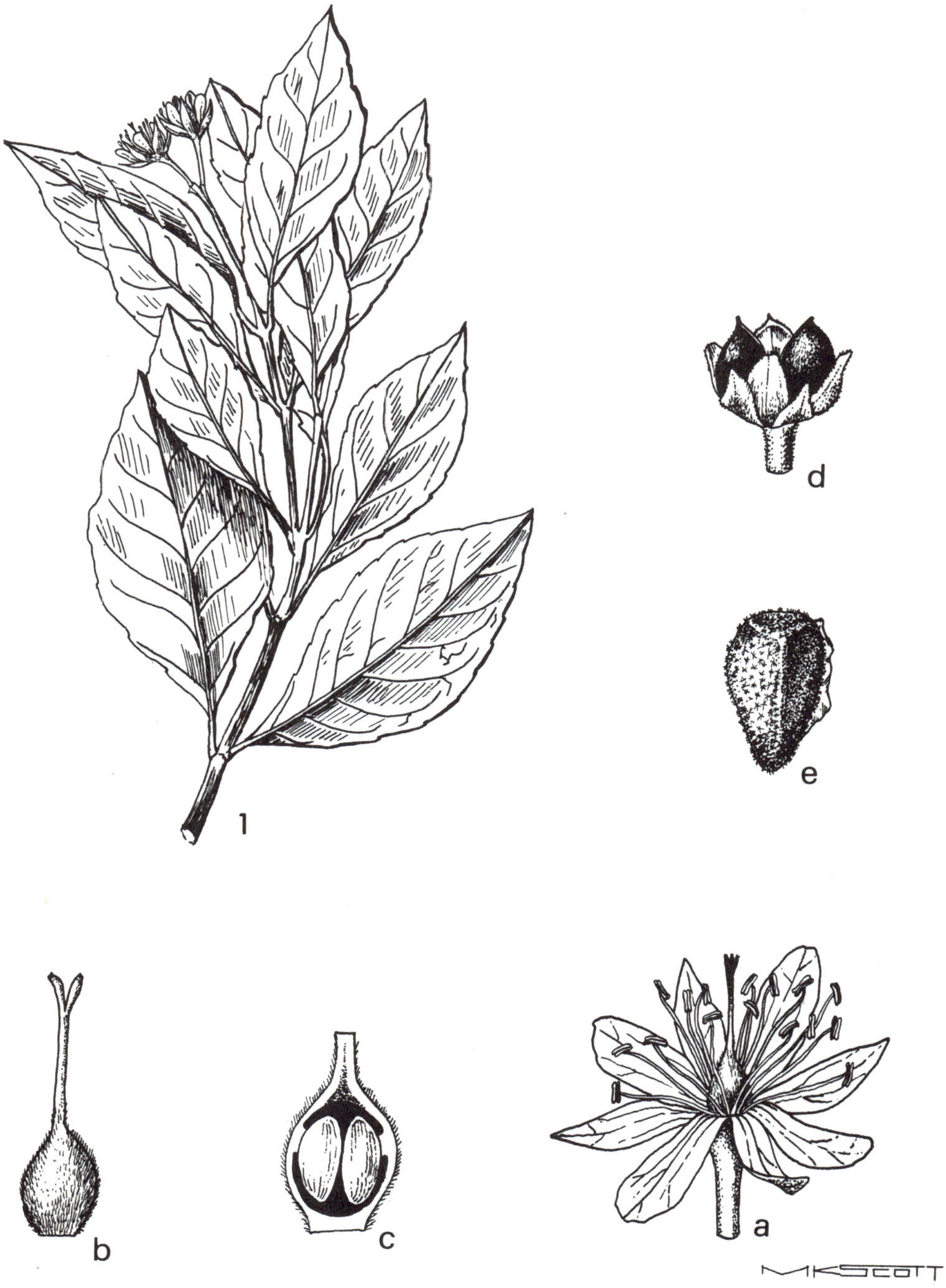

FIG. 22.—1, **Pseudoscolopia polyantha,** flowering twig, × 1 (*Marais* 696); a, flower, × 3 (*Cooper* 107); b, gynoecium, × 8 (*Cooper* 107); c, longitudinal section through ovary, × 10 (*Cooper* 107); d, fruit, × 3 (*Esterhuysen* 17946); e, seed, × 8 (*Esterhuysen* 17946).

pedicels 6–11 mm long, puberulous. *Sepals* elliptic, 6–7 mm long, 2,5–3 mm wide, apex acuminate, pubescent, margins ciliate. *Petals* subsimilar to sepals. *Stamens* many; filaments 3–4 mm long, glabrous; anthers oblong, c. 1 mm long, prominently arcuate. *Ovary* ovoid-subglobose, 1,5–2 mm long, villous; style terete, 3 mm long, glabrous; stigma 2 or 3-fid, arms c. 1,5 mm long. *Fruit* ovoid, 5 mm diam., with persistent style. *Seeds* ellipsoid, 3 mm long, stellately pubescent. FIG. 22.

A forest tree showing wide discontinuity in its distribution: one specimen, *Esterhuysen* 17946, has been collected at Rivers Kloof in the Piketberg District of the south-western Cape and the rest in Pondoland and southern Natal.

NATAL.—2831 (Nkandla): Ngoye Forest, *Ross 1863*. 2930 (Pietermaritzburg): Laager Farm, Noodsberg, *Nicholson 544* (NH). 3030 (Port Shepstone): Beacon Hill, *Strey 6530; 7226; Nicholson 817*. 3130 (Port Edward): Umtamvuna River George, *Cooper 107*.

CAPE.—3129 (Port St. Johns): Mkambati River, *Marais 969;* Ntsubane Forest, *Fraser sub PRF 3058; 3105; 3133*. 3218 (Clanwilliam): Rivers Kloof, *Esterhuysen 17946*.

With the appearance of a *Scolopia*, but differing in its opposite leaves, 2 or 3-ovuled ovary and capsular fruits. *Pseudoscolopia polyantha* bears some resemblance to *Cassipourea flanaganii* (Schinz) Alston of the family Rhizophoraceae with which it has sometimes been confused.

5312 7. GERRARDINA

Gerrardina *Oliv.* in Hook. Icon. Pl. 11: 60, t.1075 (1870); Gilg in Bot. Jahrb. 40: 488 (1908); Engl., Pflanzenw. Afr. 3,2: 578, t.256a–c (1921); Gilg in Pflanzenfam. ed. 2, 21: 425, t.191a–c (1925); Phill., Gen. ed. 2: 514 (1951); Wild in F.Z. 1: 287 (1960). Type species: *G. foliosa* Oliv.

Scrambling or erect shrubs or trees. *Leaves* alternate, simple, petiolate, stipulate; blades elliptic or ovate, sometimes acuminate, serrulate to serrate or crenato-serrate. *Inflorescence* a few-flowered axillary cyme on a long peduncle. *Flowers* bisexual. *Calyx-tube* campanulate; lobes 5, imbricate, unequal, 2 outer shorter. *Petals* 5, shorter than calyx and alternating with the calyx-lobes, inserted on the margin of the disc, deciduous. *Disc* cup-shaped, lining calyx-tube. *Stamens* 5, opposite the petals, inserted on margin of disc. *Ovary* superior, 1-locular, with 4 pendulous ovules. *Fruit* a dry capsule; seeds 1–4, ellipsoid or obovoid, testa smooth or reticulate.

A genus of 2 species, 1 occurring in Southern Africa and the other in Rhodesia, Malawi, Mozambique and Tanzania. The genus was named in honour of W. T. Gerrard, who collected in Natal about the middle of the last century.

Gerrardina foliosa *Oliv.* in Hook. Icon. Pl. 11: 60, t.1075 (1870). Gilg in Bot. Jahrb. 40: 488 (1908); Engl., Pflanzenw. Afr. 3,2: 578, t.256a–c (1921); Gilg in Pflanzenfam. ed. 2, 21: 425, t.191a–c (1925); Burtt Davy, Fl. Transv. 1: 217 (1926); Henkel, Woody Pl. Natal 94 (1934). Type: Natal, without precise locality, *Gerrard* 1513 (K, holo.!; BM!; PRE, photo.).

Shrub or tree up to 10 m high. *Branches* greyish-brown to brown, pubescent at ends. *Leaf-blade* elliptic, 2,6–6,5 cm long, 1–2,5 cm wide, apex acute to obtuse, base narrowly to broadly cuneate, margin callose-serrulate to serrate, sometimes only in upper $^2/_3$, thickened, revolute, midrib prominent below, lateral nerves 9–13, subprominent below, tertiary nerves reticulate, immersed, glabrescent, coriaceous, pale green below; petiole 2–7 mm long. *Inflorescence* an axillary cyme; peduncles erect, 1–4 cm long, slender. *Flowers* 2–6 per cyme; pedicels 1–3 mm long. *Calyx-lobes* yellowish, unequal, margins entire or glandular-denticulate, faintly keeled; 2 outer rounded, c. 1,5 mm diam.; 3 inner elliptic to obovate-spathulate, 2–3 mm long, 2 mm wide. *Petals* elliptic, 2 mm long, 1,5 mm wide, deciduous. *Stamens* 1 mm long with filaments reddish distally; anthers c. 0,7 mm long. *Disc* shallowly 5-lobed, reddish-brown. *Ovary* depressed ovoid, 8-ridged radially above, 1-locular; ovules 4, pendulous; style 1 mm long, reddish distally; stigma capitate, faintly bilobed, free. *Fruit* bright red, a dry capsule with persistent style, globose, 5–6 mm diam.; seed 1, brown, ellipsoid-obovoid, 4 mm long, 2,2 mm wide, testa smooth, brown. FIG. 23.

Occurs on rocky hillsides, krantzes and in forest in the eastern Cape, Natal, Swaziland and the eastern Transvaal.

TRANSVAAL.—2430 (Pilgrims Rest): Mamotswiri, *Junod 187; Smuts & Gillett 3573;* God's Window, *Mogg 3329*. 2531 (Komatipoort): Saddleback Range, Barberton, *Galpin 473* (K); Roses Creek, *Thorncroft 2066*.

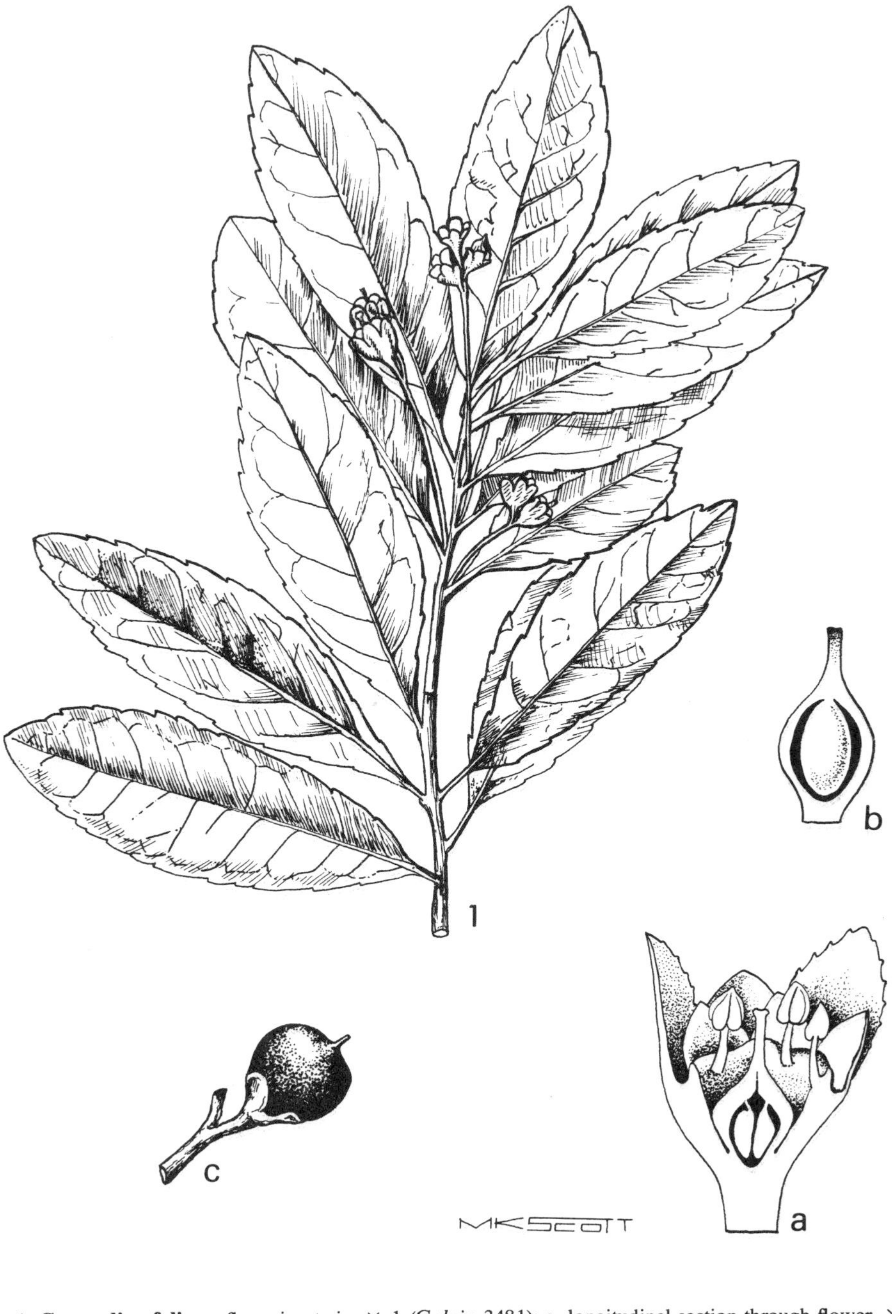

FIG. 23.—1, **Gerrardina foliosa,** flowering twig, × 1 *(Galpin* 3481); a, longitudinal section through flower, × 6 *(Strey* 3853); b, longitudinal section through ovary, × 8 *(Strey* 3853); c, fruit, × 2½ (ex colour slide, *Strey*).

SWAZILAND.—2531 (Komatipoort): King's Forest, *Compton 27817;* Havelock Mine, *Miller 5785.* 2631 (Mbabane): Mahlangatsche Mt., Mankaiana, *Miller S/268.*

NATAL.—2831 (Nkandla): Ngoye Forest, Mtunzini, *Huntley 730.* 2930 (Pietermaritzburg): 59 km from Wartburg on Tongaat Road, *Edwards 3006;* Table Mountain, *Killick 146.* 3029 (Kokstad): Murchison, *Coleman 452* (NH). 3030 (Port Shepstone): Umtwalumi, *Rudatis 584;* Oribi Gorge, *Strey 3853.* 3130 (Port Edward): Mtamvuna Forest Reserve, *Ward 7174.*

CAPE.—3129 (Port St. Johns): Umsikaba River near mouth, *Marais 1183;* top of Eagles Nest, Port St. Johns, *Schonland 4079.* 3130 (Port Edward): Umtentu River Falls, *Marais 985.*

A distinct species readily distinguishable from its tropical African congener, *G. eylesiana* Milne-Redh., by its leaves which are elliptic-acute instead of ovate-acuminate, coriaceous instead of chartaceous and the peduncles which are considerably stouter. There are two specimens of *Gerrard 1513* in Herb. Kew. The one with annotations and drawings and marked "Type specimen" is clearly the holotype.

5313 8. HOMALIUM

Homalium *Jacq.*, Enum. Syst. Pl. Ins. Carib. 5, 24 (1760); Benth. & Hook.f., Gen. Pl. 1,3: 800 (1867); Sim, For. Fl. Cape Col. 220 (1907); Engl., Pflanzenw. Afr. 3, 2: 578 (1921); Gilg in Pflanzenfam. ed. 2,21: 425 (1925); Phill., Gen. ed. 2: 514 (1951); Wild in F.Z. 1: 287 (1960); Wild & Vidigal, Fl. Mocamb. Flacourt. 31 (1973); Sleumer in Bull. Jard. Bot. Belg. 43: 239 (1973). Type species: *H. racemosum* Jacq.

Trees or shrubs. *Leaves* alternate (in African species), simple, petiolate; blade ovate, elliptic, suborbicular or oblong, sometimes large; stipules absent, minute or large. *Inflorescence* an axillary or terminal raceme or panicle. *Flowers* bisexual. *Calyx-tube* turbinate; sepals 7–9, usually narrow, persistent. *Petals* 7–9, inserted in the mouth of the calyx tube, alternating with and similar to the sepals but broader, persistent. *Stamens* solitary or in fascicles of 2–3 or more, alternating with glands which are adnate to the base of the sepals, terete; anthers small, didymous. *Ovary* semi-superior, 1-locular with 2–5 parietal placentas; styles 2–5; stigmas simple or capitate. *Fruit* a semi-superior, leathery capsule, 1- or few-seeded, opening at the apex by 2–5 valves. *Seeds* solitary or few.

A tropical and subtropical genus of about 180 species; 3 species in Southern Africa, two of which, *H. dentatum* (Harv.) Warb. and *H. rufescens* Benth., belong to subgenus *Blackwellia* Warb., section Blackwellia Benth., while the third, *H. abdessammadii* Aschers. & Schweinf. belongs to subgenus *Homalium*, section Homalium.

According to Wittstein (1856) the name *Homalium* is derived from the Greek for "similar or equal" in reference to the 21 stamens, which are arranged in 7 similar or equal bundles. In actual fact, the stamens can be more or less than 21, but when arranged in bundles the bundles are usually equal.

Stamens solitary at base of each petal:
- Leaves broadly elliptic, 5,5–9 cm long, 3,5–5,5 cm wide .... 1. *H. dentatum*
- Leaves elliptic to broadly elliptic, 1,5–5 cm long, 0,8–3 cm wide .... 2. *H. rufescens*

Stamens in 3's opposite each petal .... 3. *H. abdessammadii*

1. **Homalium dentatum** *(Harv.) Warb.* in Pflanzenfam. 3,6a: 36 (1893); Engl., Pfanzenw. Afr. 3,2: 581 (1921); Gilg in Pflanzenfam. ed. 2, 21: 427 (1925); Wild in F.Z. 1: 289 (1960); Sleumer in Bull. Jard. Bot. Belg. 43: 239 (1973); Wild & Vidigal, Fl. Mocamb. Flacourt. 31 (1973). Type: Natal, "near Port Natal", *Gerrard & McKen* 50 (TCD, holo.!; PRE, photo.).

Blackwellia dentata Harv. in F.C. 2: 585 (1862). Type as for *H. dentatum.*

Homalium subsuperum Sprague in Kew Bull. 1923: 184 (1923); Gilg in Pflanzenfam. ed. 2, 21: 427 (1925). Henkel, Woody Pl. Natal 86 (1934). Type: Cape, Ndindini Forest, Transkei, *Kaufmann* sub PRF 3225 (sub PRE 2571) (K, holo!; PRE!).

Tree 6–30 m high; branchlets glabrous or tomentulose. *Leaf-blade* broadly elliptic, 5,5–10 cm long, 3,5–7 cm wide, apex abruptly acuminate often with oblique point, obtuse or rounded, base broadly cuneate to rounded, margin crenate-serrate, midrib and lateral veins fairly prominent below, lateral veins 5–8, ½-angled, looping near margin, coriaceous, glabrous or tomentulose, often with tufts of hair in lower axils; petioles up to

2,2 cm long, glabrous or tomentulose. *Inflorescence* of axillary and terminal panicles, divaricately branched; peduncles 2,5–4 cm long, glabrous or puberulous; pedicels 0,2–2 mm long, puberulous. *Calyx-tube* puberulous; sepals 6–8, narrowly elliptic, 1,2–1,75 cm long, 0,3–0,7 mm wide, puberulous to tomentulose; glands adnate to base of sepals, round, sessile, brown, tomentulose. *Petals* 6–8, elliptic, 1,5–1,8 mm long, 0,8–1,2 mm wide, puberulous to tomentulose. *Stamens* 6–8, terete, 0,2–0,4 mm long. *Ovary* narrowly conical, 1,2 mm long, tomentulose; styles 3 or 4, joined at base, with free portion 5 mm long and tomentulose in lower half; stigmas simple. *Fruit* a leathery capsule, obovoid, 4 mm long, 2,5 mm wide, tardily dehiscent, pilose inside; seeds usually one, c. 2 mm diam.

Found in scrub and forest in the Transvaal, Natal, eastern Cape, Rhodesia, Malawi and Swaziland.

TRANSVAAL.—2330 (Tzaneen): Piesangskop, Duiwelskloof, *Scheepers 903*. 2430 (Pilgrims Rest): Erasmus Pass, *Van der Schijff 7305;* between Weltevreden Asbestos Mine and Olifants River, *Codd 10566*. 2531 (Komatipoort): Eureka City, Sheba Hills, *Scheepers SKF 4072;* near Malelane, *Codd 8257; Van der Schijff 2507;* Ship Mt., 14 km S.E. of Pretorius Kop, *Codd & De Winter 5161*. Without precise locality: Soutpansberg, *Worsdell s.n.* (K).

SWAZILAND.—? 2632 (Bela Vista): Chilobe, Lebombo Mtns., *Miller S/19;* forest bordering Tibilati stream on E. side of Lebombo Mtns., *Hornby 2835*.

NATAL.—2731 (Louwsburg): Ngome, *Gerstner 5185*. 2732 (Ubombo): Ngwalaweni Forest, *Gerstner 4017; Dutton & Tinley 23*. 2830 (Dundee): Qudeni Forest, *Edwards 2648*. 2831 (Nkandla): Nkandla Forest, *Edwards 1366*. 2832 (Mtubatuba): Hluhluwe Game Reserve, *Ward 1712* (NH). 2930 (Pietermaritzburg): Greytown, *Cooper 61*. 2931 (Stanger): Umgeni, *Hutchinson 1848;* Kildare Estates, Darnall, *Moll 2901*. 3030 (Port Shepstone): Umbogintwini, *Wood 11977* (K).

CAPE.—3129 (Port St. Johns): Mpande Forest, *Miller sub PRF 3756;* 2,4 km N. of Valley View Store, Lusikisiki, *Acocks 13416;* near Fort Donald, *Sim 2412*.

The leaves of young coppice shoots are much thinner than the adult leaves, the petioles are shorter and the hairy pockets in the leaf axils are more conspicuous. In the southern part of the Kruger National Park there is a form with densely tomentulose leaves and branchlets represented by *Codd* 5257, *Codd & De Winter* 5161, *Van der Schijff* 2507, etc. This form resembles *H. chasei* Wild from Rhodesia in its pubescence, but has the paniculate inflorescence of *H. dentatum*. According to a note on *Hutchinson* 1848 the flowers of *H. dentatum* "smell like sweet chestnut."

2. **Homalium rufescens** *Benth.* in J. Linn. Soc. 4: 34 (1 July 1859); Sim, For. Fl. Cape Col. 220 t.68, fig. 1 (1907); Wood, Natal Plants t.529 (1912); Engl., Pflanzenw. Afr. 3,2: 580 (1921); Gilg in Pflanzenfam. ed. 2, 21: 427 (1925); Henkel, Woody Pl. Natal 84 (1934); Sleumer in Bull. Jard. Bot. Belg. 43: 259 (1973). Syntypes: Cape, Enon, *Drège* s.n. (K, lecto.!; BM!; L!; S!; PRE, photo.); Natal, Port Natal, *Gueinzius* s.n. (BM!; K!; S!; PRE, photo.).

Pythagorea rufescens E. Mey. ex Arn. in Hook., J. Bot. 3: 149 (1840), nomen nudum; Presl, Bot. Bemerk. 42 (1845), nomen nudum. *P. africana* E. Mey. in Drège, Zwei Pfl. Doc. 134, 214 (1843), nomen nudum.

Blackwellia rufescens Arn. in Hook., J. Bot. 3: 149 (1840), nomen nudum; Presl, Bot. Bemerk. 42 (1845), nomen nudum; Harv. in F.C. 1: 72 (May 1860).

Tree, 2–7 m high. *Leaf-blade* elliptic to broadly elliptic, sometimes ovate or obovate, 1,5–5 cm long, 0,8–3 cm wide, apex acute to rounded, base cuneate to rounded, margin entire or irregularly crenate-serrate, wavy, midrib fairly prominent above, lateral veins 6, half-angled, dark-green and shiny above, paler below, thinly coriaceous, glabrous; petiole 3–7 mm long, channelled. *Inflorescence* of axillary, terminal panicles, divaricately branched; peduncles 2–3 mm long, puberulous; pedicels 2–5 mm long, puberulous. *Calyx-tube* pubescent; sepals 8 or 9, subulate, slightly shorter than petals, 1,4–2,4 mm long, 0,25 mm wide, pubescent, margins ciliate; glands opposite sepals, club-shaped, 0,4 mm long. *Petals* 8 or 9, elliptic-spathulate, 2–3 mm long, 0,75 mm wide, pubescent, margins white-ciliate. *Stamens* 8 or 9, terete, 2,7–3 mm long, sparsely hairy in lower half; anthers transversely elliptic, 0,4 mm long. *Ovary* cone-shaped, c. 1,5 mm long, tomentulose, with few to many ovules; styles 4 or 5, free and glabrous in upper half; stigmas simple. *Fruits* pilose inside, few-seeded. FIG. 24.

Occurring in forest and on riverbanks in the eastern Cape and Natal.

NATAL.—2930 (Pietermaritzburg): Inanda, *McKen & Buchanan 31* (K); Umzinyati Falls, *Wood 11463; 1333* (K); *McLean & Ogilvie sub PRE 28864*. 3030 (Port Shepstone): Warner Beach, *Ward 980;* Uvongo, *Mogg 13334;* Josephine Bridge, Umkomaas River, *Bayer 771*.

CAPE.—3227 (Stutterheim): banks of Nahoon River, East London, *Galpin 5717*. 3228 (Butterworth): Qora Bridge, Kentani, *Acocks 12291;* Kei Mouth, *Schlechter 22167;* confluence of Ngqageni and Qora Rivers, Willowvale, *Ward 5771*. 3325 (Port Elizabeth): Zuurberg Pass, *Archibald 5254*. 3326 (Grahamstown): Bathurst Research Station, *Brink 163*.

A clear-cut species with smaller leaves than *H. dentatum*. Sleumer (l.c.) has selected the Kew specimen of *Gueinzius* s.n. as the lectotype of *H. rufescens*. He cites the specimen as *Gueinzius* 92, but on none of the Gueinzius specimens seen by the author and certainly not the Kew specimen did the number 92 appear.

FIG. 24.—1, **Homalium rufescens,** flowering twig, × 1 *(Bayer* 771); a, flower, × 5 *(Ward* 980); b, stamen, × 7 *(Ward* 980); c, longitudinal section through ovary, × 20 *(Ward* 980).

3. **Homalium abdessammadii** *Aschers. & Schweinf.* in Sber. Ges. Naturf. Freunde Berl. 1880: 130 (1880); Gilg in Bot. Jahrb. 40: 494 (1908); Pflanzenfam. ed. 2, 21: 428 (1925); Engl., Pflanzenw. Afr. 3, 2: 581 (1921); Fernandes & Diniz in Garcia de Orta 5, 2: 252 (1957); Wild in F.Z. 1: 291 (1960); F. White, For. Fl. N. Rhod. 265 (1962); Sleumer in Bull. Jard. Bot. Nat. Belg. 43: 311 (1973). Type: Sudan, "bei Nganye", *Schweinfurth* 3954 (B, holo.†; K!; L!; P!; PRE, photo.).

H. macranthum Gilg in Bot. Jahrb. 40: 496 (1908); Pflanzenfam. ed. 2, 21: 428 (1925). Type: Mozambique, Rovumu River opposite Lissenga Mtn., *Busse* 1049 (B, holo.†; EA). *H. wildemanianum* Gilg in Bot. Jahrb. 40: 497 (1908); Pflanzenfam. ed. 2, 21: 428 (1925). Types: Zaïre, Katanga, Lukafu, *Verdick* 123 (B†; BR; lecto.!; PRE, photo.); Lofoi, *Verdick* 130 (B†; BR!). *H. rhodesicum* Dunkley in Kew Bull. 1934: 182, fig. (1934). Type: Zambia, Kafue, *Martin* 66 (K, holo.!; PRE, photo.). *H. abdessammadi* Aschers. & Schweinf. subsp. *wildemanianum* (Gilg) Wild in Bol. Soc. Brot. Sér. 2, 32: 57 (1958); Wild in F.Z. 1: 291 (1960).

Tree, 4–10 m high; branchlets glabrous to grey-pubescent. *Leaf-blade* broadly elliptic, 6–8 cm long, 4–5 cm wide, apex abruptly acuminate, acute, obtuse or rounded, base broadly cuneate to rounded, margin coarsely crenate-serrate; midrib and lateral nerves more prominent below, lateral nerves 6–8, ½-angled, looping near margin, glabrous except for few hairs on midrib, sometimes with hairy tufts or pockets in axils of nerves, coriaceous; petiole 0,8–1,5 cm long, glabrous or pubescent. *Flowers* in lax, terminal panicles up to 17 cm long, flowers solitary or 2 together, sessile or subsessile on puberulous branches of panicle. *Sepals* 5–7, subulate-triangular, 2,5–3 mm long, 1,75–2 mm wide, pubescent dorsally, glabrous ventrally, margins ciliate. *Petals* 5–7, elliptic-ovate, 3,3 mm long, 2–2,25 mm wide, pubescent dorsally, glabrous ventrally, margins ciliate. *Stamens* in 3's opposite petals, 4 mm long, glabrous or with long hairs in lower half; anthers 0,5 mm long. *Disc glands* discoid, 1 mm diam., sessile, tomentulose. *Ovary* conical, pilose outside and inside; style 2–3 mm long with diverging branches 0,3 mm long. *Fruit* capsular, woody, surrounded by persistent calyx and petals. *Seeds* usually solitary, narrowly ovoid, 1 mm long; testa thin, brown.

Found on riverbanks in tropical Africa, where it is fairly widespread, and in the Eastern Caprivi Strip of South West Africa.

S.W.A.—1724 (Katima Mulilo): Katima Mulilo, *West 3245.*

Readily distinguished from *H. dentatum* and *H. rufescens* by the stamens, which are in 3's instead of solitary. A gradation in degree of pubescence can be traced from a completely glabrous to a densely pubescent condition. For this reason, subsp. *wildemanianum* is here not accepted as a distinct taxon.

5315 9. TRIMERIA

Trimeria *Harv.*, Gen. Pl. 417 (1838); in F.C. 1: 68 (1860); Benth. & Hook.f., Gen. Pl. 1,1: 129 (1862); Harv., Gen. Pl. ed. 2: 15 (1868); Sim, For. Fl. Cape Co. 132 (1907); Engl., Pflanzenw. Afr. 3,2: 581 (1921); Gilg in Pflanzenfam. ed. 2, 21: 429 (1925); Phill., Gen. ed. 2: 514 (1951); Wild in F.Z. 1: 296 (1960); Wild & Vidigal in Fl. Mocamb. Flacourt. 38 (1973). Type species: *T. trinervis* Harv.

Monospora Hochst. in Flora 24: 660 (1841).

Renardia Turcz. in Bull. Soc. Nat. Moscou 31, 1: 466 (1858).

Shrubs or trees. *Leaves* alternate, simple, petiolate; blade ovate or orbicular, 3–9-nerved from base; stipules sometimes foliose, caducous. *Flowers* dioecious, small, in axillary racemes, spikes or panicles. *Male flowers: Sepals* 3–5, with marginal disc glands opposite each sepal. *Stamens* 9–16, inserted in 3's or 4's alternating with disc glands; anthers small, subglobose. *Female flowers: Sepals and petals* more or less as in male. *Ovary* superior, 1-locular, with 1 or 2 ovules on 3 parietal placentas; styles 3, short, persistent. *Fruit* a dry, 3-valved capsule, 1–3-seeded. *Seed* ellipsoid, testa leathery.

A genus of five species occurring in southern Africa, Rhodesia and tropical east Africa. The generic name is derived from the Greek meaning 3 parts in reference to the trimerous sepals and petals of the type species, *T. trinervis*.

Sepals and petals trimerous; leaves elliptic or obovate, 2,5–5 cm long, 1,5–2,5 cm wide .... 1. *T. trinervis*
Sepals and petals 4 or 5-merous; leaves circular or broadly obovate, 4–11 cm long, 4–10 cm wide .... 2. *T. grandifolia*

1. **Trimeria trinervis** *Harv.*, Gen. Pl. 417 (1838); in F.C. 1: 69 (1860); Sim, For. Fl. Cape Col. 133, t.8, II (1907); Engl., Pflanzenw. Afr. 3, 2: 582 (1921); Gilg in Pflanzenfam. ed. 2, 21: 430 (1925); Henkel, Woody Pl. Natal 107 (1934). Type: Cape, Van Stadens Mountains, *Zeyher* 724 (TCD, holo.!; PRE, photo.; BM!; K!; LD!; SAM!).

Tree up to 7 m high. *Branchlets* glabrous or pubescent. *Leaf-blade* elliptic or obovate, 2,5–5 cm long, 1,5–2,5 cm wide, apex acuminate, acute or rounded, base broadly cuneate to rounded, margin serrate-callose, 3–(5)-nerved at base, nerves conspicuous on lower surface, glabrous, subcoriaceous; petiole 0,4–1 cm long, glabrous or pubescent. *Inflorescence* of simple spikes, up to 2 cm long; bracts 3, broadly deltoid, 0,4 mm long, 6 mm wide, red; rhachis glabrous or puberulous. *Male flowers* sessile. *Sepals* 3, concave, oblong, 0,6 mm long, 0,7 mm wide, margin sparsely ciliate. *Petals* 3, concave, obovate-spathulate, 1,2 mm long, 1 mm wide, margin ciliate. *Stamens* 9, in 3's at base of petals, 1–2 mm long, glabrous; anthers small, subglobose. *Style* aborted, terete, 0,8–10 mm long. *Disc glands* opposite sepals, tooth-like, fleshy. *Female flowers* sessile. *Sepals* subrotund, 0,5 mm long, 0,6 mm wide, margin sparsely ciliate. *Petals* 3, subrotund, 0,5 mm diam., margin ciliate. *Disc glands* as in male flowers. *Ovary* ellipsoid, glabrous; styles 3, 0,3 mm long, slightly diverging, glabrous. *Fruits* obovoid, 4 mm long, 3,8 mm wide; seeds 1, ellipsoid-ovoid, 2,5 mm long, 1,3 mm wide; testa minutely tesselated.

A small tree occurring in forest and scrub-forest in the Transvaal, Natal and the Cape. There is a 400 km gap in the distribution of this species between Cape Town and Knysna. The fact that apparently only one specimen, *De Castelnau* 395, has ever been collected in the region of Cape Town, a well-collected area, makes this record somewhat suspect, especially when it is known that De Castelnau travelled as far east as "British Kaffraria".

TRANSVAAL.—2630 (Carolina): Mavieristad, *Pott 5115*.

NATAL.—2829 (Harrismith): Farm Glendale, Normandien, *Biggs 25*. 2929 (Underberg): above Dalton Bridge, *Wright, West & Acocks 11* (NH). 2930 (Pietermaritzburg): 8 km along Merrievale/Boston Road, *Moll 935*.

CAPE.—3028 (Matatiele): Pot River Berg, *Galpin 6575*. 3125 (Queenstown): Gwategu, *Galpin 8254*. 3225 (Somerset East): Waterkloof Valley, *Van der Walt 116*. 3227 (Stutterheim): Mount Kemp, *Killick 907*. 3318 (Cape Town): "circa Cape Town", *De Castelnau 395* (P). 3323 (Knysna): Knysna, *Bowie s.n.* (K). 3323 (Port Elizabeth): Van Stadens Mt., *Zeyher 724* (K); *3989*. 3326 (Grahamstown): Gameston, c. 20 km from Grahamstown, *Story 2626*.

Easily distinguished from *T. grandifolia* by the trimerous flowers and smaller, 3–(5) digitately nerved, elliptic leaves.

2. **Trimeria grandifolia** *(Hochst.) Warb.* in Pflanzenfam. 3, 6a: 37, f.13 H–J (1893); Dur. & Schinz, Consp. Fl. Afr. 1: 225 (1898); Burtt Davy, Fl. Transv. 1: 219 (1926); Henkel, Woody Pl. Natal 115 (1934); Killick in Bothalia 10: 568 (1972); Wild & Vidigal in Fl. Mocamb. Flacourt. 38 (1973). Type: Natal, "Port Natal", *Krauss* s.n. (B, holo.†; M, iso.!).

Monospora grandifolia Hochst. in Flora 24: 661 (1841). Type as above. *M. rotundifolia* Hochst., l.c. Type: Cape, Outeniqua and Goukamma River, *Krauss* s.n.

Antidesma alnifolium Hook., Icon. Pl. t.481 (1842). Syntypes: Eastern Cape, *Bowie* s.n. (K!; PRE, photo.); Natal, "Port Natal", *Krauss* 160 (G!; K!; PRE, photo.).

Renardia lejocarpa Turcz. in Bull. Soc. Nat. Mosc. 31, 1: 466 (1858). Type: without precise locality, *Ecklon* s.n.

Trimeria alnifolia (Hook.) Harv. in F.C. 1: 69 (1860); Sim, For. Fl. Cape Col. 132 (1907). *T. rotundifolia* (Hochst.) Gilg in Pflanzenw. Afr. 3, 2: 582 (1921); Milne-Redh. in Kew Bull. 1939: 34 (1939); Wild in F.Z. 1: 296 (1960).

Tree, 3–10 m high. *Bark* light brown, smooth or longitudinally fissured. *Branchlets* glabrous or pubescent. *Leaf-blade* circular or broadly obovate, 4–11 cm wide, apex rounded, emarginate, bilobed or apiculate, base truncate or shallowly cordate, margin serrate, digitately 5–7-nerved, nerves prominent below, glabrescent with hairs often persisting on nerves, subcoriaceous; petiole 1–3 cm long, glabrous or pubescent; stipules foliaceous, reniform, 1,2–2 cm long, 1–2,5 cm wide, caducous, pubescent. *Inflorescence* axillary, branches densely pubescent; pedicels up to 2,5 mm long, pubescent, jointed; bracts deltoid, 1 mm long. *Male inflorescences* in panicles, up to 9 cm long; female inflorescences spicate, but sometimes branched, up to 8 cm long. *Male flowers: Sepals* 4 or 5, concave, subulate, 1–1,2 mm long, 0,5 mm wide, pubescent. *Petals* 4 or 5, concave, obovate-deltoid, 1,1–1,3 mm long,

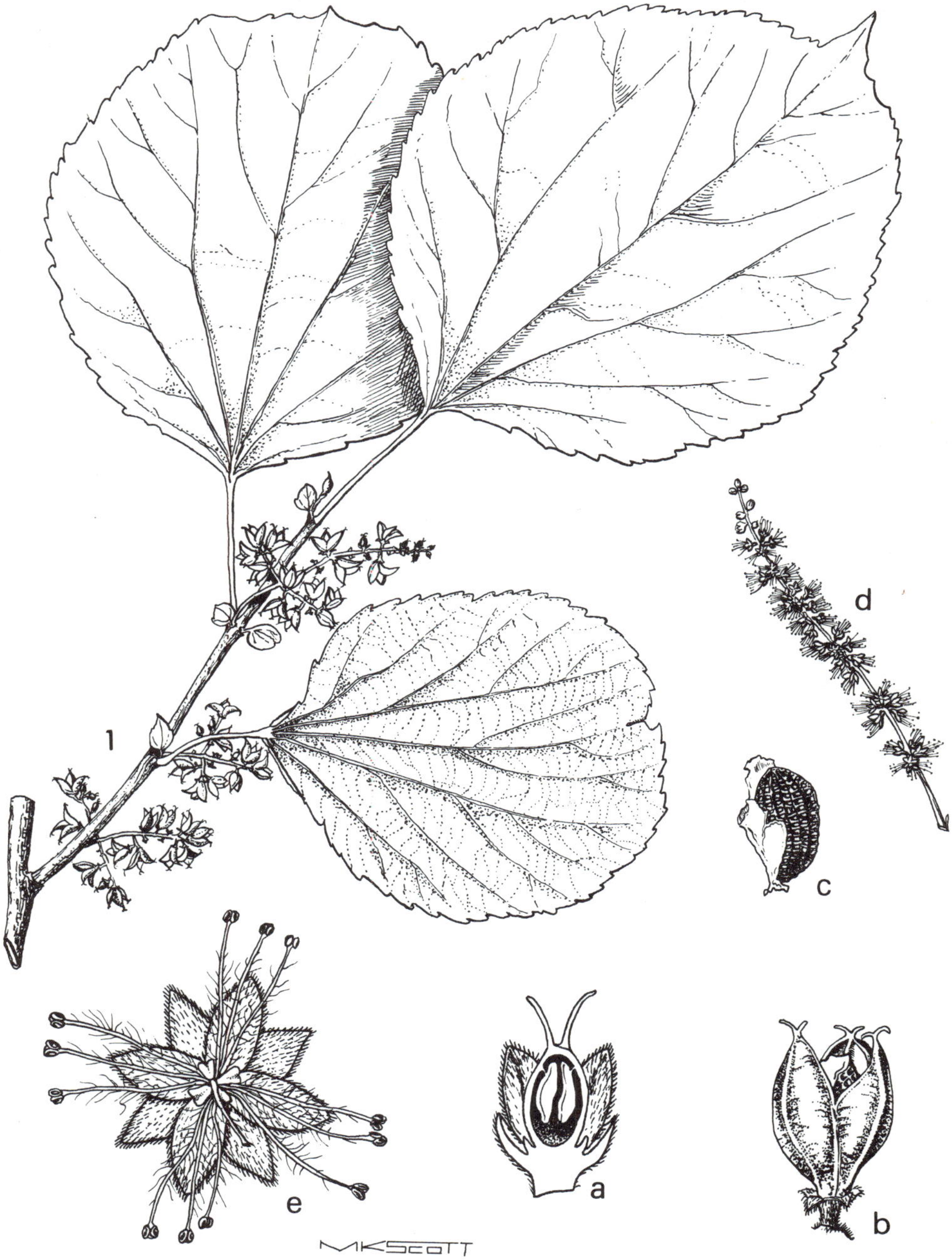

FIG. 25.—1, **Trimeria grandifolia,** twig with ♀ flowers, × 1 *(Guy* 73); a, longitudinal section through ♀ flower, × 5 *(Wells* 2136); b, fruit, × 8 *(Guy* 73); c, seed, × 8 *(Guy* 73); d, portion of male inflorescence, × 1 *(Harrison* 258); e, male flower, × 10 *(Harrison* 258).

0,6–0,8 mm wide, pubescent. *Disc* of marginal, fleshy, undulate glands opposite sepals. *Stamens* 9–16, perigynous, inserted in 3's or 4's alternating with glands; filaments 2,5 mm long, glabrous or pubescent; anthers small, subglobose; style aborted, terete, glabrous. *Female flowers: Sepals* 4, concave-deltoid, 1 mm long, 0,8 mm wide, pubescent. *Petals* as sepals. *Disc* of marginal scale-like glands. *Ovary* ellipsoid-obovoid, glabrous; styles 3, 0,5 mm long, slightly divergent. *Fruits* obovoid, 5 mm long, 3 mm wide, 3-valved; seeds 1 or 2, ellipsoid, 1,5–2 mm long, 1–1,4 mm wide, testa minutely tesselated, with axil on one side. FIG. 25.

A small tree found in forest from Knysna in the Cape eastwards to Natal and northwards to the Transvaal, Swaziland and Rhodesia.

TRANSVAAL.—2230 (Messina): Entabeni, *Obermeyer 862*. 2329 (Pietersburg): Lejuma near Louis Trichardt, *De Winter 6016*. 2330 (Tzaneen): Duiwelskloof, *Scheepers 552*. 2430 (Pilgrims Rest): Mariepskop, *Van der Schijff 4968*. 2530 (Lydenburg): Mount Anderson, *Galpin 13625*. 2531 (Komatipoort): camp between Louw's Creek and Maid of Mist Mountain, *Hutchinson 2444*.

O.F.S.—Without precise locality, *Cooper 1008*.

SWAZILAND.—2531 (Komatipoort): Piggs Peak Forest, *Compton 28220*. 2631 (Mbabane): Hlatikulu, *Compton 29499*.

NATAL.—2632 (Bela Vista): Sihadla River, *Strey & Moll 3903*. 2731 (Louwsburg): Ngotshe District, *Gerstner 5181*. 2829 (Harrismith): Mazonjwana River, Cathedral Peak, *Killick 1863*. 2830 (Dundee): Umhlumba Mt., *West 1479* (NH). 2831 (Nkandla): Ngoye Forest road, Mtunzini, *Wells & Edwards 16*. 2832 (Mtubatuba): Dukuduku East, *Moll 2708*. 2929 (Underberg): Tabamhlope, *West 1140*. 2930 (Pietermaritzburg): Byrne, *Galpin 11937*. 2931 (Stanger): 3 km from Mandini on Tugela Mouth road, *Edwards 1612*. 3030 (Port Shepstone): Ifafa, *Rudatis 498*.

CAPE.—3028 (Matatiele): Pot River Berg, *Galpin 6576* (K). 3029 (Kokstad): Insizwa Forest, *Strey 10728*. 3127 (Lady Frere): Bloemvlei, Elliot, *Van Zinderen Bakker 62*. 3129 (Port St. Johns): Lusikisiki District, *Galpin 9354* (K). 3226 (Fort Beaufort): Menziesberg, *Scully 599* (K). 3227 (Stutterheim): Kei Road, *Comins 1409*. 3228 (Butterworth): Kentani, *Pegler 746*. 3322 (Oudtshoorn): Wilderness woods, *Whellan 1511*. 3323 (Willowmore): Storms River Forest Reserve, *Dahlstrand 479;* Deepwalls Forest Reserve, *Bos 854*. 3327 (Peddie): Amalinda, *Nänni 95*.

Characterized by the large, usually circular, digitately 5–7-nerved leaves and 4-merous petals and sepals. According to a note on *West* 1140 the plant can behave as a liane. *T. grandifolia* is known to the Europeans as the Wild Mulberry or Wilde Moerbei and to the Zulus as iNdeblehlovu (ears of an elephant). According to Sim (l.c. 132), the tree is usually evergreen, but occasionally deciduous. He also states that the tree can attain a height of 16 metres in eastern Pondoland and Natal—rather higher than the heights recorded by collectors of the species. Wild & Vidigal, l.c., incorrectly attribute the combination *T. grandifolia* to Durand & Schinz.

5326 10. APHLOIA

Aphloia *(DC.) Benn.* in Benn. & R. Br., Pl. Jav. Rar. 2: 192 (1840); Benth. & Hook.f., Gen. Pl. 1, 1: 126 (1862); Perrier in Fl. Madag. Flacourt. 12 (1946); Wild in F.Z. 1: 279 (1960); Wild & Vidigal in Fl. Mocamb. Flacourt. 19 (1973). Type species: *A. theiformis* (Vahl) Benn.

Prockia sect. *Aphloia* DC., Prodr. 1: 261 (1824).
Neumannia A.Rich. in La Sagra, Hist. Fis. Cuba 10: 96 (1843).

Trees or shrubs. *Leaves* alternate, simple, petiolate; stipules minute, caducous. *Flowers* bisexual, axillary, solitary, in few-flowered fascicles or racemes. *Sepals* 4–6, free, except at base, orbicular, concave, imbricate, somewhat leathery. *Petals* 0. *Stamens* numerous, free, inserted outside or on a flat disc. *Ovary* sessile or shortly stipitate, superior, 1-locular, with few to several ovules on a single parietal placenta; style very short; stigma broadly peltate with a median furrow. *Fruit* a fleshy berry with c. 6 seeds. *Seeds* obovoid or globose, somewhat flattened, incurved testa crustaceous with thin endosperm; embryo incurved; cotyledons ovate.

About 6 species occurring from Tanzania, southwards to Natal and in Madagascar, the Mascarene and Comoro Islands and Seychelles. The generic name is derived from the Greek for "without a corolla (petals) or bark" (Wittstein).

Aphloia theiformis *(Vahl) Benn.* in Benn. & R. Br., Pl. Jav. Rar. 2: 192 (1840); Bak.f. in J. Linn. Soc. (Bot.) 40: 23 (1911); Perrier in Fl. Madag. Flacourt. 13 (1946); Wild in F.Z. 1: 279 (1960); Wild & Vidigal in Fl. Mocamb. Flacourt. 20 (1973). Type: Reunion Island, *Commerson* s.n. (C, holo.; PRE, photo.).

Lightfootia theiformis Vahl, Symb. Bot. 3: 69 (1794). Type as above.

Prockia theiformis (Vahl) Willd., Sp. Pl. 2, 2: 1214 (1799). Type as above.

Neumannia theiformis (Vahl) A.Rich. in La Sagra, Hist. Fis. Cuba 10: 97 (1845); Dur. & Schinz, Consp. Fl. Afr. 1, 2: 218 (1898); Gilg in Bot. Jahrb. 40: 503 (1908); Pflanzenfam. ed. 2, 21: 437, t.200 (1925); Engl., Pflanzenw. Afr. 3,

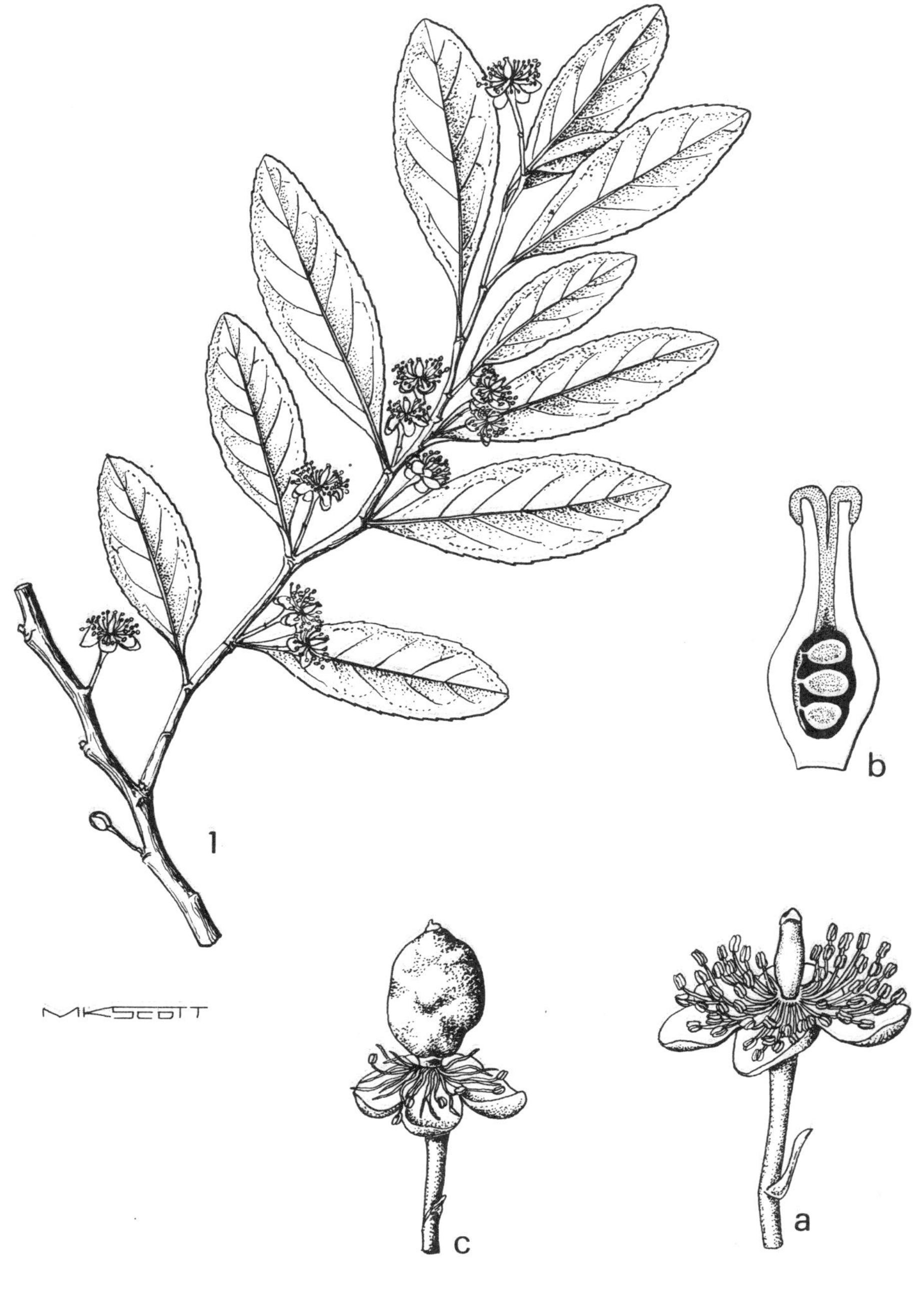

FIG. 26.—1, **Aphloia theiformis,** flowering twig, × 1 *(Scheepers* 679); a, flower, × 5 *(Compton* s.n.); b, longitudinal section through ovary, × 10 *(Strey* 7750); c, fruit, × 1½ *(Obermeyer* 1107).

2: 584, t.260 (1921). *N. myrtiflora* (Galpin) Th. Dur. in Dur. & Schinz, Consp. Fl. Afr. 1, 2: 218 (1898). Type as for *Aphloia myrtiflora* Galpin.

Aphloia myrtiflora Galpin in Kew Bull. 1895: 142 (1895); Burtt Davy, Fl. Transv. 1: 215 (1926); Brenan, Checklist Tang. Terr. 2: 229 (1949). Type: Transvaal, United Ivy Reef, Moodies near Barberton, *Galpin* 1082 (PRE, holo.!; BOL!; K!; NH!; SAM!).

Tree or shrub up to 10 m high. *Branchlets* brown, longitudinally striate, with narrowly decurrent wings arising from outer extremities of stipular cushions, glabrous. *Leaf-blade* elliptic, 3–8 cm long, 1–3 cm wide, apex acuminate, acute or rounded, base cuneate to broadly cuneate, margin serrulate, often upper $^2/_3$'s only, lateral nerves about 10 pairs, inconspicuous, glabrous, chartaceous; petiole 2–4 mm long. *Flowers* axillary, solitary or in fascicles of 1–3; pedicels greenish, up to 2 cm long, with trifid bracteoles up to 1,8 mm long. *Sepals* white, somewhat leathery, broadly elliptic to circular, concave, 2,5–3,5 mm diam., connate at base. *Stamens*: filaments 3 mm long, glabrous; anthers orbicular, 0,7 mm diam. *Ovary* ellipsoid, sometimes shortly stipitate, 3–4 mm long; stigma subsessile, peltate, with a median groove. *Fruit* a white, fleshy berry, c. 8 mm diam., with persistent stigma on top. *Seeds* several, subglobose-obovoid, somewhat compressed, 2,5–3 mm diam. FIG. 26.

A widely distributed species occurring in Tanzania, Rhodesia, Mozambique, Transvaal, Swaziland, Natal, Madagascar, the Comoro and Mascarene Islands and Seychelles.

TRANSVAAL.—2230 (Messina): Entabeni Forest, *Galpin 9558; Hutchinson & Gillett 4226* (K). 2329 (Pietersburg): De Hoek, *Keet 1185.* 2330 (Tzaneen): Westfalia Estate, Duiwelskloof, *Scheepers 679;* Debengeni Falls, *Vahrmeijer 468.* 2430 (Pilgrims Rest): Mariepskop, *Van der Schijff 6366.* 2531 (Komatipoort): Ida Doyer Nature Reserve, *Edwards 4124;* United Ivy Leaf, Moodies, near Barberton, *Galpin 1082.*

SWAZILAND.—2531 (Komatipoort): Piggs Peak, *Compton 28228;* Kings Bush, *Miller 7475.*

NATAL.—2930 (Pietermaritzburg): Zwati, Ndwedwe, *Strey 7526; 7750a; 7750b.*

A smallish tree or shrub with longitudinally striate branchlets, elliptic, serrulate leaves and smallish apetalous flowers on longish pedicels. A variable species with numerous infraspecific taxa described from Madagascar. Perrier (l.c. 18) considers the South African plant to be *A. theiformis* (Vahl) Benn. subsp. *madagascariensis* (Clos.) Perrier var. *closii* Tul. In the present treatment, *A. theiformis* is accepted in a broad sense without recognition of infraspecific taxa.

5327 11. FLACOURTIA

Flacourtia *L'Hérit.*, Stirp. Nov. 3: 59, t.30, 30 β (1786); Benth. & Hook.f., Gen. Pl. 1, 1: 128 (1862); Sim, For. Fl. P.E. Afr. 13 (1909); Engl., Pflanzenw. Afr. 3,2: 585 (1921); Gilg in Pflanzenfam. ed. 2,21: 438 (1925); Phill., Gen. ed. 2: 515 (1951); Wild in F.Z. 1: 285 (1960); Bamps in F.C.B. Flacourt. 48 (1968); Wild & Vidigal in Fl. Mocamb. Flacourt. 27 (1973). Type species: *F. ramontchi* L'Hérit.

Trees or shrubs, often spiny. *Leaves* alternate, simple, petiolate, mostly crenate. *Flowers* dioecious, rarely bisexual, in short axillary racemes or solitary. *Calyx* 4–7-partite or sepals free and imbricate. *Petals* 0. *Male flowers: Stamens* numerous; filaments terete; anthers small, elliptic or somewhat quadrate, sometimes subdidymous. *Female flowers: Stamens* 0 or few. *Ovary* surrounded by annular or interrupted disc, usually 4–8-locular with 2 superposed ovules in each loculus; styles 4–8, short, thick, more or less radiating, persistent; stigma retuse or marginate. *Fruit* a berry, 5–16-locular. *Seed* obovoid or subcompressed; testa bony, rough; embryo straight; cotyledons flat, ovate; endosperm present.

About 15 species found in south-east Asia, Malaysia, Polynesia, Madagascar, tropical and southern Africa. The genus was named in honour of Etienne de Flacourt (1607–1660), General Director of the French East India Company, Governor of Madagascar and author of a history of that island.

Flacourtica indica *(Burm.f.) Merr.*, Interpr. Rumph. Amboin. 377 (1917); Gilg in Pflanzenfam. ed. 2,21: 440, t.201 (1925); Brenan, Checklist Tang. Terr. 2: 231 (1949); Sleumer in Fl. Males. Flacourt. 76 (1954); Wild in F.Z.: 286 (1960); Bamps in F.C.B. Flacourt. 48 (1968); Wild & Vidigal in Fl. Mocamb. Flacourt. 28 (1973). Type: Java, Doerÿ Roekan, *Burmann* collection (G, holo.!; PRE, photo.).

Gmelina indica Burm.f., Fl. Ind. 132, t.39, fig. 5 (1768).

Flacourtia ramontchi L'Hérit., Stirp. Nov. 3: 59, t.30, 30 β (1786); Oliv. in F.T.A. 1: 120 (1868); Sim, For. Fl. P.E.

FIG. 27.—1, **Flacourtia indica,** flowering twig, × 1 (*Smuts & Gillett* 3415); a, fruit, × 1 (*Acocks* 12875); b, male flower, × 6 (*Codd & De Winter* 5057); c, bisexual flower, × 6 (*Codd & De Winter* 5057); d, female flower, × 6 (*Edwards* s.n.); e, longitudinal section through ♀ flower, × 6 (*Edwards* s.n.).

Afr. 13 (1909); Bak.f. in J. Linn. Soc. (Bot.) 40: 23 (1911); R.E. Fr., Schwed. Rhod.-Kongo-Exped. 1: 157 (1914); Perrier, Fl. Madag. Flacourt. 9 (1946). Syntypes: Madagascar, *Poivre* s.n.; *Commerson* s.n. (P!). *F. hirtiuscula* Oliv., l.c. 121 (1868); Sim, l.c. 13 (1909); Bak. f., l.c. 23 (1911); Gilg, l.c. 440 (1925); Burtt Davy, Fl. Transv. 1: 215 (1926). Type: Moçambique, near Sena, *Kirk* s.n. (K, holo.!; PRE, photo.).

Tree or shrub, 1,5–16 m high. *Bark* smooth, stippled white (teste *De Winter* 7748) or rough and yellowish or orange-brown, occasionally silvery on young branches (Wild l.c.). *Branches* with or without axillary straight spines, sometimes with branching spines up to 12 cm long on the trunk near the base or with very spiny coppice shoots, glabrous or pubescent. *Leaf-blade* elliptic, ovate or obovate, 5–9,5 cm long, 3,5–6 cm wide, apex acuminate, acute or rounded, base cuneate to rounded, margins crenate, crenate-serrate or entire, veins 4–6 pairs, more prominent on lower surface, chartaceous, glabrous to densely pubescent; petiole 4–10 mm long, channelled above, glabrous or pubescent. *Flowers* dioecious or bisexual, greenish. *Calyx* segments or sepals often unequal, narrowly to broadly ovate, 2–3 mm long, 1,5–3 mm wide, ciliate. *Male flowers: Stamens* very numerous, surrounded by disc of fleshy free glands; filaments terete, 2,2 mm long. *Female flowers: Ovary* globose c. 3 mm diam., surrounded by an annular disc; styles 4–8, channelled above; stigmas retuse. *Bisexual flowers* as female flowers, but with up to 10 stamens. *Fruit* red, globose, up to 3 cm diam., fleshy with persistent styles. *Seeds* c. 10, obovoid, subcompressed, c. 7 mm diam.; testa beige, hard, rough. FIG. 27.

Occurring in scrub, woodland and forest, sometimes riverine, in the Transvaal. Also found in south and eastern tropical Africa, Madagascar, south-east Asia, Polynesia and Malaysia.

TRANSVAAL.—2229 (Waterpoort): northern end of Wyllies Poort, *De Winter 7748.* 2230 (Messina): Makonde Mission Station, *Codd 6851.* 2328 (Baltimore): 37 km E. of Potgietersrust on road to Swartwater, *Van der Schijff 5323.* 2329 (Pietersburg): Louis Trichardt, *Gerstner 5680.* 2330 (Tzaneen): Elim, *Obermeyer 541.* 2427 (Thabazimbi): upper slopes of Krantzberg, *Prosser 1717* (K). 2428 (Nylstroom): between Nylstroom and Warmbaths, *Pole Evans 188.* 2429 (Zebediela): Potgietersrust to Zebediela, *Pole Evans 3104 (23).* 2431 (Acornhoek): 5 km N.E. of Skukuza on Tshokwane Road, *Codd & De Winter 5057.* 2526 (Zeerust): Enselberg camp, *Pole Evans 2225.* 2527 (Rustenburg): Jacksontuin, Brits, *Mogg 14979.* 2529 (Witbank): Loskop Dam, *Theron 923.* 2530 (Lydenburg): Lowveld Botanic Garden, *Buitendag 815.* 2531 (Komatipoort): Pretoriuskop, *Van der Schijff 3924.*

A small tree with or without spines, elliptic, ovate or obovate leaves with usually crenate margins and red, globose fruits up to 3 cm in diam. An extremely variable species as regards leaf shape and degree of pubescence. The plant is variously known as the Madagascar, Mauritius, Governor's or Batoko Plum and is cultivated in the tropics of both the Old and New Worlds, because of its edible fruit. The rather acid fruits may be eaten raw or cooked to make jam and preserves. The wood is used for making agricultural implements and the plant is used medicinally in tropical Africa.

5328 12. DOVYALIS*

Dovyalis *E. Mey. ex Arn.* in Hook., J. Bot. 3: 251 (1841); Drège, Cat. Pl. Afr. Austr. 3 (1840); Zwei Pfl. Doc. 125 (1843), p. 180, as *Doryalis,* nom. nud.; Endl., Gen. Pl. Suppl. 2: 91 (1842); Sond. in Linnaea 23: 12 (1850); Harv. in F.C. 1: 69 (1860); Benth. & Hook.f., Gen. Pl. 1: 128 (1862); Harv., Gen. Pl. ed. 2: 16 (1868); Warb. in Pflanzenfam. 3,6a: 44 (1893), as *Doryalis*; Sim, For. Fl. Cape Col. 129 (1907); For. Fl. P.E. Afr. 13 (1909); Engl. in Bot. Jahrb. 40: 504 (1908), as *Doryalis*; Gilg in Pflanzenfam. ed. 2, 21: 440 (1925); Burtt Davy, Fl. Transv. 1: 216 (1926); Phill., Gen. ed. 2: 515 (1951); Wild in F.Z. 1: 281 (1960); Sleumer in Bot. Jahrb. 92: 64 (1972). Type species: *D. zizyphoides* E. Mey. ex Arn.

Aberia Hochst. in Flora 27: 2 (1844); Sond. in Linnaea 23: 9 (1850); Harv. in F.C. 1: 70 (1860); in F.C. 2: Addenda 584 (1862); Benth. & Hook.f., Gen. Pl. 1: 128 (1862); Oliv. in F.T.A. 1: 121 (1868); Harv., Gen. Pl. ed. 2: 16 (1868). Type species: *A. verrucosa* Hochst.

Trees or shrubs, evergreen or deciduous, dioecious or rarely polygamous, often armed with spines; branches lenticellate. *Leaves* exstipulate, alternate or sometimes fascicled, petiolate; blade 3–5-veined at or near the base, margins entire, crenate or serrate. *Male flowers* pedicellate, in fascicles of 2–10, or 5–10 flowers in groups of 3; pedicels surrounded at the base by small hairy

---

* By J. E. Langenegger.

scales; calyx (tepals of Sleumer l.c.) 2–6-lobed; corolla 0; stamens 10–many; filaments surrounded at the base by sparsely to densely hairy nectaries which form a honeycomb structure; anthers bilocular, dehiscing by means of longitudinal slits. *Female flowers* pedicellate, axillary, solitary or in fascicles of 2 or 3; pedicels surrounded at the base by small hairy scale-like bracts; calyx (tepals) deeply 3–7-lobed; lobes sometimes revolute and sometimes fringed with stipitate glands; corolla 0; disc annular, lobed, glutinous, hairy, sparsely hairy or glabrous; ovary unilocular or sometimes incompletely 2 or 3-locular; placentas 2–7, parietal, each placenta with 1 or 2 ovules; styles channelled, as many as placentas. *Fruit* a subglobose to oblong berry, minutely papillose, glabrous or hairy. *Seeds* embedded in fleshy pulp, elliptic to broadly elliptic, glabrous to densely woolly; testa leathery; embryo straight; cotyledons flat, elliptic, sometimes thin and pinnately veined; endosperm present.

A genus of about 20 species, widespread in Africa and extending to Sri Lanka and New Guinea; 6 species recognized in Southern Africa.

The spelling of the generic name has been the subject of some confusion. It first appeared as a *nomen nudum* in Drège, Cat. Pl. Afr. Austr. (1840), spelt *Dovyalis* E. Mey. A year later, in 1841, Arnott provided a description of *Dovyalis*. In Drège's Zwei Pfl. Doc. (1843) the generic name was spelt *Dovyalis* in the text and *Doryalis* in the index. It seems very probable that the latter spelling was intended, because (a) it is placed in the correct alphabetical sequence in the index, and (b) it is no doubt derived from the Greek "dory", a spear, in allusion to the conspicuous spines found on most plants. However, although Warburg in 1895, restored *Doryalis*, the original spelling *Dovyalis* used by Arnott must be adopted according to Article 73 of the International Code of Botanical Nomenclature (1972).

Ovary with 5–7 placentas, each placenta with 2 ovules; styles 5–7; fruit with many seeds; fruit 3–4 cm in diam.; ultimate branchlets often abbreviated, up to about 2 cm long or forming cushion-like structures; leaves fascicled or, on young shoots, alternate .... 1. *D. caffra*

Ovary with 2 or 3 placentas, each placenta with 1 ovule; styles 2 or 3; fruit with 1–3 seeds; fruit 0,6–2 cm in diam.; ultimate branchlets often abbreviated, but not forming cushion-like structures; leaves not fascicled:

- Exocarp of fruit glabrous:
  - Testa of seed glabrous .... 2. *D. rotundifolia*
  - Testa of seed hairy:
    - Leaves ovate to elliptic-ovate; leaf base cordate to truncate; the secondary nerves tending to spread towards the leaf margin .... 3. *D. rhamnoides*
    - Leaves obovate; leaf base cuneate; the secondary nerves tending to ascend to the leaf tip .... 4. *D. longispina*
- Exocarp of fruit hairy or papillose:
  - Calyx (tepal) lobes of ♀ flowers 4–6 mm long in anthesis, accrescent .... 5. *D. zeyheri*
  - Calyx (tepal) lobes of ♀ flowers 1,5–2 mm long in anthesis, not accrescent .... 6. *D. lucida*

1. **Dovyalis caffra** *(Hook.f. & Harv.) Hook.f.* in Harv., Gen. Pl. ed. 2: 16 (1868); Warb. in Pflanzenfam. 3, 6a: 44 (1893), as *Doryalis*; Sim, For. Fl. Cape Col. 129 (1907); For. Fl. P. E. Afr. 13 (1909); Engl. in Bot. Jahrb. 40: 604 (1908), as *Doryalis*; Marloth, Fl. S. Afr. 2: 195 (1925); Gilg in Pflanzenfam. ed. 2, 21: 440 (1925); Burtt Davy, Fl. Transv. 1: 216 (1926); Wild in F.Z. 1: 285 (1960); Sleumer in Bot. Jahrb. 92: 68 (1972). Syntypes: Cape, eastern districts, *R. Hallack* s.n. (TCD!; PRE, photo.); Kaffirland, *Holland* s.n.

Aberia caffra Hook.f. & Harv. in F.C. 2: 584 (1862); T. Anders. in J. Linn. Soc. 7: 68 (1862); Fourcade, Rep. Natal Forests 94 (1889). *A. edulis* T. Anders. l.c. 68, nom. nud.

Tree or shrub up to 6 m tall, evergreen, dioecious or rarely polygamous, armed with spines up to 6 cm long; bark ashen-white to brown with longitudinal ridges and fissures; ultimate branchlets often much abbreviated, up to about 2 cm long or forming cushion-like structures. *Leaves* fascicled or, on young shoots, alternate; blade soft to coriaceous, obovate or sometimes elliptic-rhomboid, 2–5,5 cm long, 1–3 cm broad, glabrous or rarely puberulous, with prominent venation on both sides, 3–5-veined from the base, apex acute to obtuse or emarginate, base usually cuneate, margin entire, slightly revolute; petiole 2–4 mm long, glabrous. *Male flowers* light green, fascicled, with 5–10 flowers in groups of 3; calyx 2–5-lobed; lobes 3 mm long, elliptic, pubescent; stamens numerous; filaments 5 mm long; nectaries hairy. *Female flowers* light green, solitary or in fascicles of 2–3 on abbreviated shoots; pedicels 4–10 mm long; calyx deeply 5–7-lobed, lobes 3 mm long, elliptic-lanceolate, puberulous, acuminate, somewhat recurved; disc sparsely hairy; ovary 5–7-lobed, unilocular; placentas 5–7, each placenta with 2 ovules; styles 5–7. *Fruit* subglobose, 3–6 cm in diam., minutely velvety, bright yellow. *Seeds* many, woolly, 1 cm long. FIG. 28: 2; 29: 1.

A common constituent of mixed scrub, riverine bush, open woodland and forest; distributed on a variety of soils from the Bathurst district in the eastern Cape northwards along the coast and in river valleys to Natal, eastern, central and northern Transvaal, and extending to Rhodesia. It has been introduced into California and Australia.

TRANSVAAL.—2231 (Pafuri): Shipudza area, Punda Milia, *Van Wyk 4741*. 2329 (Pietersburg): on tropic of Capricorn, *Story 4840*. 2330 (Tzaneen): Elim, *Obermeyer 752*. 2430 (Pilgrims Rest): Ohrigstad, *Burtt Davy 7314*; Nooitgedacht Mountain, *Young A 657*. 2531 (Komatipoort): Pretoriuskop, *Van der Schijff 3902*.

NATAL.—2732 (Ubombo): Mkuze Game Reserve, *Ward 3831*. 2830 (Dundee): Keats Drift road from Muden, *Edwards 913*. 2831 (Nkandla): Eshowe, *Lawn 1270* (NH); Hlabisa, *Ward 4783*.

CAPE.—3129 (Port St. Johns): Tabankulu, 6 km from Welsh Bridge on Ceveraland road, *Story 4218*. 3228 (Butterworth): Kei River Valley, *Flanagan 467*.

Dovyalis caffra, the Kei-apple, is an attractive, usually evergreen tree which, from December to about February, is covered with masses of yellow apricot-like fruits. The fruits are edible and are used to make jelly and other preserves. To counter the rather acidic taste of the fruits due to a high malic acid content, the fruits are often mixed with a sweet fruit, e.g. grapes. The plant is spinous and makes an excellent impenetrable hedge. It has the added virtue that it is fairly drought resistant.

2. **Dovyalis rotundifolia** *(Thunb.) Thunb. & Harv.* in F.C. 1: 70 (1860); Warb. in Pflanzenfam. 3, 6a: 44 (1893), as *Doryalis*; Sim, For. Fl. Cape Col. 132 (1907); Gilg in Bot. Jahrb. 40: 504 (1908); Pflanzenfam. 2, 21: 440 (1925); Burtt Davy, Fl. Transv. 1: 216 (1926); Sleumer in Bot. Jahrb. 92: 82 (1972). Type: Cape, *Thunberg* s.n. (UPS, holo.).

Celastrus rotundifolius Thunb., Prodr. 42 (1794); Willd., Sp. Pl. 1: 1129 (1798); Thunb., Fl. Cap. ed. Schult. 221 (1823).

Prockia rotundifolia (Thunb.) Eckl. & Zeyh., Enum. 16 (1834).

Dovyalis celastroides Sond. in Linnaea 23: 9 (1850), nom. illegit.; Sim, For. Fl. P.E. Afr. 13 (1909); Marloth, Fl. S. Afr. 2: 194 (1925); Henkel, Woody Pl. Natal 102 (1934).

Tree or shrub up to 6 m tall, dioecious, armed with spines up to 8 cm long; bark ashen-white with longitudinal ridges and fissures, strongly lenticellate. *Leaves* alternate; blade coriaceous, sub-orbicular to obovate, 2,2–3 cm long, 1,5–2 cm broad, glabrous, with prominent venation on both sides, 3–5-veined from the base, apex obtuse to emarginate, base obtuse to cuneate, margin entire; petiole 2–3 mm long, glabrous. *Male flowers* light green, in fascicles of 2–10; calyx 4–6-lobed; lobes 1,5–2,5 mm long, lanceolate to obovate-rhomboid, pubescent; stamens 9–20, filaments 3 mm long; nectaries hairy. *Female flowers* yellow-green, solitary or in fascicles of 2 or 3; pedicels 0,5–1,5 mm long; calyx 4–6-lobed; lobes 1–2 mm long, obovate, pubescent, acuminate, somewhat recurved; disc sparsely hairy; ovary 2-lobed, unilocular; placentas 2, each placenta with 1 ovule; style 2. *Fruit* oblong, 0,6–1 cm in diam., glabrous, red when ripe. *Seeds* 2, glabrous, 0,2–0,6 mm long. FIG. 29: 4.

Found in coastal bush on sand dunes in the eastern Cape Province from Port Elizabeth to East London.

CAPE.—3227 (Stutterheim): Bonza Bay, *Comins 1068*. 3325 (Port Elizabeth): Amanzi, *Wells 3178*. 3326 (Grahamstown): Port Alfred, *Tyson sub TRV 17246*. 3327 (Peddie): Great Fish River, *MacOwan 825* (NBG).

3. **Dovyalis rhamnoides** *(Burch. ex DC.) Burch. & Harv.* in F.C. 1: 69 (1860); Fourcade, Rep. Natal Forests 94 (1889); Warb. in Pflanzenfam. 3, 6a: 44 (1895), as *Doryalis*; Sim, For. Fl. Cape Col. 131 (1907); Engl. in Bot. Jahrb. 40: 504 (1908), as *Doryalis*; Gilg in Pflanzenfam. ed. 2, 21: 440 (1925); Burtt Davy, Fl. Transv. 1: 216 (1926); Verdoorn, Edible Wild Fr. Transv. 33 (1938); Sleumer in Bot. Jahrb. 92: 80 (1972). Type: Cape, George, Uitenhage and Albany, *Burchell* 6012 (PRE!).

Flacourtia rhamnoides Burch. ex DC., Prodr. 1: 256 (1824); Eckl. & Zeyh., Enum. 16 (1834).

Dovyalis zizyphoides E. Mey. ex Sond. in Linnaea 23: 12 (1850); E. Mey. in Drège, Cat. Pl. Afr. Austr. 3 (1840), nom. nud.; E. Mey. ex Arn. in Hook., J. Bot. 3: 251 (1841), comb. superfl.; Pappe, Fl. Cap. Med. Prodr. 4 (1859); Marloth, Fl. S. Afr. 2: 194 (1925). Syntypes: Cape, Ruigtevalei, *Drège* s.n. (PRE!; S!); Krakakamma and Olifantshoek, *Ecklon* 115 (S!; PRE, photo.); Van Stadens Mts, *Zeyher* s.n. (S!; PRE, photo.); Port Natal, *Gueinzius* 58 (S!; PRE, photo.).

Tree or shrub up to 5 m tall, dioecious, armed with spines up to 6,2 cm long; bark white to brown with shallow longitudinal ridges and fissures, young twigs and spines pubescent.

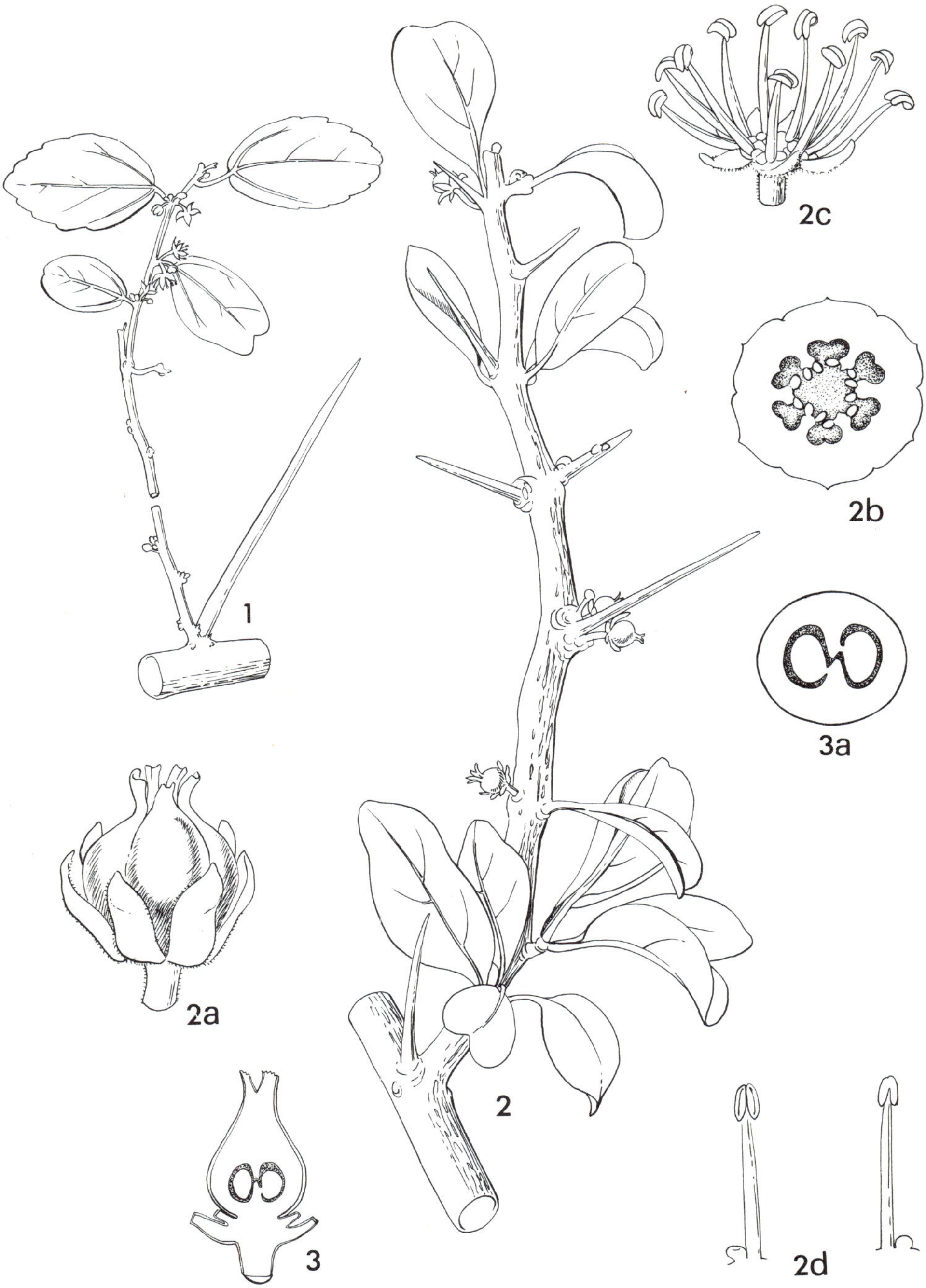

FIG. 28.—1, **Dovyalis rhamnoides,** branch with male flowers, × $^2/_3$ (*Wells* 3912). 2, **D. caffra,** branch with female flowers, × $^2/_3$; 2a, female flowers, × $5^1/_3$; 2b, cross-section of ovary, × $5^1/_3$; 2c, male flower, × 6; 2d, stamen, × $5^1/_3$ (*Thom* 1 and 2). 3, **D. longispina,** longitudinal section of female flower, × $5^1/_3$; 3a, cross-section of ovary, × $5^1/_3$ (*Bourquin* 557).

Leaves alternate; blade thin and somewhat membranous, ovate to elliptic-ovate, 2–4,5 cm long, 1–2,5 cm broad, glabrous or sometimes puberulous especially on the veins, veins prominent on both surfaces, 3–5-veined from the base, apex acute to obtuse, base truncate to cordate or sometimes cuneate, margin entire to crenate; petiole 2–3 mm long, glabrous or puberulous. *Male flowers* light green, in fascicles of 3–6; calyx 4–5-lobed, lobes 2–3 mm long, ovate, pubescent; stamens 15, filaments 1,5 mm long; nectaries hairy. *Female flowers* light green, solitary or in fascicles of 2 or 3; pedicels 2–3 mm long; calyx deeply 5-lobed, lobes 1–5 mm long, linear, acuminate, erect, puberulous and usually fringed with stipitate glands; disc hairy; ovary 2-lobed, unilocular, with 2 much-reduced placentas, each placenta with 1 ovule; styles 2. *Fruit* ellipsoid to ovoid, 1 cm in diam., glabrous, orange to scarlet red when ripe. *Seeds* 2–3, 9 mm long, slightly woolly. FIG. 28: 1; 29: 5.

Found in dune forests, on riverbanks, in temperate ravine forest, scrub forest and thornveld. Recorded from the eastern Cape, Natal, Zululand, Swaziland and the Transvaal.

TRANSVAAL.—2431 (Acornhoek): Lothian, *Strey 3555*.

SWAZILAND.—2631 (Mbabane): Hlatikulu, *Compton 29496*.

NATAL.—2731 (Louwsburg): Wendelane Kloof, *Gerstner 4674*. 2831 (Nkandla): Mtunzini, *Edwards 1582*. 2832 (Mtubatuba): Hluhluwe Game Reserve, *Ward 1690*. 2930 (Pietermaritzburg): Mt. Ashley, Lions River, *Moll 1308*. 2931 (Stanger): Groutville, *Acocks 11820*. 3030 (Port Shepstone): Umpanyoni River, *Rudatis 1989* (STE).

CAPE.—3227 (Stutterheim): Stutterheim Commonage, *Acocks 9720*. 3228 (Butterworth): Komga, *Flanagan 374* (NBG). 3322 (Beaufort West): Homtini River gorge, *Taylor s.n.* (Saasveld Herb.). 3325 (Port Elizabeth): Van Staadens Kloof, *Taylor 834*. 3326 (Fort Beaufort): Kowie, *Britten 768*. 3422 (Mossel Bay): Gouwkamma River, *Keet 611* (STE). 3423 (Knysna): Knysna River, *Schonland 3431* (GRA). 3424 (Humansdorp): Humansdorp, *Galpin 4577*.

According to Smith (1966), *D. rhamnoides* was known as Wynbessie, because of the pleasantly vinous taste of the ripe fruits from which brandy and vinegar were made in the early settler days of the eastern Cape. The wood of this species was formerly used for making yokes, wagons and agricultural implements.

4. **Dovyalis longispina** *(Harv.) Warb.* in Pflanzenfam. 3, 6a: 44 (1893), as *Doryalis*; Gilg in Bot. Jahrb. 40: 506 (1908), as *Doryalis*; Pflanzenfam. ed. 2, 21: 441 (1925). Syntypes: Natal, near Durban, *Gerrard & McKen* 542; 541 (TCD, holo.!; PRE, photo.).

Aberia? longispina Harv. in F.C. 2, App.: 585 (1862).
Dovyalis celastroides sensu Sim, For. Fl. P.E. Afr. 13, t.17fB (1909).

Tree or shrub up to 11,5 m high, dioecious, armed with spines up to 8 cm long; bark ashen-white with longitudinal ridges and fissures, strongly lenticellate. *Leaves* alternate; blade soft to coriaceous, obovate to rhomboid-elliptic, seldom orbicular, 2,2–9,5 cm long, 1,5–5,5 cm broad, glabrous, sometimes puberulous on veins, venation very prominent on both sides, 3–5-veined from the base, apex acute to emarginate, base cuneate to obtuse, margin entire to sub-undulate; petiole 2–6 mm long, glabrous. *Male flowers* light green, in fascicles of 2–10; calyx 4–6-lobed; lobes 1,5–2,5 mm long, obovate-rhomboid, pubescent; stamens 9–20; filaments 3 mm long; nectaries hairy. *Female flowers* yellow-green, solitary or in fascicles of 2 or 3; pedicel 3–5 mm long; calyx 4–6-lobed; lobes 3–5 mm long, elliptic-lanceolate to obovate, pubescent, sometimes fringed with gland-tipped segments, acute, erect or somewhat recurved; disc hairy; ovary 2-lobed, unilocular, placentas 2, each with 1 ovule; styles 2. *Fruit* oblong, 1,5 cm in diam., glabrous, orange or red with white spots when ripe. *Seeds* 2, woolly, 0,9 cm long. FIG. 28: 3; 29: 2.

Found in coastal bush on sand dunes and in low-lying areas near mangroves. Further inland, it occurs in mixed scrub and scrub forest, on sandy soil. Recorded from the northern coastal areas of Natal, northwards to Mozambique.

NATAL.—2732 (Ubombo): between Sordwana Bay and Jozini, *Van der Schijff 6577*. 2831 (Nkandla): Umlalazi Nature Reserve, *Fakude 18*. 2832 (Mtubatuba): Hluhluwe Game Reserve, *Ward 4782*. 2930 (Pietermaritzburg): near Isipingo River, *Ward 493*. 2931 (Stanger): Hlogweni Forest, *Moll 3621*.

5. **Dovyalis zeyheri** *(Sond.) Warb.* in Pflanzenfam. 3, 6a: 44 (1893), as *Doryalis*; Gilg in Bot. Jahrb. 40: 506 (1908), as *Doryalis*; Pflanzenfam. ed. 2, 21: 440 (1925); Burtt Davy, Fl. Transv. 1: 216 (1926); Wild in F.Z. 1: 284 (1960); Killick in Flow. Pl. Afr. 39: t.1546 (1969); Sleumer in Bot. Jahrb. 92: 84 (1972). Type: Transvaal, Crocodile River, *Zeyher* 52 (S, holo.!; TCD!; PRE, photo.).

Aberia zeyheri Sond. in Linnaea 23: 10 (1850); Harv. in F.C. 1: 70 (1860); Szyszyl., Polypet. Thalam. Rehm. 112 (1888). —var. *velutina* Szyszyl., l.c. (1888). Syntypes: Transvaal, Wonderboompoort, *Rehmann* 4557 (B†, Z); Aapiespoort, *Rehmann* 4097 (B†; BM; K, lecto.); *Rehmann* 4098 (B†; Z). *A. tristis* Sond. in Linnaea 23: 9 (1850); Harv. in F.C. 1: 70 (1860). Type: Cape, Phillipstown, Kat River, *Ecklon & Zeyher* s.n. (S, holo.!; SAM!; PRE, photo.).
Dovyalis tristis (Sond.) Warb. in Pflanzenfam. 3, 6a: 44 (1893), as *Doryalis*; Sim, For. Fl. Cape Col. 129 (1907); Gilg in Bot. Jahrb. 40: 506 (1908), as *Doryalis*; Pflanzen-

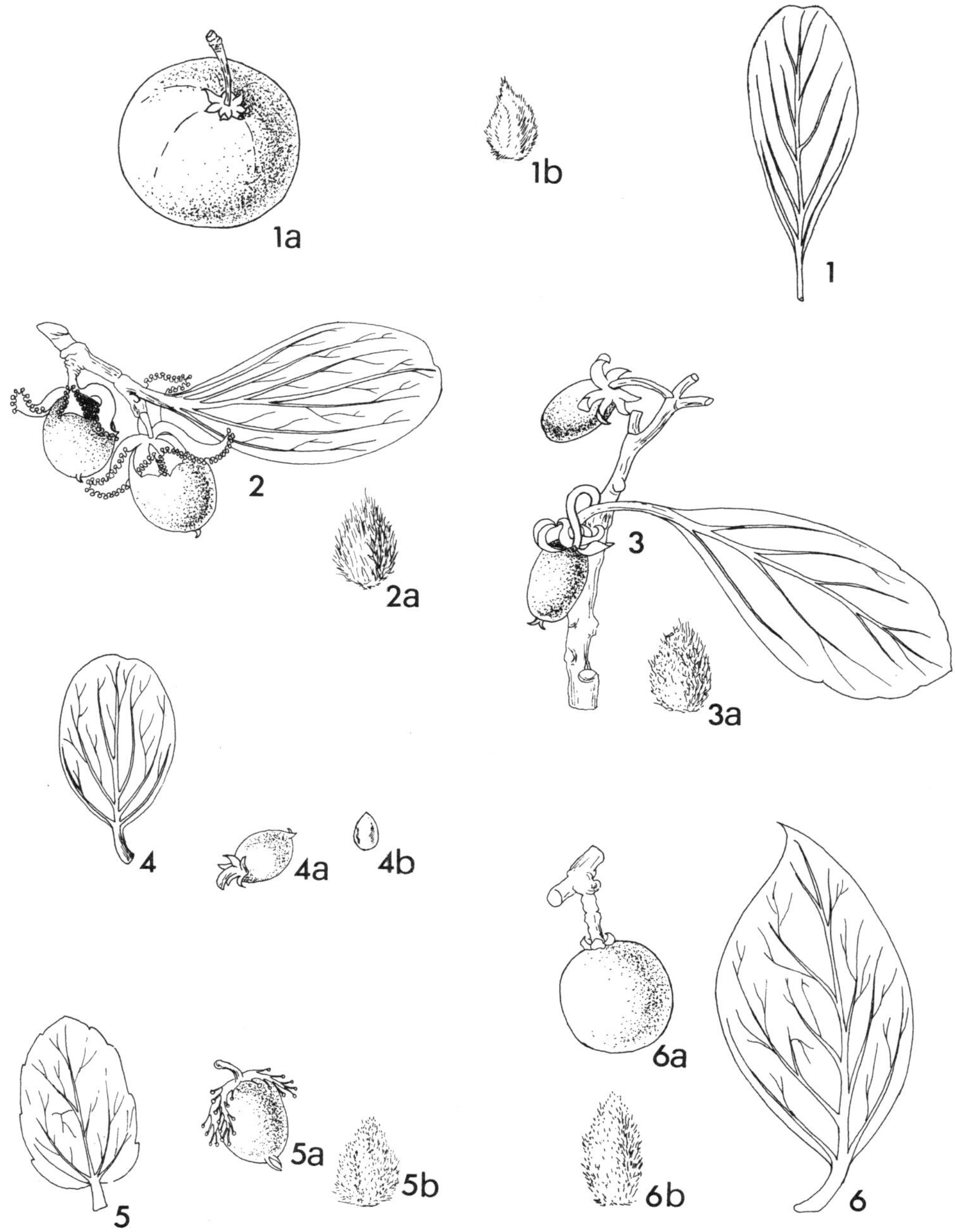

FIG. 29.—1, **Dovyalis caffra,** leaf, × $^2/_3$; 1a, fruit, × $^2/_3$; 1b, seed, × $1^1/_3$ (*Edwards 5328*). 2, **D. longispina,** fruiting twig, × $^2/_3$; 2a, seed, × $1^1/_3$ (*De Winter & Vahrmeijer* 8426). 3, **D. zeyheri,** fruiting twig, × $^2/_3$; 3a, seed, × $1^1/_3$ (*Letty* 477). 4, **D. rotundifolia,** leaf, × $^2/_3$; 4a, fruit, × $^2/_3$; 4b, seed, × $1^1/_3$ (*Tyson* 186). 5, **D. rhamnoides,** leaf, × $^2/_3$; 5a, fruit, × $^2/_3$; 5b, seed, × $1^1/_3$ (*De Winter* 7650). 6, **D. lucida,** leaf, × $^2/_3$; 6a, fruit, × $^2/_3$; 6b, seed, × $1^1/_3$ (Forester, Entabeni, December 1969).

fam. ed. 2, 21: 441 (1925); Marloth, Fl. S. Afr. 2: 194 (1925); Burtt Davy, Fl. Transv. 1: 216 (1926). —var. *depauperata* Sim, For. Fl. Cape Col. 131 (1907). Type: Cape, Keiskama, about Line Drift, *Sim* 2570 (NU, holo.!). *D. revoluta* Thom in Bothalia 10: 434 (1971). Type: Natal, False Bay, *Moll* 5112 (PRE!).

Tree or shrub up to 11 m tall, dioecious, often armed with pubescent or glabrous spines up to 3,5 cm long; bark ashen-white with longitudinal ridges and shallow fissures; ultimate branchlets often abbreviated; young twigs pubescent. *Leaves* alternate; blade thin and soft, sometimes coriaceous when mature, obovate to rhomboid-elliptic, 1,5–6 cm long, 1–4 cm broad, pubescent to glabrescent, 3-veined from the base, apex acute to obtuse, or emarginate, base cuneate, margin entire to crenate; petiole 2–5 mm long, pubescent or sometimes glabrous. *Male flowers* yellow-green, in fascicles of 3–6; calyx 4–6-lobed, lobes 4 mm long, narrowly ovate, pubescent; stamens 24–48, filaments 2,5–3 mm long; nectaries densely hairy. *Female flowers* yellow-green, solitary or rarely in pairs; pedicels 4–7 mm long; calyx deeply 4–7-lobed; lobes 4–6 mm long, lanceolate-obovate, acute, tomentose, somewhat recurved; disc densely hairy; ovary 2 or 3-lobed, almost completely 2 or 3-locular, placentas 2 or 3, each placenta with 1 ovule; style 2–3. *Fruit* oblong, 1,1–1,5 cm in diam., densely and shortly hairy, orange to red when ripe. *Seeds* 2–3, hairy, 7 mm long, with a coffee smell. FIG. 29: 3.

Occurs in several veld types including forest margins, open woodland and streambanks, often among rocks on a variety of soil types. Recorded from the coastal regions of the eastern Cape, inland to as far as Queenstown, extending to Natal, Zululand, Swaziland, and the southern, eastern and northern Transvaal. Also recorded from southern Rhodesia.

TRANSVAAL.—2230 (Messina): Makonde Mission Station, *Codd 6817*. 2328 (Baltimore): Leipzig, *Bremekamp & Schweickerdt 56*. 2329 (Pietersburg): Dorps River, Louis Trichardt, *Gerstner 5697*. 2330 (Tzaneen): Letaba, *Scheepers 999*. 2429 (Zebediela): Farm Magnet Heights, *Barnard & Mogg 864*. 2430 (Pilgrims Rest): Sekukuniland, *Barnard 423*. 2527 (Rustenburg): Hartebeespoortdam, *Vahrmeijer 413*. 2528 (Pretoria): *Liebenberg 3*. 2529 (Witbank): Loskopdam Nature Reserve, *Mogg 30385*. 2531 (Komatipoort): Barberton, *Pott 17749*. 2627 (Potchefstroom): Witpoortjie, *Rogers 18538*.

SWAZILAND.—2631 (Mbabane): Tulwane, *Compton 32173*.

NATAL.—2731 (Louwsburg): 21 km N. of Nongoma, *Acocks 13020*. 2829 (Harrismith): Collings Pass, *Acocks 13859*. 2830 (Dundee): Roodedraai, *Acocks 10482*. 2832 (Mtubatuba): False Bay, *Edwards 3199; Moll 5109; 5112*. 2929 (Underberg): Cathedral Peak, *Killick 1901*. 2930 (Pietermaritzburg): Albert Falls, *Comins 466*.

CAPE.—3128 (Umtata): Mqanduli, *Pegler 592*. 3225 (Somerset East): Bosberg, *Van der Walt 354*. 3226 (Fort Beaufort): Kei River Valley, *Galpin 8074*. 3227 (Stutterheim): Cata Forest, *Wells 3229*. 3327 (Peddie): *Sim 20440*.

The species varies a great deal in its pubescence. The unarmed form with relatively glabrous leaves has in the past been separated as *D. tristis*, while *D. zeyheri* is described as being spiny and having pubescent leaves. Intermediates are, however, so common that the step taken by Wild (l.c.) to merge the two species is supported. Following Sleumer (l.c.), *D. revoluta* is treated here as a local form of *D. zeyheri* with glabrous leaves and papillose-puberulous ovaries and fruit.

6. **Dovyalis lucida** *Sim*, For. Fl. Cape Col. 131, t.6 (1907); Wild in F.Z. 1: 282 (1960); Sleumer in Bot. Jahrb. 92: 87 (1972). Type: Eastern Cape, Plate 6, For. Fl. Cape Col. (1907), no specimen found.

Tree or shrub up to 10 m tall, dioecious, rarely armed with spines up to 3,5 cm long, bark yellow-white to brown, strongly lenticellate. *Leaves* alternate; blade soft to coriaceous, obovate to rhomboid, rarely ovate, 2–7,5 cm long, 1,2–4 cm broad, glabrous, shining and dark green above, paler and dull below, veins prominent on both sides, 3–5-veined from the base; apex acuminate, rarely emarginate, base cuneate, rarely obtuse, margin entire or serrate; petiole 3–5 mm long, glabrous. *Male flowers* light green, in fascicles of 3–10, rarely in a small raceme; pedicels 1–3 mm long; calyx 3–5-lobed, lobes 3 mm long, ovate-elliptic, pubescent; stamens 11–27, filaments 3 mm long; nectaries densely hairy. *Female flowers* yellow-green, solitary or in fascicles of 2 or 3; pedicels 3–4 mm long; calyx 3–5-lobed, lobes 1,5–2 mm long, obovate, obtuse, pubescent, somewhat recurved; disc densely hairy; ovary 2 or 3-lobed, unilocular, with 2 or 3 placentas, each placenta with 1 ovule; styles 2 or 3. *Fruit* ellipsoid, 1,5 cm in diam., minutely papillose, orange to red when ripe. *Seeds* 2 or 3, hairy, 1,1 cm long. FIG. 29: 6.

Found in scrub forest and forest from Port Elizabeth in the Eastern Cape, northwards through Natal to the Soutpansberg in the Northern Transvaal, extending to the eastern districts of Rhodesia.

TRANSVAAL.—2230 (Messina): Entabeni Forest Reserve, *Leistner, Thom & Gillham 3310*. 2329 (Pietersburg): Farm Seodin, *Leistner, Thom & Gillham 3279*. 2330 (Tzaneen): Magoebaskloof, *Codd 1687*. 2430 (Pilgrims Rest): 1,6 km S. of Graskop, *Leistner, Thom & Gillham 3322*.

NATAL.—2830 (Dundee): Qudeni Forest, *Edwards 2633*. 2831 (Nkandla): Inhlazatje Mt., *Acocks 11686*. 2929 (Underberg): Ntambamhlope, *Edwards 695*. 2930 (Pietermaritzburg): Umgeni Poort Forest, *Moll 861*.

CAPE.—3128 (Umtata): Engcobo, *Acocks 13815*. 3226 (Fort Beaufort): Hogsback, *Story 369*. 3227 (Stutterheim): King William's Town, *Comins 1024*. 3325 (Port Elizabeth): Van Stadens Kloof, *Taylor 837*. 3326 (Grahamstown): Albany, *Wells 3016*.

5343 13. CASEARIA

Casearia *Jacq.*, Enum. Syst. Pl. Ins. Carib. 4, 21 (1760); Benth. & Hook.f., Gen. Pl. 1,3: 796 (1867); Gilg in Bot. Jahrb. 40: 510 (1908); Engl., Pflanzenw. Afr. 3,2: 589 (1921); Gilg in Pflanzenfam. ed. 2, 21: 451 (1925); Phill., Gen. ed. 2: 516 (1951); Wild in F.Z. 1: 293 (1960); Sleumer in Bull. Jard. Bot. Nat. Belg. 41: 397 (1971); Wild & Vidigal in Fl. Mocamb. Flacourt. 35 (1973). Type species: *C. nitida* (L.) Jacq.

Trees or shrubs. *Leaves* alternate, simple, petiolate, entire, serrate or with subspinose teeth; stipules small. *Inflorescence* an umbel or flowers fascicled or rarely solitary. *Flowers* bisexual. *Calyx-tube* short or subelongated or 0; sepals 4–6, suborbicular, concave, imbricate. *Petals* 0. *Stamens* 6–15, rarely 20–40, free or connate, hypogynous or inserted on the calyx-tube or at the base of calyx; filaments linear, connective of anthers sometimes penicillate at the apex; staminodes alternating with the stamens, sometimes oblong and hairy at the apex. *Ovary* superior, 1-locular with few to many ovules on 3 or 4 parietal placentas; style short; stigma capitate or peltate or stigmas 3. *Fruit* a subfleshy or dry capsule, opening by 3 or 4 valves, few to many-seeded. *Seeds* sometimes angled, with a fleshy aril; testa chartaceous; embryo straight; cotyledons flat, elliptic or suborbicular; endosperm present.

Species about 160 in tropical countries of the world. *Casearia* was named in honour of Johann Casearius, a Dutch missionary in Indo-China in the 1700's, who was responsible for the first part of the Hortus Malabaricus (Wittstein).

Casearia gladiiformis *Mast.* in F.T.A. 2: 493 (1871); Gilg in Bot. Jahrb. 40: 510 (1908); Engl., Pflanzenw. Afr. 3, 2: 589 (1921); Gilg in Pflanzenfam. ed. 2, 21: 454 (1925); Wild in F.Z. 1: 293 (1960); Sleumer in Bull. Jard. Bot. Nat. Belg. 41: 423 (1971); Wild & Vidigal in Fl. Mocamb. Flacourt. 35 (1973). Type: Mozambique, Shupanga, Zambesi River, *Kirk* s.n. (K, holo.!; PRE, photo.).

C. junodii Schinz in Mém. Herb. Boiss. 10: 52 (1900); Gilg in Bot. Jahrb. 40: 513 (1908); Engl., Pflanzenw. Afr. 3, 2: 590 (1921); Gilg in Pflanzenfam. ed. 2, 21: 454 (1925); Henkel, Woody Pl. Natal 87 (1934). Type: Mozambique, Delagoa Bay, *Junod* 351 (Z, holo.!; K!; PRE, photo.).

Tree up to 25 m high; young branches puberulous, becoming glabrous; bark grey, smooth. *Leaf-blade* elliptic or ovate, 7–18 cm long, 3–7 cm wide, apex acuminate, acute or obtuse, cuneate and oblique at base, margins usually entire but sometimes serrate-crenate especially young and coppice shoot leaves, coriaceous, glabrous or puberulous on midrib when young, with circular and linear pellucid dots, lateral nerves 7–9, ½-angled, curved, more prominent on lower than upper side; petiole 0,1–2 cm long, channelled above, puberulous or glabrous, dark green and shiny. *Flowers* greenish, in dense, axillary fascicles on a cushion of minute, brown, membranous, ovate bracteoles with ciliate margins; pedicels 1–3 mm long, puberulous. *Sepals* 5, broadly elliptic, 3 mm long, 2 mm wide, puberulous dorsally. *Staminal tube* united for about 0,3–0,5 mm and then dividing into 10 fertile stamens and 10 alternating staminodes; filaments 1 mm long, glabrous; anthers ovate, 0,8 mm long, apiculate; staminodes 1 mm long, villous at apex. *Ovary* ovoid, 1,75 mm long, 1 mm wide, glabrous or with long hairs in upper $^1/_3$; style 0,5 mm long; stigma capitate. *Fruit* yellowish-orange, angular, ellipsoid capsule, splitting from apex into 3–4 longitudinal valves. *Seeds* c. 10, ellipsoid or ovoid, 4,5 mm long, 2 mm wide, with pale beige testa clasped by fleshy aril. FIG. 30.

Occurring in forest, chiefly along the coast, in Natal and the eastern Cape. Also found in Kenya, Tanzania, Malawi and Mozambique.

NATAL.—2632 (Bela Vista): Mangusi Forest, *Venter 6220;* Lake Nhlange, Kosi System, *Vahrmeijer & Tölken 953.* 2732 (Ubombo): Lake Sibayi, *Stephen 461.* 2831 (Nkandla): Lower Nkandla Forest, *Edwards 1446.* 2832 (Mtubatuba): Charter's Creek, *Ward 3029.* 2930 (Pietermaritzburg): Howick, *Cooper 3;* Farm "The Start", Lions River, *Moll 3366.* 3029 (Kokstad): Mt. Ingeli, *Nicholson 760.* 3030 (Port Shepstone): Paddock, *Strey 5995.*

CAPE.—3128 (Umtata): Mpane Forest, Elliotdale, *Miller sub PRF 5824.* 3129 (Port St. Johns): Qokama, Ngqeleni District, *Acocks 12820;* Egossa Forest, *Strey 8870.* 3228 (Butterworth): Manubi Forest, Kentani, *Marais 481.*

A tree with elliptic or ovate leaves varying considerably in size with margins usually entire, but sometimes serrate-crenate and with circular and linear pellucid dots scattered over the blade. The flowers are small and arranged in dense, axillary fascicles. The terminology relating to the structure of the flowers is not consistent, thus Phillips (l.c.) and Hutchinson (1967) refer to "calyx lobes", Wild (l.c.) to "sepals" and Sleumer (l.c.) to "tepals". Similarly Phillips, Hutchinson and Wild refer to "staminodes", while Sleumer refers to "disc lacinia".

According to Palmer and Pitman (1973) the ashes of the bark of *C. gladiiformis* are used by the Bantu as snuff.

FIG. 30.—1, **Casearia gladiiformis,** flowering twig, × 2/3 (*Moll* 3290); a, flower, × 10 (*Moll* 3290); b, fruit, × 1 (*Dutton* 22).

TURNERACEAE

by A. A. OBERMEYER

Herbs, shrubs or trees, usually pubescent. *Leaves* alternate, simple, variously incised or entire; stipules small or 0. *Flowers* regular, bisexual, in racemose or cymose, axillary or terminal inflorescences or solitary, opening in the morning, fading at night; pedicels free or rarely adnate to petiole; bracts small. *Calyx* tubular, 5-dentate or sepals 5, fused near the base, quincuncial. *Petals* 5, free, erect or rotate, contorted in bud, usually unguiculate with the claw inserted on the calyx-tube or near its base, sometimes with a ligule or small thin corona on upper part of claw. *Stamens* 5, hypogynous or inserted on the calyx-tube, filaments free, of equal or unequal length, anthers bilocular, introrse, opening lengthwise, basifixed. *Ovary* superior, 1-locular, with 3 parietal or, very rarely, basal placentas; styles 3, usually free, slender, often heterostylous; stigmas multipartite; ovules many to few (–1). *Capsule* ovoid or silique-like (resembling Brassicaceae fruits), 3-valved, with 1- to multi-seriate placentas, dehiscent; seeds cylindrical, straight or curved, reticulate or striate, often pitted, on long funicles and enveloped by a white, leaf-like aril.

Genera 9, species about 90, predominantly South American; also in Africa, the Mascarenes and south-east Asia. Four genera with 10 species recorded from the northern parts of South Africa and South West Africa.

Literature: Urban in Jahrb. K. Bot. Gart. Mus. Berl. 2 (1883).

Annuals or chamaephytes, 10–60 cm tall; flowers in axillary inflorescences or solitary; petals orange-red to pale yellow, claw not dark-coloured; pedicels free:
- Flowers on long racemes; capsule resembling a siliqua of the Brassicaceae, long, thin, terete, rostrate; seeds straight, cylindrical .... 1. **Wormskioldia**
- Flowers solitary or on short, few-flowered racemes; capsule ovoid; seeds curved:
  - Calyx tubular; petals not clawed; capsule erect, narrowly ovoid, beaked, thin-walled; leaves without 1–2 pairs of sessile glands below .... 2. **Streptopetalum**
  - Calyx of 5 sepals fused in lower third; petals clawed, alternating with sepals; capsule recurved, ovoid, woody; leaves with 1–several pairs of round sessile glands below on lower half of leaf .... 3. **Piriqueta**

Shrubs up to 2 m tall with axillary, solitary flowers; petals orange or yellow with a dark brown or purplish black claw; pedicels fused to petiole .... 4. **Turnera**

5355 1. WORMSKIOLDIA

Wormskioldia *Thonn.* in Schum. & Thonn., Beskr. Guin. Pl. 165 (1827); Urb. in Jahrb. K. Bot. Gart. Mus. Berl. 2: 48 (1883); Gilg in Pflanzenfam. ed. 2, 21: 464 (1925); Burtt Davy, Fl. Transv. 1: 118 (1926); Phill., Gen. ed. 2: 517 (1951); Lewis in F.T.E.A. Turneraceae: 9 (1954); Hutch., Gen. Flow. Pl. 2: 352 (1967). Type species: *W. pilosa* (Willd.) Schweinf. ex Urb.

Annuals or perennials (often chamaephytes), variously hispid and often glandular-pubescent. *Leaves* alternate, linear to narrowly ovate, usually serrate or lobate to deeply incised; stipules small or 0. *Racemes* simple, secund, few- to many-flowered with the apical part recurved in bud, accrescent and erect with age, the flowers developing one after another, opening in the morning, fading at night; scape long and thin; pedicels short, somewhat accrescent and often recurved in fruit; bracts small; a few sterile bracts often present on the rhachis. *Calyx* tubular, 5-dentate, usually setiferous on the outside, pubescent in lower half inside. *Corolla* rotate, yellow, orange-red or white; petals obovate, unguiculate, the claw appressed to the calyx-tube in upper half, fused to it below; with a ligule or small pustule on upper part of claw. *Stamens* 5, hypogynous, reaching to the throat, filaments filiform, sometimes narrowly winged, occasionally 3 shorter; anthers oblong. *Ovary* ellipsoid, styles 3, erect, heterostylous, stigmas apical, small, multipartite. *Capsule* resembling a silique of Brassicaceae, terete, usually long and thin, often constricted between the uniseriate seeds, rostrate; seeds straight, terete, obtuse, reticulate, usually with 2 pores in each cavity, aril unilateral, short.

An African genus with about 11 species in tropical and subtropical regions; recorded from the northern parts of South Africa and South West Africa.

Annuals:
With bulbous-based setae on stems; exstipulate:
Capsule erect, hairy; leaves coarsely serrate .... 1. *W. glandulifera*
Capsule patent or reflexed, glabrous; leaves pinnatisect, discolorous, soft and thin .... 2. *W. tanacetifolia*
Without bulbous-based setae on stems, hairs simple, short, sparse; leaves lobate in lower half, rarely entire; stipules auriculate, fused to base of leaf .... 3. *W. lobata*
Perennials (chamaephytes) forming a woody basal complex and a long thick taproot; the parts above ground dying back annually; without bulbous-based setae:
Leaves narrowly linear, sessile, entire, glabrous .... 4. *W. mossambicensis*
Leaves broader, petiolate or sessile, margin variously incised, pubescent:
Capsule c. 3 cm long, few-seeded, not constricted between the seeds; plants glandular-pubescent but without long setae; leaves coarsely and irregularly double-serrate .... 5. *W. schinzii*
Capsule c. 5–8 cm long, many-seeded, constricted between the seeds; plants without glandular pubescence but with long dark red or white setae on the stems; leaves subentire to laciniate:
Stems with long patent red to dark brown bristles, other parts glabrescent; leaves greyish-green, subentire to dentate, rarely laciniate; racemes 6–12-flowered .... 6. *W. longipedunculata*
Stems, leaves, peduncles and capsules echinulate; bristles on stems not reddish; leaves yellow-green, laciniate; racemes 4–7-flowered .... 7. *W. lacerata*

1. **Wormskioldia glandulifera** *Klotzsch* in Peters, Reise Mossamb. Bot. 1: 146, t.26 (1864); Mast. in F.T.A. 2: 503 (1871); Urb. in Jahrb. K. Bot. Gart. Mus. Berl. 2: 49 (1883); Gilg in Pflanzenfam. ed. 2, 21: 463 (1925); Burtt Davy, Fl. Transv. 1: 119 (1926); A. & R. Fernandes in Mem. Junta Invest. Ultram. ser. 2, 34: 16 (1962). Type: Mozambique, Tete, *Peters* (B, holo.†; K; EA; PRE, photo.).

Small glandular-pubescent annuals up to 30 cm tall, usually few-branched. *Stems,* petioles, veins, scapes and pedicels bearing erect multicelled setae which are swollen and dark basally, the capitate glands usually lost early on, interspersed with simple, soft hairs. *Leaves* petiolate, broadly linear-acuminate, 4–8 cm long, 8–20 mm broad, coarsely and irregularly double-serrate, drying a dark brown colour, lamina sparsely and softly puberulent. *Flowers* c. 3 on each axillary raceme; peduncle 5–10 cm long, wiry, arcuate-erect; pedicels up to 5 mm in fruit; with minute bracts. *Calyx* tubular, c. 1 cm long, puberulent; lobes short, acute. *Petals* c. 1,5 cm long, yellow or orange. *Stamens* unequal. *Ovary* cylindrical, hairs erect; styles heterostylous. *Capsule* erect, overtopping the leaves, up to 6 cm long, puberulous; seeds narrowly clavate, cream, brown or black, the ridges minutely knobbed, aril tongue-shaped. FIG. 31: 1.

Recorded from Mozambique, Rhodesia, South West Africa, Transvaal and Natal; behaving as a weed.

S.W.A.—1819 (Karakuwisa): Cigarette, N.E. of Karakuwisa, *Maguire 2462.* 1820 (Tarikora): Omuramba Khaudum at Tamso Camp, *De Winter & Marais 4672.* 1918 (Grootfontein): near Tsebeb, *Schoenfelder S536.*

TRANSVAAL.—2229 (Waterpoort): Dongola Hill, *Codd & Dyer 3907;* Liliput, *Strey 3486;* Waterpoort, *Bremekamp & Schweickerdt 297.* 2230 (Messina): Messina, *Rogers 19405.* 2231 (Pafuri): Pafuri, *Van der Schijff 3568;* 5 km N. of Kloppersfontein, *Acocks 16782.* 2327 (Ellisras): 4 km N.E. of P.O. Tom Burke, *Codd 6615.* 2330 (Tzaneen): Hans Merensky Nature Reserve, *Gilliland 780.* 2331 (Phalaborwa): Kruger National Park, near Letaba Camp, *Munro s.n.* 2530 (Lydenburg): 16 km N. of Abel Erasmus Pass on road to Tzaneen, *Mauve 4328.* 2531 (Komatipoort): 16 km from Gorge, *Van der Schijff 2283.*

NATAL.—2731 (Louwsburg): Pongola Dam at Magut, *Gerstner 3174.*

2. **Wormskioldia tanacetifolia** *Klotzsch* in Peters, Reise Mossamb. Bot. 1: 147 (1864); Urb. in Jahrb. K. Bot. Gart.Mus. Berl. 2: 51 (1883); Burtt Davy, Fl. Transv. 1: 119 (1926); A. & R. Fernandes in Mem. Junta Invest. Ultram. 2, 21: 16 (1962). Type: Mozambique, between Boror and Querimba, *Peters* 23 (B, holo.†; K).

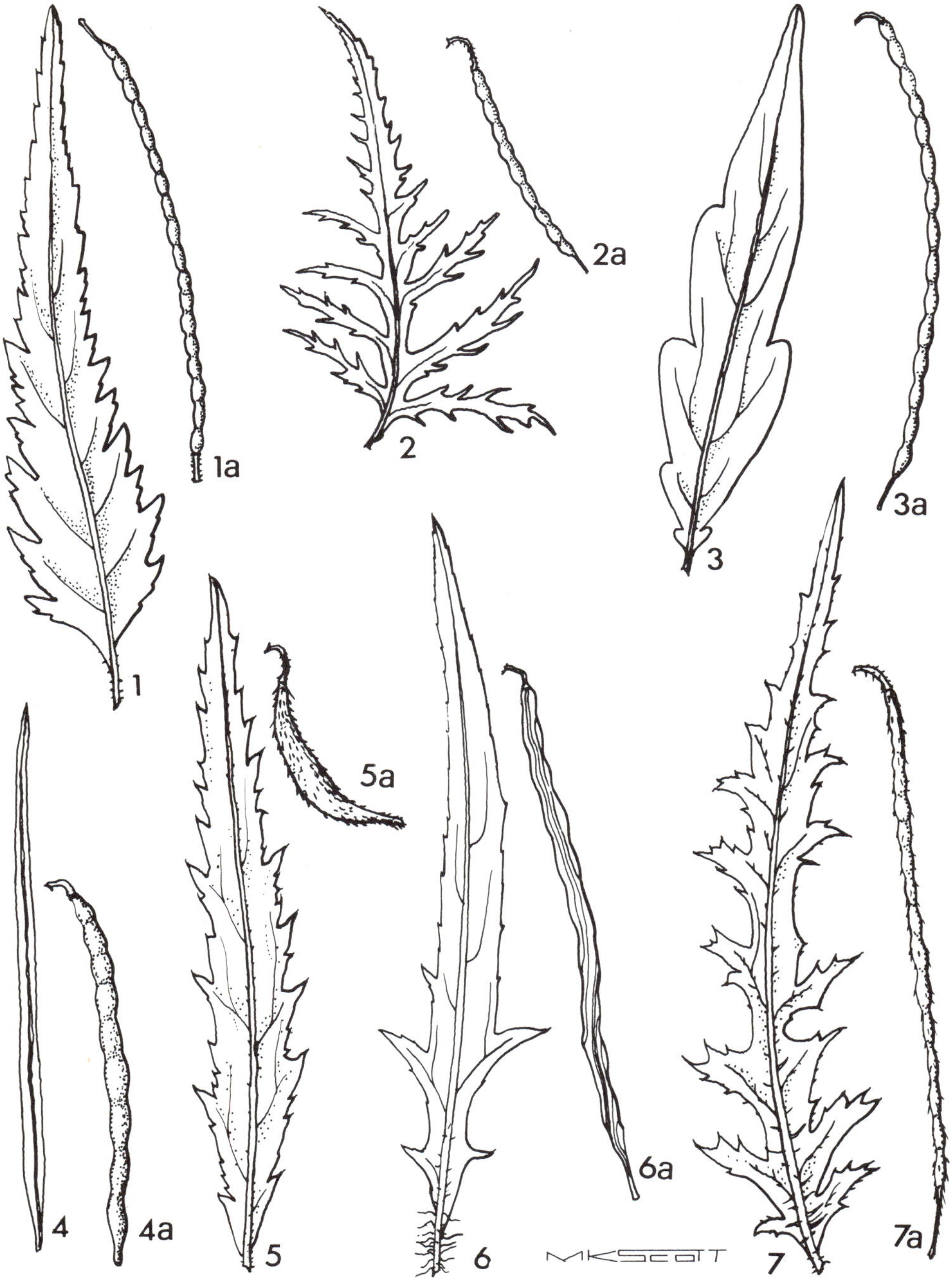

FIG. 31.—Leaves and capsules of **Wormskioldia** spp., all × 1: 1, **W. glandulifera,** leaf; 1a, capsule (*Van der Schijff* 1885); 2, **W. tanacetifolia,** leaf; 2a, capsule (*Rogers* 20830); 3, **W. lobata,** leaf; 3a, capsule (*Welwitsch* 2493); 4, **W. mossambicensis,** leaf; 4a, capsule (*Vahrmeijer* 1057); 5, **W. schinzii,** leaf; 5a, capsule (*Mostert* 1239); 6, **W. longipedunculata,** leaf; 6a, capsule (*Codd & Dyer* 4698, *Rogers* 25503); 7, **W. lacerata,** leaf; 7a, capsule (*Rogers* 22578).

Annuals up to 60 cm tall, malodorous. *Stems*, petioles, veins, scapes and pedicels bearing erect multicelled setae which are swollen and dark basally, the apical capitate glands usually early deciduous; with some soft, white woolly hairs on the stems. *Leaves* with petioles 1–3 mm long; lamina somewhat ovate-acuminate in outline, c. 6 cm long and 4 cm broad, deeply bipinnatipartite, thin, soft, discolourous, nearly glabrous. *Flowers* in c. 3-flowered patent racemes, longer than leaves, bracts small, circular. *Calyx* c. 16 mm long, glabrous. *Petals* yellow, c. 27 mm long. *Stamens* unequal with 2 longer than the others. *Ovary* glabrous. *Capsule* recurved, c. 3 cm long; seed oblong, brown or yellow. FIG. 31: 2.

Recorded from the eastern Transvaal, Mozambique and Rhodesia. Like *W. glandulifera* it behaves as a weed; often in sandy disturbed ground, emitting an unpleasant scent.

TRANSVAAL.—2530 (Lydenburg): Nelspruit, *Liebenberg 2465; Buitendag 402;* Kaapse Hoop, *Rogers 20830.* 2531 (Komatipoort): Storiwane, S.E. of Pretoriuskop, *Van der Schijff 1927;* 12 km W. of Malelane, *Codd 5246;* Louws Creek, *Thorncroft 2044.*

3. **Wormskioldia lobata** *Urb.* in Jahrb. K. Bot. Gart. Mus. Berl. 2: 52 (1883); Thonner, Flow. Pl. Afr. t.106 (1908); Lewis in F.T.E.A. Turneraceae: 10 (1954); A. & R. Fernandes in Mem. Junta Invest. Ultram. ser. 2, 28: 14 (1961); ibid. 34: 15 (1962); Schreiber in F.S.W.A. 88: 4 (1968). Syntypes: several from tropical Africa including Angola, Pungo Andongo, *Welwitsch* 2493, 2494 (B†; BM; K; LISC; PRE!).

Annual up to 60 cm tall, minutely tawny pubescent with short and long hairs and with scattered small pyramidal, attenuate, multicellular glands. *Stems* woody, few-branched. *Leaves* with the primary small, attenuate below, petiolate, exstipulate; following sessile, broadly linear, up to 12 cm long and 2 cm broad, apex acute, lower half usually expanded into 2 large unequal lobes, occasionally with a few shallow indentations above, margin shallowly serrate, soft, discolourous; stipules small, circular, fused to lamina below. *Flowers* 2–6 on long thin scapes overtopping the leaves. *Calyx* tubular, c. 15 mm long, lobes short, acute, setose. *Petals* orange (or pale yellow in specimens from S.W.A.) up to 2 cm long. *Stamens* narrowly winged. *Ovary* glabrous. *Capsule* patent to reflexed, up to 8 cm long, glabrous. FIG. 31: 3.

Recorded from South West Africa. Widespread in tropical Africa, in sandy habitats.

S.W.A.—1718 (Kuring-kuru): near Mupini Mission, *De Winter & Marais 5433;* Musese Camp W. of Lupala Mission, *De Winter & Marais 4975.* 1721 (Mbambi): 64 km W. of Andara, *Merxmüller & Giess 2074.* 1724 (Katima Mulilo): Ngoma area, *Killick & Leistner 3005.* Grid uncertain: Kaokoveld, Kunene River bank, *Barnard 592.*

4. **Wormskioldia mossambicensis** *A. & R. Fernandes* in Bol. Soc. Brot. Sér. 2, 35: 158, t.5 (1961); in Mem. Junta Invest. Ultramar. ser. 2, 34: 18 (1962). Type: Mozambique, Inhambane, between Quissico and Chiducuane, *Mendonça* 3308 (LISC, holo.).

Chamaephyte, slender, fastigiate, glaucous, glabrous to slightly woolly pubescent. *Stems* c. 40 cm tall, thin, hard, striate. *Leaves* narrowly to broadly linear, variable, 8–15 cm long, 2–8 mm broad, margin entire or shallowly denticulate (the teeth consisting of glands), sessile; stipules minute, pectinate, sparsely lanate inside axil. *Flowers* on long erect scapes, 2–7-flowered. *Calyx* c. 1 cm long with a few short ovoid-attenuate glands, lobes with hyaline minutely fimbriate margins. *Petals* orange, c. 12 mm long. *Stamens* equal or unequal. *Ovary* glabrous. *Capsule* c. 7 cm long, reflexed. FIG. 31: 4.

Recorded from Mozambique and Natal in Palm veld; grassy sandy localities.

NATAL.—2632 (Bela Vista): Tongaland, north of Lake Sibayi, *Vahrmeijer & Tölken 274;* 5 km W. of Maputa, *Strey & Moll 3788.* 2732 (Ubombo): S. of Maputa on road to Lake Sibayi, *Gerstner 4114;* Manzengwenya, *Moll 4717;* Bazwana, *Gerstner 3557; 707; Strey 5115.* 2831 (Nkandla): Hlabisa Bird Sanctuary, *Gerstner 5027.* 2832 (Mtubatuba): W. of Charters Creek, *Ward 2832;* 26 km N.E. of Mtubatuba, *Acocks 13085;* 21 km N. of bridge over St. Lucia estuary on road to Cape Vidal, *Codd 10168.*

5. **Wormskioldia schinzii** *Urb.* in Bot. Jahrb. 15: 159 (1892); in Warb., Kunene-Samb. Exped. 310 (1903); A. & R. Fernandes in Mem. Junta Invest. Ultramar. ser. 2, 28: 15 (1961). Type: Angola, Gambos, Calculovar River, *Newton* 26 (Z, holo.!; COI; PRE, photo.); incorrectly stated by Urban as collected in Mozambique.

W. juttae Dinter & Urb. in Feddes Repert. 13: 153 (1914); Schreiber in F.S.W.A. 88: 4 (1968). Syntypes: S.W. Africa, between Neitsas and Otjituo, *Dinter* 652 (B†; SAM!), 652a (B†). *W. rehmii* Suesseng. in Mitt. Bot. Staatssamml. München 1: 55 (1950). Type: S.W. Africa, farm Rotenfels, *Rehm* s.n. (M, holo.).

Small scabrid and glandular-pubescent, spreading chamaephytes c. 30 cm tall. *Stems* densely leafy, woody and brittle. *Leaves* linear, 9–18 cm long (rarely shorter), 12–25 mm broad, apex acuminate, base acute, margin coarsely double-serrate, the teeth diminishing in size towards the apex, scabrid, midrib raised, often winged below; petiole short or 0; stipules 0. *Flowers* in few-flowered racemes. *Calyx* 16 mm long. *Petals* c. 35 mm long, bright orange red. *Ovary* densely covered with erect bristles. *Capsule* c. 3–4 cm long, few-seeded, not constricted between seeds, glandular-pubescent. FIG. 31: 5.

Recorded from South West Africa, Botswana, northern Cape, Transvaal and Angola. In sandy soil.

S.W.A.—1718 (Kuring-kuru) or 1818 (Tsitsib): on road from Tsintsabis to Kuring-kuru, *Schoenfelder S528*. 1917 (Tsumeb): Farm Kameeldoorn, GR 158, *Seydel 2146*. 1918 (Grootfontein): Neitsas, *Dinter 652*. 2017 (Waterberg): near Okakarara, *Liebenberg 4730*. 2119 (Epukiro): Farm Hekel, G0415, *Giess 10667*. 2218 (Gobabis): Gobabis district, farm Gemsbokfontein, *Merxmüller & Giess 1180*.

TRANSVAAL.—2326 (Mahalapye): Buffelsdrift, *Vahrmeijer 1297*.

CAPE.—2523 (Pomfret): Bray, *Mostert 1239*.

6. **Wormskioldia longipedunculata** *Mast.* in F.T.A. 2: 502 (1871), as *longepedunculata;* Urb. in Jahrb. K. Bot. Gart. Mus. Berl. 2: 53 (1883); Burtt Davy, Fl. Transv. 1: 119 (1926) excl. f.11a; Lewis in F.T.E.A. Turneraceae: 14 (1954); A. & R. Fernandes in Mem. Junta Invest. Ultram. ser. 2, 34: 18 (1962); Letty, Wild Flow. Transv. 224, t.111 (1962). Type: Malawi, Manganja Hills, *Meller* (K, holo.; PRE, photo.).

Small chamaephytes up to 40 cm tall, covered with a short white woolly pubescence and scattered short to long hairs; upper part of stem armed with c. 3 mm long dark red to purple, coarse patent hairs. *Leaves* close together, sessile, variable, linear to oblong, 8–15 cm long, 2–20 mm broad, apex usually obtuse, margin subentire to serrate to coarsely pinnatifid in lower half, glaucous, minutely white woolly pubescent and with some short scattered hairs, becoming glabrescent with age. *Raceme* 7–15-flowered. *Calyx-tube* 15 mm long, setulose, *Petals* up to 24 mm long, bright orange-red. *Stamens* unequal in length. *Ovary* with a few scattered setae, probably heterostylous. *Capsule* 5–8 cm long, indistinctly moniliform, many-seeded. FIG. 31: 6.

Recorded from South West Africa, the Transvaal, Rhodesia, Malawi, Mozambique, Zambia and Botswana, in dry bushveld, road verges and lands.

S.W.A.—1723 (Singalamwe): On the road from Sibinda to Katima Mulilo, *Pienaar & Vahrmeijer 224*. 1724 (Katima Mulilo): about 8 km from Katimo Mulilo to Linyanti, *Killick & Leistner 3101*.

TRANSVAAL.—2228 (Maasstroom): road from Steilloop Bridge to Tolwe, *Ihlenfeldt 2118a*. 2230 (Messina): Sibasa, *Junod 4547;* Makonde, *Codd 6871*. 2231 (Pafuri): Punda Milia, *Lang sub TRV 32178; Schlieben 9283*. 2328 (Baltimore): Blouberg, *Schlechter 4645*. 2329 (Pietersburg): Lejuma Mt., *Strey 7991;* Sand River near Mara, *Schlieben & Strey 8290;* Soutpansberg, 8 km N. of Louis Trichardt, *Van Vuuren 1664;* Spelonken (Cave Mts.), *Nelson 122;* near Pietersburg, *Bolus 10895*. 2330 (Tzaneen): Elim, *Obermeyer sub TRV 29324;* Tshakhuma, *Van Warmelo 5337/11;* Mbayinbayi, *Lang sub TRV 32254;* Mokeetsi, *Van Dam sub TRV 25021;* Westfalia Estate near Duiwelskloof, *Scheepers 24;* 13 km S.W. of Leydsdorp, *Codd & Dyer 4698*. 2428 (Nylstroom): Moorddrif, *Leendertz sub PRE 7316*. 2429 (Zebediela): Naboomfontein, *Galpin 133156;* 13 km N.E. of Zebediela, *Story 5355*. 2430 (Pilgrims Rest): Shilouvane, *Junod 4220;* Marouvuni, *Junod 85*.

Because of its soboliferous habit and deep hard woody roots it is able to persist in ploughed lands and on road verges, being considered a pest by farmers and attractive to passersby who admire the bright red flowers. It is said to be poisonous to cattle but the unpleasant odour acts as a repellant.

7. **Wormskioldia lacerata** *Oberm.* in Bothalia 11: 288 (1974). Type: Transvaal, Newington, *Rogers* 22578 (PRE, holo.; NH).

W. schinzii sensu Burtt Davy, Fl. Transv. 1: 119 (1926) incl. f.11a.

Chamaephytes with a dense short echinulate pubescence, c. 30 cm tall. *Stems* usually short, firm, erect or ascending. *Leaves* close together, sessile, linear-lacerate, 8–12 cm long, 2–3,5 cm broad, the lobes patent, irregular, dentate, decreasing in size upwards, apex acuminate, echinulate along the broad midrib, veins and margin, with a white woolly pubescence on the midrib and lateral veins above. *Flowers* 4–7 on long straight or slightly arcuate scapes, ultimately exceeding the leaves; bracts small. *Calyx* c. 15 mm long, with erect setae and some sparse white woolly pubescence. *Petals* deep orange, the blade broadly obovate, c. 2 cm long. *Stamens* 8 mm long in short-styled flowers, 4 mm long in long-styled flowers. *Ovary* densely pubescent with erect hairs, heterostylous, the styles 9 mm long and exserted above the stamens or 4 mm long, reaching only about half way. *Capsule* reflexed, c. 7 cm long, echinulate. FIG. 31: 7.

Recorded from the eastern Transvaal and Swaziland, in open spaces in subtropical woodland or in disturbed places.

TRANSVAAL.—2330 (Tzaneen): Hans Merensky Nature Reserve, *Oates 77;* 5 km S. of Gravelotte, *Leach 11559; Werdermann & Oberdieck 1922.* 2430 (Pilgrims Rest): 10 km from Mica on road to Hoedspruit, *Ihlenfeldt 2303;* 16 km from Mariepskop on road to Klaserie, *Coetzer 151.* 2431 (Acornhoek): Klaserie, *Killick & Strey 2511;* Newington, *Rogers 22578; Buitendag 909;* Skukuza, *Cholmondeley s.n.* 2531 (Komatipoort): Pretoriuskop, *Van der Schijff 1075; 1217; Codd & De Winter 4918;* Malelane, *Lang sub TRV 31644;* Kaapmuiden, *Rogers 25034;* De Kaap Valley, *Thorncroft 54;* Barberton, *Galpin 747; Thorncroft sub Rogers 19171.*

SWAZILAND.—2531 (Komatipoort): Tshaneni, near Sand River reservoir, *Edwards 213.* 2631 (Mbabane): Ranches Ltd., *Compton 24704.*

5356 2. STREPTOPETALUM

Streptopetalum *Hochst.* in Flora 24: 665 (1841); Urb. in Jahrb. K. Bot. Gart. Mus. Berl. 2: 55 (1883); Gilg in Pflanzenfam. ed. 2, 21: 463 (1925); Phill., Gen. ed. 2: 517 (1951); Lewis in F.T.E.A. Turneraceae: 15 (1954); Hutch., Gen. Flow. Pl. 2: 352 (1967). Type species: *S. serratum* Hochst.

Small glandular-pubescent annuals or perennials. *Stems* woody. *Leaves* alternate, exstipulate. *Flowers* in few- to many-flowered axillary racemes. *Calyx* narrowly tubular, hairy on the outside and within in lower half. *Corolla* rotate, petals 5, convolute in bud, inserted in the throat of the calyx, spathulate, yellow or orange. *Stamens* 5, free, perigynous, equal, or unequal in length, reaching to the mouth of the tube. *Ovary* oblong-ovoid, pubescent, with 3 parietal placentas; ovules pluriseriate; styles filiform just over-topping the stamens. *Capsule* ellipsoid, 3-valved, splitting to the base, glandular-hispid; seeds allantoid, testa reticulate, pits with 2 pores, aril unilateral, as long as, or shorter than the seed, with a wavy margin.

Species 3, African; 1 species recorded from Ethiopia to the Transvaal and South West Africa.

Urban and others observed 5 large tubercles on the calyx-tube above the insertion of the stamens but in the (unfortunately) few flowers dissected, these appeared to be absent.

Streptopetalum serratum *Hochst.* in Flora 24: 666 (1841); Urb. in Jahrb. K. Bot. Gart. Mus. Berl. 2: 56 (1883); Lewis in F.T.E.A. Turneraceae: 17 (1954); Schreiber in F.S.W.A. 88: 2 (1968). Type: Ethiopia, Tsellenti, Takkaze River, *Schimper* 1260 (B, holo.†; K; BM).

Small bushes up to 40 cm tall, annual or possibly sometimes perennial, sparsely to densely hispid with short simple hairs and glanduliferous setae with swollen bases, many-celled. *Leaves* sessile, narrowly ovate, up to 10 cm long and 2,5 cm broad, attenuate below and above, serrate, thin, yellow-green. *Inflorescences* with the primary racemes axillary, simple, several-flowered, secund; the inner serial axillary bud later develops into a strong side branch which in its turn produces much abbreviated side branches bearing reduced leaves and a few flowers towards the top. *Calyx* narrowly tubular, c. 1 cm long, lobes 2,5 mm long, hairy outside and inside below. *Petals* spathulate, c. 7 mm long, yellow or orange, alternating with sepals. *Stamens* 5, attached just below the middle of the tube, free above, reaching to the throat. *Ovary* ovoid, densely covered with glandular setae which are swollen at the base; styles 3, terete, reaching to the throat; stigmas apical, obtuse, slightly swollen. *Capsule* ovoid, c. 12 mm long, rostrate, scabrid; seeds allantoid, 2,5 mm. FIG. 32.

Recorded from northern and central Transvaal, Botswana, South West Africa and further north to Ethiopia; rare but often locally abundant, in warm dry bushveld, usually in sandy soil, often along road sides.

S.W.A.—1917 (Tsumeb): on sandy road from Tsumeb to Heidelberg, *Dinter 7530;* Tsumeb, *Dinter 3034, 1714.*

TRANSVAAL.—2228 (Maasstroom): Kremetartberg, *Mara Research Station 475.* 2230 (Messina): Messina, *Rogers 20820.* 2231 (Pafuri): 16 km E. by N. of Punda Milia, *Acocks 16775.* 2329 (Pietersburg): Blouberg, *Mogg 24548.* 2428 (Nylstroom): Magabene, *Schlechter 4671.* 2529 (Witbank): Rooikop, *Smuts & Gillett 2536.*

5357 3. PIRIQUETA

Piriqueta *Aubl.*, Hist. Pl. Guiane Fr. 1: 298, t.117 (1775); Urb. in Jahrb. K. Bot. Gart. Mus. Berl. 2: 57 (1883); Gilg in Pflanzenfam. ed. 2, 21: 463 (1925); Phill., Gen. ed. 2: 517 (1951); A. & R. Fernandes in Mem. Junta Invest. Ultram. sér. 2, 34: 22 (1962). Type species: *P. villosa* Aubl.

FIG. 32.—1, **Streptopetalum serratum,** habit, × 1; a, longitudinal section showing attachment of filaments, × 6; b, ovary and styles, × 4; c, capsule, × 3; d, seed with aril, × 14 (*Mara Research Station* 475).

Erblichia Seem., Bot. Voy. Herald 130 t.27 (1853) pro minore parte quoad species austro-africanae; Hutch., Gen. Flow. Pl. 2: 353 (1967).

Annuals, shrubs or trees, variously pubescent with simple and/or compound hairs. *Leaves* alternate, sessile or petiolate, stipules minute or 0, lamina simple, serrate, with or without large sessile circular glands on lower surface. *Flowers* axillary, solitary or rarely cymose. *Calyx* forming a short to long tube. *Petals* inserted in the throat, obovate, clawed, alternating with the calyx-lobes and together forming a fimbriate annulus in the throat. *Stamens* hypogynous with long or short filaments. *Ovary* with 3 parietal placentas, ovules many, uni- to multi-seriate; styles 3, heterostylous. *Capsule* globose, 3-valved, hairy; seed straight or slightly curved, reticulate, aril unilateral, as long as or shorter than seed.

Species about 20, mostly South American. One species in Transvaal, Natal, Swaziland and Mozambique.

Hutchinson places *Piriqueta capensis* in the genus *Erblichia* in his Gen. Flow. Pl. 2: 353 (1967), but without validating the new combination. Urban and Gilg regard *Erblichia* as a section of *Piriqueta.* Four species are involved; the type species, *E. odorata* from Central America and 2 indigenous in Madagascar, all 3 of which fall outside the scope of this work. Our species, the fourth, forms a short but distinct calyx-tube and a fimbriate annulus in the throat, characters which Urban considers of importance for *Piriqueta.* Harvey overlooked this annulus for it is small and unobtrusive. The differences enumerated by Hutchinson in some cases do not agree with those observed by Urban. There appears to be no reason why it should not be placed under *Piriqueta.*

Piriqueta capensis *(Harv.) Urb.* in Jahrb. K. Bot. Gart. Mus. Berl. 2: 78 (1883); Gilg in Nat. Pflanzenfam. ed. 2, 21: 464 (1925); Fernandes in Bol. Soc. Brot. Sér. 2, 35: 160, t.7 (1961); in Mem. Junta Invest. Ultram. sér. 2, 34: 22 (1962). Type: Transvaal, Apies River, *Burke & Zeyher* (TCD, holo.! K, SAM!, PRE, photo.).

Turnera capensis Harv. in F.C. 2: 599 (1862); Thes. Cap. 2: 25, t.140 (1862).

Small woody, sericeous, aromatic shrublets (chamaephytes) 15–30 cm high, with a long thick woody taproot and gnarled basal stems covered with thick corky bark, producing annual shoots. *Leaves* subsessile, stipules reduced to 1–2 pairs of minute, subulate glands, lamina narrowly ovate, 2–5 cm long, 7–14 mm broad, apex acute to obtuse, base cuneate, margin serrate, with 1–2 pairs of sessile circular glands irregularly placed on lower surface near margin, densely pubescent with coarse simple hairs. *Flowers* axillary, solitary, opening in the morning, fading at night, on short to long pedicels which become pendulous in fruit, with a pair of minute bracts. *Sepals* fused in lower third forming a cup, free above, narrowly ovate-acuminate, apiculate, c. 12 mm long, 3 mm broad, thin, dorsally somewhat hairy, 3-nerved, deciduous. *Petals* about as long as sepals, yellow, narrowly obovate, clawed, fused to the sepaloid cup below, alternating with the sepals and together producing a small erect, annular fringe at the mouth of the cup, deciduous. *Stamens* 5, free, hypogynous, filaments subulate, glabrous, long or short, anthers versatile, introrse. *Ovary* ovoid, densely covered with erect hairs; ovules 4–8, more or less uniseriate on each parietal placenta; styles 3, long or short (heterostylous), terete; stigmas spreading-fimbriate. *Capsule* ovoid, 8 mm, the valves splitting to the base; seeds oblong often somewhat curved, 3 mm, yellow or black, reticulate, with a short areole.

Recorded from the Transvaal, Swaziland, northern Natal, Mozambique and Rhodesia, usually on stony hillsides in dry bushveld. FIG. 33.

TRANSVAAL.—2229 (Waterpoort): between Waterpoort and Wylliespoort, *Obermeyer, Schweickerdt & Verdoorn 339.* 2327 (Ellisras): farm Alfred, 4 km W. of Monte Christo P.O., *Codd 6609;* near Elmeston, about 64 km from Vaalwater, *Meeuse & Strey 10432.* 2330 (Tzaneen): Sandrivier, *Schlechter 4597.* 2428 (Nylstroom): between Warmbad and Nylstroom, *Bremekamp & Schweickerdt 3; Sidey 1312; Bolus 11890.* 2429 (Zebediela): 9 km N.W. of Marble Hall, *Codd & Verdoorn 10371.* 2528 (Pretoria): 7 km N.E. of Hammanskraal on road to Rust de Winter, *Codd 3619.* 2529 (Witbank): Loskop Dam on road to "The Hell", *Codd & Verdoorn 10361.*

SWAZILAND.—2631 (Mbabane): Palata near Stegi, *Compton 31209.*

NATAL.—2732 (Ubombo): Ingwavuma hills, *Strey 10263.*

FIG. 33.—1, **Piriqueta capensis,** habit, × ¾; a, leaf-base, lower surface showing glands, × 2; b, flower laid out, × 3; c, seed with aril, × 13 (*Strey* 10263).

5360 4. TURNERA

Turnera *L.*, Sp. Pl. 1: 271 (1753); Benth. & Hook.f., Gen. Pl. 1: 806 (1862); Urb. in Jahrb. K. Bot. Gart. Mus. Berl. 2: 81 (1883); Gilg in Pflanzenfam. ed. 2, 21: 464 (1925); Hutch., Gen. Flow. Pl. 2: 352 (1967). Type species: *T. ulmifolia* L.

Annual or perennial herbs, shrubs or trees, pubescent. *Leaves* alternate, sessile or petiolate, entire, toothed or subpinnatifid, often with 2 basal glands on the lower side at the base, pubescence simple or stellate; stipules small or 0. *Flowers* axillary, solitary, rarely forming inflorescences, the peduncle sometimes fused to the petiole, bracteate. *Calyx* forming a short tube, 5-partite. *Petals* 5, inserted in the calyx-throat, shortly clawed. *Stamens* inserted below the petals, sometimes hypogynous. *Ovary* ovoid to cylindrical, ovules pluriseriate; styles 3, spreading; stigmas fimbriate. *Capsule* globose to ovoid, seeds globose to obovoid or cylindrical and curved, reticulate, arillate.

Species about 100, mostly South American; one species endemic in South West Africa and Angola, another in tropical Africa. One South American species naturalized in the Seychelles, Bourbon and Indo-Malaysia.

Turnera oculata *Story* in Bothalia 7: 493 (1961). Type: S.W. Africa, Kaokoveld, Kunene River at Otjinungua, *De Winter & Leistner 5770* (PRE, holo.; WIND).

Shrub up to 2 m tall, with slender long erect branches and short side branches, covered by a silvery (turning tawny when dry) velvet pubescence of stellate and simple sericeous hairs or hair-covering more sparse; internodes often shortened above. *Leaves* obovate, up to 3 cm long and c. 12 mm broad, apex obtuse to acute, base narrowed into a petiole, margin shallowly to deeply dentate, velvety pubescent to sparsely pubescent with short stellate hairs and long simple sericeous hairs, pale green, nerves distinct below, with 2 (rarely 1 or 0) conspicuous sessile circular, rimmed glands, 1 mm in diam. at junction of petiole and blade; stipules minute. *Flowers* solitary, axillary on the upper abbreviated side branchlets; pedicels fused to petioles; bracts 2, filiform, c. 10–18 mm long. *Calyx* tubular, c. 12 mm long, 9 mm in diam., 10-nerved, lobes linear-acuminate, somewhat longer than tube, pubescent with a membranous glabrous margin. *Petals* convolute, inserted on the calyx throat, spreading above, broadly obovate, c. 2,5 cm long and 1,8 cm broad, apex obtuse, apiculate, base cuneate, bright yellow above, dark red-brown below forming the dark heart of the flower. *Stamens* 5, exserted, filaments 3 cm long, arcuate, inserted 3 mm below throat with the shortly winged margins adhering to the calyx-tube below, anthers basifixed, the filaments inserted in a dorsal pit, introrse. *Ovary* cylindric, obtuse, pubescent, ovules pluriseriate, numerous; styles 3, longer than stamens, spreading; stigmas fimbriate. *Capsule* 3-valved, ovoid, c. 9 mm long, pubescent and pustulate; seed cylindrical, somewhat curved, 4 mm long, cream, reticulate, aril unilateral, membranous, lobed, longer than seed.

Recorded from the Kaokoveld in northern South West Africa and from southern Angola. Two varieties are recognized.

Pubescence velvety, dense; leaf margin with shallow teeth; bracts c. 10 mm long ....(a) var. *oculata*

Pubescence more sparse; leaf margin with distinct sharp teeth; bracts 10–19 mm long ....(b) var. *paucipilosa*

(a) var. **oculata.**

Turnera oculata Story in Bothalia 7: 493 (1961); Fernandes in Mem. Junta Invest. Ultram. ser. 2, 34: 22 (1962). Schreiber in F.S.W.A. 88: 3 (1968). Type: S.W. Africa, Kaokoveld, Kunene River at Otjinungua, *De Winter & Leistner* 5770 (PRE, holo.; WIND).

Plants with a densely velvety pubescence hiding the epidermis; teeth on leaf-margin small and shallow; bracts up to 10 mm long. FIG. 34: 1.

Recorded from the Kaokoveld in northern South West Africa and southern Angola.

S.W.A.—1712 (Posto Velho): 1 km S. of Kunene River on the mountain E. of Otjinungua, *Giess & Wiss 3290; De Winter & Leistner 5770;* Otjihipa Mountain, *Davies, Thompson & Miller 77.* 1812 (Sanitatas): 20 km S. of Kunene River on way to Anabib (Orupembe), *Story 5778.*

(b) var. **paucipilosa** *Oberm.* in Bothalia 11: 290 (1974). Type: S.W. Africa, near Otjitanda, *Giess* 9377 (PRE, holo.; WIND).

Plants sparsely and minutely pubescent with stellate and simple hairs not hiding the epidermis; teeth on leaf margin usually deeply incised; bracts 12–18 mm long. FIG. 34: 2.

Recorded from one area in the Kaokoveld, around Etengua and Otjitanda.

S.W.A.—1712 (Posta Velho): Otjinungua, 12 km S. of Kunene River, *De Winter & Leistner 5751.* 1713 (Swartbooisdrift): 32 km W. of Etengua near Otjitanda, *Giess 9377;* 8 km W. of Otjitanda, *Kotze 118.*

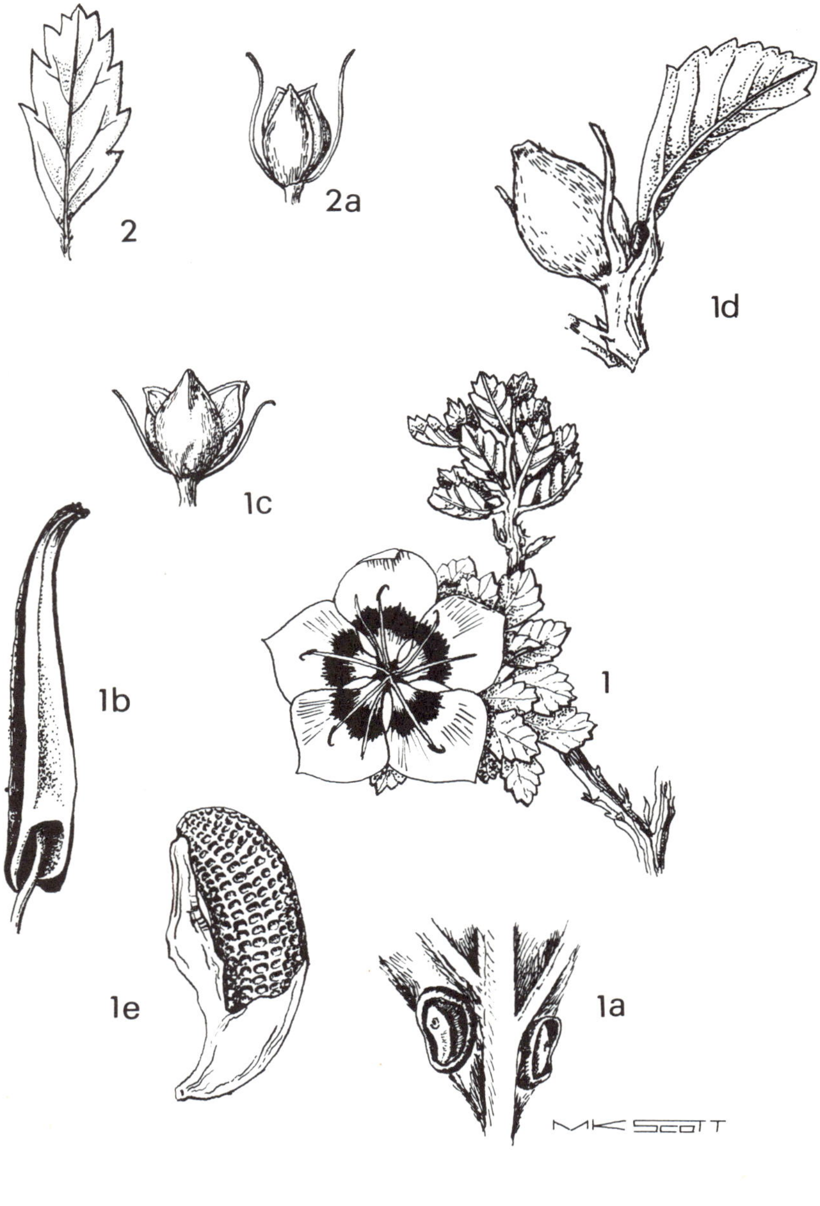

FIG. 34.—1, **Turnera oculata** var. **oculata,** branch with flower, × ½; 1a, base of leaf, lower surface showing glands, × 4; 1b, anther, × 10; 1c, capsule and bracteoles, × 1½; 1d, capsule and leaf showing fusion of petiole and pedicel, × 2; 1e, seed with aril, × 10 (*Story* 5778). 2, **T. oculata** var. **paucipilosa,** leaf, × 1; 2a, capsule and bracteoles, × 1 (*Giess* 9377).

PASSIFLORACEAE

by W. J. J. O. DE WILDE*

Climbers or sometimes (erect) herbs, mostly provided with axillary tendrils (tribe Passifloreae), or erect shrubs or trees (tribe Paropsieae); glabrous or hairy, rarely thorny. *Leaves* (mostly) alternate, simple or compound, entire or lobed, often with glands on petiole and blade; stipules small, sometimes caducous. *Inflorescences* axillary, either cymose (Passifloreae), ending in (a) tendril(s) or not, or racemose (Paropsieae); bracts and bracteoles usually small. *Flowers* hermaphrodite or (functionally) unisexual (and then plants often dioecious or polygamous); stipe articulate to pedicel; hypanthium saucer-shaped to tubiform. *Sepals* 4 or 5 (or 6), imbricate, free or partially connate (*Adenia* p.p.), often persistent. *Petals* (3–4 or 5 (–6), imbricate, rarely absent. *Corona* extra-staminal, inserted on the hypanthium, various, composed of hairs, or of one or more whorls of thread-like processes or scales, or tubiform or cup-shaped, or absent (*Adenia* p.p.). *Disc* mostly extra-staminal, annular or composed of 5 mostly strap-shaped parts (*Adenia*), or absent. *Stamens* 4–many, in S. Africa 4 or 5 (–8), inserted on the hypanthium or on an androgynophore, if few alternipetalous, free or partially connate; anthers 2-celled, basifixed to dorsifixed, versatile or not, sometimes apiculate, opening lengthwise. *Ovary* superior, sessile or on a gynophore or androgynophore, 1-celled, 3–5 (–6)-carpellate, with 3–5 (–6) parietal placentas; ovules mostly numerous, anatropous, with 2 integuments; styles 1 or 3–5, very short to distinct, free or partially united; stigma(s) capitate to subglobose, sometimes much divided (*Adenia*). *Fruit* a loculicidally 3–5-valved capsule, or berry-like. *Seeds* mostly compressed, enveloped by a membranous or mostly pulpy aril; funicles often distinct, testa crustaceous, mostly pitted; endosperm horny; embryo large, straight, with foliaceous cotyledons.

A pantropical family of about 18 genera and about 500 species; *Passiflora*, the largest, mainly in America.

The family can be divided into two tribes, viz. tribe Paropsieae, and tribe Passifloreae. Tribe Paropsieae (all African, 1 species of *Paropsia* in the Caprivi-area) comprises the erect, arborescent species, destitute of tendrils. Tribe Passifloreae consists of the mainly climbing species, provided with tendrils; of this tribe 3 genera are indigenous, whereas *Passiflora* is cultivated as an ornamental or for the flavoured edible fruit (*P. edulis* Sims), but sometimes escaped and running wild. Two species (*P. foetida* L. and *P. suberosa* L.) are locally naturalized, and may behave as weeds.

Shrubs or trees, without tendrils; vegetative ramification through the axillary bud; corona tomentose, composed of threads arranged in 5 bundles (Tribe Paropsieae) ....1. **Paropsia**

Mostly tendril climbers, rarely low herbs without tendrils, or shrub-like, sometimes provided with thorns; vegetative ramification through the accessory bud (Tribe Passifloreae):

- Flowers (mostly) unisexual; corona ill-developed, or absent; disc consisting of 5 strap-shaped parts, or absent; stigmas divided or distinctly papillate ....2. **Adenia**
- Flowers hermaphrodite; corona well-developed; disc annular or absent; stigmas capitate or subglobose:
  - Styles single, stigma 3 or 4-lobed ....3. **Schlechterina**
  - Styles 3 or 4 (or 5), free or partially connate:
    - Flowers small, up to 1,5 cm diam.; androgynophore absent, stamens inserted inside in the inner corona ....4. **Basananthe**
    - Flowers usually much larger; androgynophore distinct ....5. **Passiflora**

*Rijksherbarium, Schelpenkade 6, Leiden, Netherlands.

5369a

1. PAROPSIA

Paropsia *Noronha ex Thouars*, Hist. Vég. Isles Austr. Afr., ed. 1, 59, tab. 19 (1805); ed. 2, 59, tab. 19 (1806); Gilg in Pflanzenfam. ed. 2, 21: 414 (1925); A. & R. Fernandes in Garcia de Orta 6: 244, 667 (1958); Sleumer in Bull. Jard. Bot. Nat. Belg. 40: 50 (1970). Type: *P. edulis* Thouars.

Trichodia Griff. in Not. 4: 570 (1854).
Hounea Baill. in Bull. Soc. Linn. Paris 1: 301 (1881).

Shrubs or trees, mostly hairy, without tendrils. *Leaves* alternate, ± distichous, not lobed, ± oblong, petiolate, margin subentire to mostly distinctly serrate-dentate or repand-dentate; glands at the tops of the teeth, and on the blade margin (rarely on the surface) towards the base; stipules small, mostly caducous. *Inflorescences* 1–many-flowered, subfasciculate, axillary, peduncled or not, yellow to reddish-brown hairy, sometimes forming terminal panicles by reduction or abscission of sustaining leaves; bracts linear, caducous. *Flowers* hermaphrodite, pedicellate, often fragrant; hypanthium shallowly cup-shaped; sepals (4 or) 5, imbricate, persistent; petals (4 or) 5, imbricate. *Corona* composed of hairy threads in one series, ± free or connate at base, or these collected into 5 bundles opposite the petals. *Stamens* 5, opposite the sepals; filaments glabrous or pilose at base, faintly connate at base, inserted at base of ovary or on short gynophore; anthers oblong to linear, ± sagittate, subdorsifixed. *Gynophore* short or absent; ovary ovoid, placentas 3–5, many-ovuled; styles (2 or) 3(–5), free, sometimes pilose; stigmas subreniform or capitate. *Fruit* a 3(–5)-valved capsule, subglobose, sometimes shortly stipitate. *Seeds* ovoid, compressed, 5–7 mm, arillate; testa crustaceous, scrobiculate.

A genus of 11 species, 1 in S.E. Asia (Malesia), 6 species in Madagascar, and 4 in tropical Africa, 1 of which enters our Flora area.*

Paropsia brazzeana *Baill.* in Bull. Soc. Linn, Paris 1: 611 (1886); Warb. in Pflanzenfam. 3, 6a: 27 (1893); Gilg in Bot. Jahrb. 40: 472 (1908); in Engl., Pflanzenw. Afr. 3, 2: 572, fig. 253 D–F (1921); Gilg in Pflanzenfam. ed. 2, 21: 415 (1925); Exell in J. Bot. Lond. 67 (Suppl.): 191 (1929); Gossw. & Mendonça, Carta Fitogeogr. Angola 111, 161 (1939); Gossw., Agron. Angol. 7: 274 (1953); A. & R. Fernandes in Garcia de Orta 6: 653 (1958) as *brazzaeana*; F. White, For. Fl. N. Rhod. 268 (1962); Sleumer in Bull. Jard. Bot. Nat. Belg. 40: 70 (1970). Type: Congo Republic, *De Brazza* 48 (P, holo.!).

P. reticulata Engl. in Bot. Jahrb. 4: 391 (1892); ibid. 15: 586 (1893); Warb. Pflanzenfam. 3, 6a: 27, fig. 9D–F (1893); Gilg. in Warb., Kunene–Samb. Exped. 309 (1903); Pflanzenfam. ed. 2, 21: 414, fig. 183D–F (1925). Type: Angola, *Mechow* 541 (B†; K, photo.). —var. *ovatifolia* Engl. in Bot. Jahrb. 14: 14 (1891). Type: Zaïre, *Pogge* 951 (B†). —var. *proschii* Briq. in Annu. Cons. Jard. Bot. Genève 6: 1 (1902). Type: Zambia, *Prosch* 35 (G). *P. argutidens* Sleumer in Fedde Repert. 45: 13 (1938). Type: Zambia, *Milne-Redhead* 726 (K, holo.!; BR).

Shrub or shrublet 1–2,5 (–3) m; stem up to 5 cm across, mostly branched from the base; young twigs yellowish-tomentose. *Leaves* oblong-elliptic or ovate-oblong, rarely broadly elliptic or ovate, apex shortly acuminate, rarely obtuse, base broadly attenuate to rotundate, hard-chartaceous, when young woolly yellowish-tomentose all over, later on glabrescent and shiny above, margin regularly glandular sinuate-dentate or subserrate (teeth 1–4 mm, 2–5 (–7) mm spaced), (3–) 6–10 (–13) × (1,5–) 2,5–4 (–6,5) cm (in sterile shoots somewhat larger), lateral nerves 5–8 pairs, upward curved, conjunct, distinctly prominent beneath; petioles (3–) 5–7 mm long, c. 2 mm thick. *Inflorescences* 1–2 (–3) (rarely–5)-flowered fascicles, axillary to normal leaves; bracts ovate, c. 2 mm. *Flowers* subsessile or stipe (pedicel) up to 5 mm, c. 1 mm thick; sepals oblong, greenish-whitish, outside (as the pedicel) yellowish-tomentose, inside finely tomentose, 10–12 × (3–) 4–5 mm, slightly enlarging with age; petals oblong, white, especially outside puberulous, resembling the sepals but slightly narrower; corona c. 3 mm high, split up into 5 rather

*After going to press specimen *Wiens & Van Wyk* 5334 from Pafuri (Grid 2231 AD) was received. The specimen is immature and has been provisionally identified as *P. brazzeana* Baill. by the author. In the editor's opinion the specimen probably belongs to *P. braunii* Gilg, a species known from Mozambique and areas further north.

distinct bundles; threads connate at base, outside tomentose, connate part inside glabrous; filaments c. 5 mm; anthers oblong-cordate, c. 2 mm; ovary sessile, reddish-brown tomentose; styles hardly 1 mm. *Fruit* ovoid-subglobose, red-brown tomentose, 1,5–2 × 1,2–1,8 cm. FIG. 35.

Distributed mainly in Equatorial Africa, viz. Cameroon, Central African Republic, Congo Brazzaville, Zaïre, Angola, Zambia, Rhodesia, Botswana; 1 collection from Caprivi; found in forest and thickets, gallery- and secondary forest; rather common on clay and Kalahari sand; 0–1000 (in Angola up to 1600) m.

S.W.A.—1723 (Singalamwe): Caprivi area, Singalamwe to Katima Mulilo, *Killick & Leistner 3281*.

5370 2. ADENIA

Adenia *Forsk.*, Fl. Aegypt.-Arab. 77 (1775); Engl. in Bot. Jahrb. 14: 374 (1891); Harms in Pflanzenfam. ed. 2, 21: 488 (1925); Burtt Davy, Fl. Transv. 1: 220 (1926); Liebenberg in Bothalia 3: 513 (1939); A. & R. Fernandes in Garcia de Orta 6: 255, 656 (1958); C.F.A. 4: 217 (1970); Phill., Gen. ed. 2: 518 (1951); De Wilde in Meded. Landbouwhogeschool Wageningen 71–18: 1–281 (1971); Ross, Fl. Natal 251 (1972). Type: *A. venenata* Forsk.

Modecca Lam., Encycl. 4: 208 (1797); Harv. in F.C. 2: 499 (1862); Thes. Cap. 2: 43 (1863); Gen. Pl. ed. 2: 121 (1868); Benth. & Hook.f., Gen. Pl. 1: 813 (1867); Mast. in F.T.A. 2: 512 (1871).
Kolbia P. Beauv., Fl. d'Oware 2: 91, t.120 (1807).
Blepharanthes J. E. Smith, Gram. Bot. 188 (1821), nom. illegit.
Paschanthus Burch., Trav. 1: 543 (1822).
Microblepharis (Wight & Arn.) Roem., Syn. Mon., 2 Pepon. 200 (1846).
Erythrocarpus Roem., Syn. Mon., 2 Pepon. 204 (1846).
Clemanthus Klotzsch in Peters, Reise Mossamb. Bot. 143 (1862).
Ophiocaulon Hook.f. in Benth. & Hook.f., Gen. Pl. 1: 813 (1867); Harv., Gen. Pl. ed. 2: 121 (1868); Mast. in F.T.A. 2: 517 (1871); Engl. in Bot. Jahrb. 14: 385 (1891).
Machadoa Welw. ex Benth. & Hook.f., Gen. Pl. 1: 814 (1867).
Keramanthus Hook.f. in Bot. Mag. t.6271 (1876).
Jäggia Schinz in Verh. Bot. Ver. Prov. Brandenb. 30: 253 (1888).
Echinothamnus Engl. in Bot. Jahrb. 14: 383 (1891).

(Sub)ligneous to herbaceous perennial climbers with tendrils, sometimes erect herbs or shrublets mostly without tendrils, often with a rootstock or tuber, or a pachypodous main stem; sometimes thorny or spiny; glabrous or sometimes pubescent. *Leaves* either simple, entire or lobed, or palmately parted or pseudo-compound; glands (0 or 1) or 2 at the blade-base, at or near the apex of the petiole, and with or without glands elsewhere on the lower surface or margin of the blade; stipules minute, narrowly triangular or reniform; tendrils axillary. *Inflorescences* axillary, cymose, the middle (or the first three) flower(s) often replaced by (a) tendril(s); bracts and bracteoles minute, triangular to subulate. *Flowers* dioecious or rarely monoecious, bisexual or polygamous, campanulate or urceolate to tubular or infundibuliform, mostly greenish or yellowish, always glabrous, stipe articulate at base. *Hypanthium* saucer- or cup-shaped, or tubular. *Sepals* (4–) 5 (–6), free or partially connate into a calyx-tube, imbricate, persistent. *Petals* (4–) 5 (–6), free, included in the calyx, sometimes adnate with the calyx-tube, mostly fimbriate or laciniate. *Corona* annular, or consisting of 5 cup-shaped parts, or of a laciniate rim or membrane, or composed of hair-like processes, or absent. *Disc-glands* 5, ligulate or strap-shaped, inserted at or near the base of the hypanthium, alternating with the petals, or absent. *Male flowers: Stamens* (4–) 5 (–6), hypogynous or perigynous (variably inserted in the hypanthium), free or partially connate into a tube; anthers basifixed, oblong to linear, often apiculate, 2-locular, opening introrsely to latrorsely; vestigial ovary minute. *Female flowers* mostly smaller than the male flowers, with smaller petals; staminodes ± subulate. *Ovary* superior, shortly stipitate or subsessile, globose to oblong; placentas 3 (–5), ovules usually numerous; styles 3(–5) free or partially united, sometimes styles very short; stigmas mostly subglobose, laciniate or plumose or densely woolly-papillate.

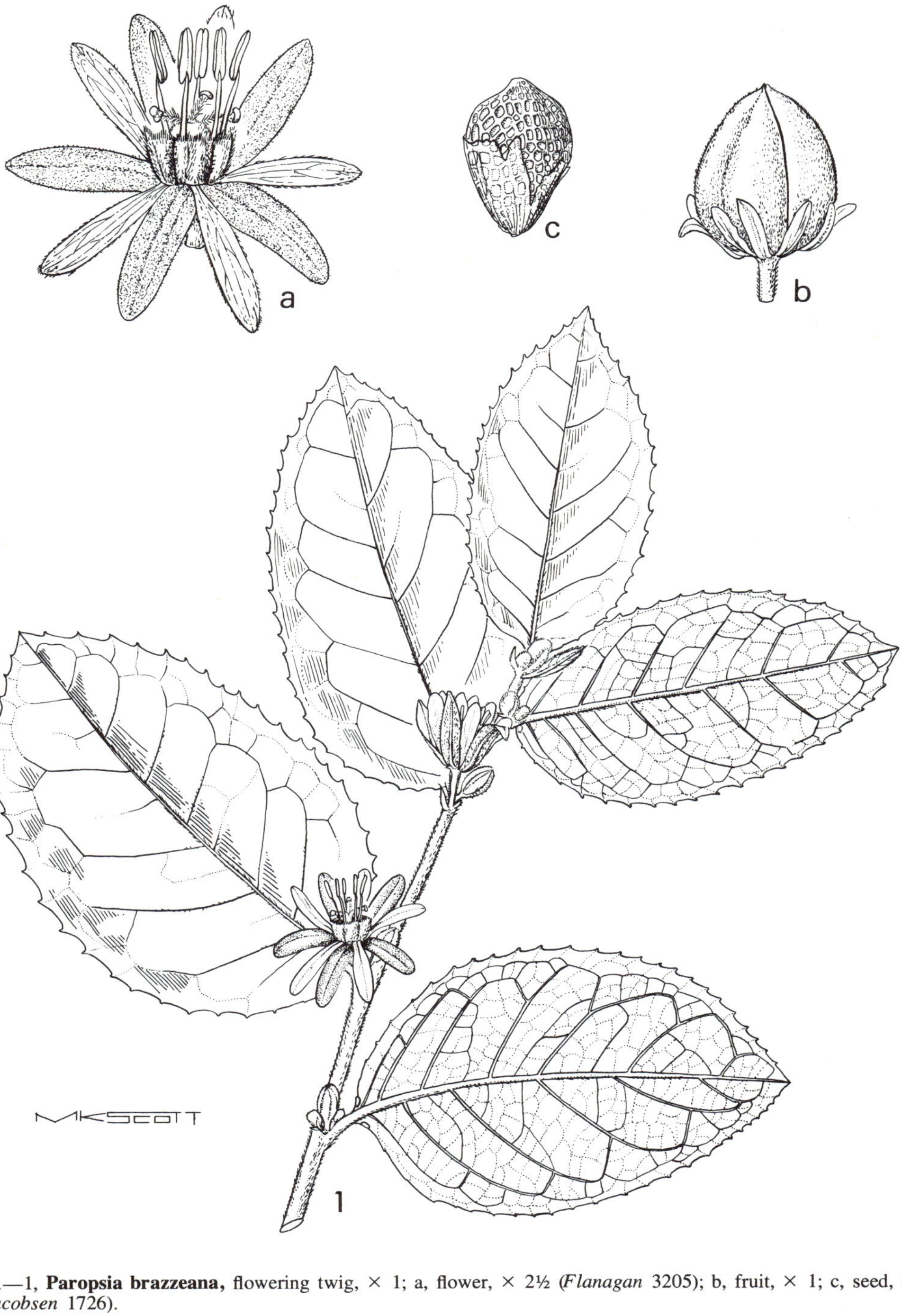

FIG. 35.—1, **Paropsia brazzeana,** flowering twig, × 1; a, flower, × 2½ (*Flanagan* 3205); b, fruit, × 1; c, seed, × 3 (*Jacobsen* 1726).

Fruit a stipitate 3 (–5)-valved capsule; pericarp (woody-)coriaceous to rather fleshy (and hence fruit ± berry-like), greenish to yellow or bright red. *Seeds* ± compressed, with crustaceous pitted testa, enclosed in a membranous to pulpy (juicy) aril.

A genus of 93 species, divided into 6 sections, in tropical Africa (with a few species in temperate Southern Africa), Madagascar, and S.E. Asia, Malesia, and N. Australia. 10 species occur in our area.

Gland at blade-base single, on a single median hemispherical to spathulate appendage; corona absent; disc-glands absent; sepals free; stigmas (sub)sessile. Sect. Ophiocaulon .... 10. *A. gummifera*

Gland(s) at blade-base (0 or) 1 or 2, (sub)sessile, never on a single distinct median appendage with narrow insertion; corona absent or membranous or laciniate, or composed of (fine) hairs; disc-glands present or not; sepals free or partially connate into a calyx-tube; styles distinct:

- Sepals free or nearly so; sepals and petals inserted at or at about the same level; anthers wholly or largely extending beyond the hypanthium; corona present; disc-glands present. Sect. Microblepharis:
  - Male flowers including stipe usually more than 10 mm long; plants mostly provided with tendrils, or with thorns:
    - Plant unarmed:
      - Gland at blade-base 1; leaves entire or 3–5-foliolate; hypanthium wider than long, 2,5–4 mm wide .... 1. *A. fruticosa*
      - Glands at blade-base 2, though approximate; leaves deeply 5-lobed; hypanthium longer than or about as long as wide, 2–3 mm wide .... 2. *A. glauca*
    - Plant thorny .... 3. *A. spinosa*
  - Male flowers small, including stipe 8–10 (–14) mm long; stems without tendrils, erect, short, hard-shrubby, arising from a swollen lumpy base; glands at blade-base 2, sessile .... 4. *A. pechuëlii*
- Sepals partially connate into a calyx-tube which extends wholly or partially above the insertion of the petals, hence calyx-lobes and petals not inserted at the same level, but if approximate because of adnation of the petals to the calyx-tube, then stamens (anthers) largely or entirely enclosed in the calyx-tube:
  - Midrib not ending in a gland; stamens inserted at the base of the hypanthium; corona present; disc-glands present. Sect. Blepharanthes:
    - Leaves (deeply) lobed or entire, not digitately compound:
      - Hypanthium wider than long, 5–12 mm wide, base flattish, about as wide as the calyx-tube; leaves 3–5-lobed .... 5. *A. natalensis*
      - Hypanthium about as long as, or longer than wide, 3–6 mm wide, base tapering; leaves entire .... 6. *A. hastata*
    - Leaf-blade digitately dissected to the base, the lobes or leaflets sometimes more or less stalked:
      - Climbing herbs 1–5 m tall, provided with tendrils; leaves palmately (3–) 5-parted or -foliolate .... 7. *A. digitata*
      - Erect herbs, up to 50 cm tall, without tendrils; leaves palmately (5–) 7-parted .... 8. *A. wilmsii*
  - Midrib ending in a (sub)apical gland; stamens inserted well above the base in the flower tube; corona absent; disc-glands absent. Sect. Paschanthus .... 9. *A. repanda*

1. **Adenia fruticosa** *Burtt Davy,* Fl. Transv. 1: 36; 221 (1926); Bremekamp, Vegetationsbilder 23, 3: 6, pl. 18 (1932); Liebenberg in Bothalia 3: 538, 528, 532, pl. 1, 2 (1939); Letty, Wild Flow. Transv. 225 (1962); De Wilde in Meded. Landbouwhogeschool Wageningen 71–18: 69 (1971). Type: Transvaal, *Pole Evans* H15723 (K, holo.!; PRE!).

Shrub or shrubby tree to 6 m, glabrous; main stem mostly thick, soft-woody, branched or not, up to 2 × 0,6 m; branches shrubby or lianoid, provided with tendrils. *Leaves* some grey-green or glaucous, punctate or not beneath, simple or 3–5-foliolate, in outline ovate to suborbicular, base (sub)cordate, 1–8 × 1–8 cm, (1–) 3–5-plinerved; petiole (0,3–) 1–5 cm; leaflets (or lobes) orbicular, ovate or obovate, base rounded to attenuate, apex subacute to broadly rounded, rarely retuse, 1–6 × 1–4 (–6) cm, ± penninerved; margin entire; petiolule 0–5 (–7) mm. *Gland* at blade-base 1, on a fleshy ± upward curved median lobelet at the top of the petiole; no other glands present. *Stipules* narrowly triangular, c. 0,5 mm. *Inflorescences* either solitary in the leaf-axils, or grouped in short-shoots, peduncled up to c. 1 cm, 2–5-flowered in ♂, 1–3-flowered in ♀; tendrils 0; sterile tendrils 3–12 cm, sometimes breaking off and leaving a thorn-like structure. *Male flowers* campanulate, including the 0,5–5 mm long stipe 9–17 mm; hypanthium cup-shaped, 1–2,5 × 2,5–4,5 mm; calyx-tube 0; sepals lanceolate, obtusish, 7–9 (–10) mm, (sub)entire; petals oblong to lanceolate, acute, 4–8 mm, 3-nerved, serrulate; filaments 1,5–4 mm, connate for 0,5–2 mm, inserted at the base of the hypanthium; anthers 3–6,5 mm, obtuse; septa 1–2 mm high; corona consisting of fine hairs 0,2–0,5 (–1) mm, sometimes partly membranous; disc-glands c. 0,5 mm, or absent. *Female flowers* campanulate, including the 0,5–1 mm long stipe 7–9 mm; hypanthium 1–1,5 × 2,5–3,5 mm; calyx-tube 0; sepals lanceolate, obtuse, 5–7 mm, entire; petals lanceolate, acute, 3–4 mm, 1-nerved, ± serrulate; staminodes 2–4 mm; septa 1 (–2) mm high; corona consisting of fine hairs 0,2–0,5 mm, or nearly absent; disc-glands 0–0,5 mm; gynophore 1,5–3 mm; ovary subglobose 2,5–4 mm; styles connate for 0,5–0,7 mm, style arms 0,5–1 mm; stigmas subglobose, papillate, each 1–1,5 mm diam. *Fruit* 1–2 per inflorescence, subglobose (to ellipsoid), excluding the 2–5 mm long gynophore 1–2 × 0,8–1,8 cm; pericarp coriaceous; seeds 3–6 per capsule, broad ovate to orbicular, 6–6,5 mm.

A species of dry savanna and bushveld, on sandy or granitic soils; 100–1400 m. Distributed in Transvaal and Natal, also in Rhodesia.

Three more or less allopatric subspecies are recognized.

Leaves 3–5-foliolate; petiolule of leaflets 2–5 (–7) mm; anthers c. 3 mm . . . . (a) subsp. *fruticosa*
Leaves simple or 3 (or 4)-foliolate; leaflets sessile; anthers 4–5,5 mm:
- Hypanthium of ♂ flowers broadly cup-shaped, ± 5-saccate; corona hairs 0,5–1 mm; disc-glands present; leaves simple or 3-foliolate . . . . (b) subsp. *simplicifolia*
- Hypanthium of ♂ flowers cup-shaped, tapering, not saccate; corona hairs up to 0,5 mm, or partly absent; disc-glands absent; leaves 3 (or 4)-foliolate . . . . (c) subsp. *trifoliolata*

(a) subsp. **fruticosa.**

De Wilde in Meded. Landbouwhogeschool Wageningen 71–18: 70 (1971).

Leaves 3–5-foliolate, 2–8 × (2–) 2,5–8 cm; petiole 1–5 cm; leaflets suborbicular to obovate, base attenuate to rounded, apex broadly rounded, rarely retuse, 1–6 × 1–4 (–6) cm; petiolules 2–5 (–7) mm. *Male flowers* including the c. 0,5 mm long stipe c. 9 mm; hypanthium cup-shaped, not distinctly saccate, c. 1 × 2,5–3 mm; sepals 7–8 mm; petals c. 4 mm; filaments c. 1,5 mm; anthers c. 3 mm; corona partly consisting of fine hairs up to 0,5 mm, partly membranous; disc-glands c. 0,2 mm. *Female flowers* including the 0,5–1 mm long stipe 7–9 mm, hypanthium c. 1,5 × 2,5–3,5 mm; sepals 5–7 mm; petals 3–4 mm; staminodes 3–4 mm; corona consisting of fine hairs up to 0,2 mm; disc-glands absent; pistil 6,5–8 mm; gynophore 2–3 mm; ovary 3–4 × 3–3,5 mm; styles connate for 0,7 mm, style-arms 0,5–0,7 mm.

Found in bushveld, on dry stony slopes, dry sandy loam soil and granite in the eastern Transvaal, not elsewhere; 800–1400 m. Flowers in Aug. and Sept., fruits in Oct. and Nov.

TRANSVAAL.—2429 (Zebediela): Chuniespoort, *Obermeyer & Verdoorn 10;* Sekukuniland, *Barnard 454.* 2430 (Pilgrims Rest): 8 km N. of Burgersfort, *Codd & Dyer 7732;* Kasparsnek, *Strey 3693.*

(b) subsp. **simplicifolia** *De Wilde* in Meded. Landbouwhogeschool Wageningen 71–18: 71 (1971). Type: Rhodesia, *Chase* 1321 (SRGH, holo.!; BM; K; LISC).

Leaves simple, ovate, or (2–) 3-foliolate, 1–6 × 0,7–5 cm; petiole (0,3–) 0,5–2,5 cm; leaflets suborbicular to elliptic, or ovate, or rhomboid, base attenuate to rounded, apex (sub)obtuse, 1–5 × 1–3 (–4) cm; petiolules ± 0. *Male flowers* including the 3–4 mm long stipe 13–15 mm; hypanthium broadly cup-shaped, 5-saccate, c. 2 × 4–4,5 mm; sepals 8–10 mm; petals 7–8 mm; filaments 3–4 mm; anthers 5–5,5 mm; corona of densely set fine hairs 0,5–1 mm; disc-glands 0,5–0,7 mm. *Female flowers* including the 0,5–1 mm long stipe 7–8 mm; hypanthium 1–1,5 × 2,5–3,5 mm; sepals 5–6 mm; petals 3–4 mm; staminodes 2–2,5 mm; corona hairs c. 0,5 mm; disc-glands 0,2–0,5 mm; pistil 5–6,5 mm; gynophore 1,5–2 mm; ovary 1,5–3 × 1,5–2,5 mm; styles connate for 0,5–0,7 mm; style-arms 0,5–1 mm.

Confined to a restricted area in Transvaal and southern Rhodesia, in thorny bushveld, found in rocky (basalt) places and on sandy soils, also near hot springs; 400–1000 m. Flowers and fruits Sept.–Nov.

TRANSVAAL.—2229 (Waterpoort): Dongola Reserve, *Pole Evans 3747.*

The fruits are reported as being green berries whitish striped.

(c) subsp. **trifoliolata** *De Wilde* in Meded. Landbouwhogeschool Wageningen 71–18: 72 (1971). Type: Natal, *Acocks* 12937 (PRE, holo.!; P).

Leaves 3- (or 4)-foliolate, 1,5–3 × 1,5–2,5 cm; petiole 0,5–1,5 cm; leaflets suborbicular, base and apex rounded to subacute, 0,5–2 × 0,5–2,5 cm; petiolules ± 0. *Male flowers* including the c. 5 mm long stipe c. 14–17 mm; hypanthium cup-shaped, tapering, not saccate, c. 2,5 × 4–5 mm; sepals 6,5–8 mm; petals c. 5 mm, filaments 3–3,5 mm; anthers 4–4,5 mm; corona hairs up to 0,5 mm, or corona partly membranous or partly absent; disc-glands 0. *Female flowers* not known. *Fruit* subsessile, ovoid, c. 1,2 cm long, 2–3-seeded.

An endemic of Natal and Zululand, in sandy bushveld on sandstone; locally frequent; 100–900 m. Flowers and fruits Sept.–Dec.

NATAL.—2732 (Ubombo): Ubombo Mt., *Venter 1751*. 2831 (Nkandla): Umfolozi Game Reserve, *Ward 4378;* near Ntonjaneni, *Acocks 12937*. 2832 (Mtubatuba): Hluhluwe Game Reserve, *Ward 4475*.

Fresh fruits are recorded as: ''almost globose, mottled green when immature, darker towards base, becoming orange-green when ripe'', or: ''very dark green with light green veins''.

Gerstner 5241 is annotated as having an underground tuber, and not an overground stem.

2. **Adenia glauca** *Schinz* in Bot. Jahrb. 15, Beibl. 33, 1: 1 (1892); Engl., Pflanzenw. Afr. 3, 2: 605 (1921); Burtt Davy, Fl. Transv. 1: 222 (1926); Steyn, Tox. Pl. S. Afr. 314, fig. 43B, 46, 47B (1934); Liebenberg in Bothalia 3: 539, 523, 532, fig. 6–8, pl. 4 & 5 (1939); Letty, Wild Flow. Transv. 225, tab. 112 (1962); Watt & Breyer-Brandwijk, Med. Pois. Pl. ed. 2: 828 (1962); De Wilde in Meded. Landbouwhogeschool Wageningen 71–18: 73 (1971). Type: Transvaal, *Rehmann* 4799 (Z, lecto.!).

Modecca glauca Schinz in Bot. Jahrb. 15, Beibl. 33, 1: 1 (1892); ex Engl. in Bot. Jahrb. 14: 393 (1891), nom. nud.

Climber, sometimes shrub-like, up to 3,5 m, the basal part thickened, tapering, ''bottle-shaped'', up to c. 1 × 0,4 m, ± fleshy, smooth, greenish, glabrous. *Leaves* greyish or glaucous or purplish-grey, sometimes punctate beneath, 5-parted to the base, suborbicular in outline, 2–12 × 2,5–12 cm, base cordate, 5-plinerved; petiole (0,5–) 1–5 cm; leaf-parts (ob)ovate to oblong, ± attenuate at base, apex (sub)obtuse, sometimes retuse, 1–7 × 0,5–3,5 (–4,5) cm; margin entire. *Glands* at blade-base 2, contiguous, situated on two ± connate thickish auricles making the blade slightly peltate; blade-glands 1–3 (–5), (sub)apical on the lobes. *Stipules* narrowly triangular, acute, 1–1,5 mm. *Inflorescences* peduncled up to 1 (–2) cm, 2–5-flowered in ♂, 1–3-flowered in ♀; tendril (0 or) 1, 2–6 cm; sterile tendrils up to 10 cm. *Male flowers* (tubiform to) infundibuliform, including the 3–7 mm long stipe (13–) 15–30 mm; hypanthium cup-shaped, (1–) 2–4 × 2–4 mm; calyx-tube 0; sepals lanceolate, obtuse, (8–) 10–18 (–20) mm, entire; petals lanceolate, apex acute to obtuse, 6–11 mm, 1–3-nerved, serrulate; filaments (3–) 4,5–7 mm, connate for (1,5–) 2,5–4 mm, inserted at the base of the hypanthium; anthers 3—5,5 mm subacute; septa (1–) 2,5–4 mm high; corona consisting of either sparse thick hairs c. 0,5 mm or of thinner hairs 0,5–1 mm; disc-glands absent. *Female flowers* tubiform to infundibuliform, including the 1,5–4 mm long stipe (9–) 10–15 mm; hypanthium cup-shaped, 1,5–2 × 2–3 mm; calyx-tube 0; sepals lanceolate, obtuse, 7–9 mm, (sub)entire; petals lanceolate, acute, 4–5 mm, 1-nerved, (sub)serrulate; staminodes 2–3,5 mm; septa 1–2 mm high; corona hairs sparse, thick, 0,5–1 mm; disc-glands 0; gynophore 1–2 mm; ovary subglobose, 2,5–4,5 mm; styles connate up to 0,5 mm, style-arms 1–1,5 mm; stigmas subglobose, papillate, each c. 1,5 mm diam. *Fruit* 1–2 per inflorescence, (subglobose to) ellipsoid, excluding the 3–5 mm long gynophore 1,8–2,5 × 1,5–2 cm; pericarp coriaceous; seeds 3–5 per capsule, orbicular to heart-shaped, 5,5–7,5 mm.

Found in a restricted area in Transvaal and S.E. Botswana, in dry bushveld, in rocky places or on sandy soil; locally common; 1 000–1 600 m. Flowers mainly Aug.–Jan., fruits Oct. and Nov.

TRANSVAAL.—2427 (Thabazimbi): 17 km S. of Matlabas, *Codd 4433*. 2428 (Nylstroom): near Naboomspruit, *Galpin 11605;* near Warmbaths, *Burtt Davy 2622; Acocks 13923*. 2528 (Pretoria): Premier Mine, *Rogers 25027;* Magaliesberg, *C. A. Smith 6271*.

Said to be poisonous to cattle.

Liebenberg (l.c.) mentions occasional 4-merous ovaries and fruits, and once a male flower with 6 petals and stamens.

3. **Adenia spinosa** *Burtt Davy,* Fl. Transv. 1, 36: 222 (1926); Liebenberg in Bothalia 3: 533, 523, fig. 9, pl. 3 (1939); Letty, Wild Flow. Transv. 225 (1962); De Wilde in Meded. Landbouwhogeschool Wageningen 71–18: 76 (1971). Type: Transvaal, *Rogers* 19299 (K, holo.!; PRE; Z).

Thorny shrub up to 1,5 (–6?) m high, arising from an irregularly shaped ± fleshy trunk up to 0,5 m tall, up to 2,5 m wide. *Leaves* ± coriaceous, grey-glaucous, sometimes punctate beneath, entire, broadly ovate to elliptic, base cordate to rounded, apex obtuse, sometimes retuse, 1–3,5 × 0,7–3 cm, 3-plinerved to ± penninerved; margin entire; petiole 0,2–0,7 cm. *Gland(s)* at blade-base 1 or 2, sessile at the top of the petiole; blade-gland 1, (sub)apical. *Stipules* triangular, acute, c. 0,5 mm. *Thorns* 0,5–4 cm, patent, acute, when young ending in a tendril. *Inflorescences* solitary or in short-shoots, peduncled up to 0,5 cm, 2–6-flowered in ♂, 1–3-flowered in ♀; tendril (0 or) 1, strong, 4–6 cm, mostly replaced by a thorn; sterile tendrils simple, 4–8 cm. *Male flowers* tubiform-campanulate, including the 1,5–3 mm long stipe (10–) 12–24 mm; hypanthium (narrowly) cup-shaped, 1–3 × 2,5–4 mm; calyx-tube 0; sepals lanceolate, obtuse, (8–) 9–18 mm, (sub)entire; petals lanceolate, acute, (6–) 8–10 mm, 1–3-nerved, serrulate; filaments 2,5–5,5 mm, connate for 1–3 mm, inserted at or near the base of the hypanthium; anthers 4–5 mm, obtuse; septa 1–3 mm high; corona consisting of a few stiff hairs 0,2–1 mm, mainly near the insertion of the petals; disc-glands absent or inconspicuous. *Female flowers* tubiform-campanulate, including the 1–1,5 mm long stipe 7–10 mm, hypanthium 1–1,5 mm, calyx-tube 0; sepals lanceolate, obtuse, 5–8 mm, entire; petals lanceolate, acute, 2,5–3,5 mm, 1-nerved, (sub)entire; staminodes c. 1,5 mm; septa 0,5–0,7 mm; corona absent or consisting of a few stiff hairs c. 0,2 mm; disc-glands 0; gynophore 1–2 mm; ovary subglobose, 2–4 (–5) mm; styles connate for c. 1 mm, style-arms c. 0,5 mm; stigmas subglobose, papillate, each c. 1 mm diam. *Fruit* 1–2 per inflorescence, subglobose (to ellipsoid), excluding the 3–4 mm long gynophore 1,4–2,2 × 1,2–1,8 cm; pericarp coriaceous; seeds 3–6 per capsule, suborbicular, c. 6,5–7 mm.

Occurs in a restricted area in the Transvaal, also in southern Rhodesia. It is locally common in open bushland and scrub, in dry rocky (basalt) places; 200–1000 m. Flowers in Jan., May, July, Sept. and Nov., fruits in Sept. and Nov.

The thorns and tendrils are homologous with the inflorescences as found in other Adenias.

The flowers resemble those of *A. glauca*. The mostly early deciduous leaves often much resemble the simple leaves of *A. fruticosa* subsp. *simplicifolia*.

TRANSVAAL.—2220 (Waterpoort): Soutpan, *Obermeyer, Schweickerdt & Verdoorn 137; Galpin 15140;* 6 km N.W. of Wylliespoort, *Codd 2997.* 2230 (Messina): near Messina, *Rogers 19299; 21664.* 2329 (Pietersburg): Naauwpoort, *Bremekamp & Schweickerdt 469;* Vivo, *Mogg 24453.* 2331 (Phalaborwa): Kruger National Park, near The Gorge Rest Camp, *Codd 6187.* 2429 (Zebediela): Molsgat, near Chuniespoort, *Obermeyer sub TRV 34667.*

4. **Adenia pechuëlii** *(Engl.) Harms* in Pflanzenfam. 3, 6a, Nachtr. 1: 255 (1897); ibid., ed. 2, 21: 490, fig. 216 (1925); Engl., Pflanzenw. Afr. 3, 2: 595, 600, 601, 608, fig. 267 (1921); Schreiber in F.S.W.A. 89: 2 (1968); De Wilde in Meded. Landbouwhogeschool Wageningen 71–18: 78 (1971). Syntypes: S.W. Africa, *Pechuël-Loesche* s.n.; *Gürich* 18 (B†).

Echinothamnus pechuëlii Engl. in Bot. Jahrb. 14: 383, fig. 9 (1891).

Bushy shrub up to 1,5 m, mostly leafless, without tendrils; pruinose, glabrous; the branches up to 60 cm, thorny, growing from lumpy subglobose to cylindrical main stem up to 1 m diam. *Leaves* coriaceous, grey-green, minutely papillate beneath, entire or shallowly 3-lobed, ovate-oblong to lanceolate, base attenuate to rounded, apex (sub)acute to obtuse, 1–6 × 0,2–2 (–2,5) cm, penninerved; margin entire; petiole up to 0,2 cm. *Glands* at blade-base 2, lateral at the very base of the midrib; blade-glands 0–4, one pair at about $^1/_3$–½, and one pair $^2/_3$ from the base. *Stipules* narrowly triangular, acute, 0,5–1 mm. *Inflorescences* peduncled up to 0,2 cm, 1–2 (–3)-flowered in ♂ and ♀; tendrils 0. *Male flowers* campanulate, including the c. 3 mm long stipe 8–10 (–14) mm; hypanthium cup-shaped, 0,7–2 × 2,5–4 (–6) mm; calyx-tube 0; sepals elliptic to oblong, obtuse, 3–5,5 (–7) mm, entire; petals lanceolate, acute, 3–5 mm, 1-nerved, subserrulate; filaments 1–2 mm, connate for 0,5–1 mm, inserted at the base of the hypanthium; anthers 2,5–4,5 mm, obtuse; septa 0,5–1 mm high; corona hairs fine, 0,5–1 mm; disc-glands c. 0,2 mm. *Female flowers* campanulate, including the 2–3 mm long stipe 7–8 mm; hypanthium (broadly) cup-shaped, 1–1,5 × 2,5–3 mm; calyx-tube 0–0,5 mm, sepals oblong, obtuse, c. 4 mm, entire; petals lanceolate, (sub)acute, c. 2,5 mm, 1-nerved, ± serrulate; staminodes 1,5–2 mm; septa 0,5–0,7 mm high; corona hairs c. 0,5 mm; disc-glands 0; gynophore c. 1 mm, ovary subglobose to ellipsoid, c. 2 mm; styles connate for c. 0,5 mm, style-arms c. 0,5 mm; stigmas subglobose, papillate, c. 1 mm diam. *Fruit* 1 per inflorescence, subglobose to ellipsoid, excluding the 1–2 mm long gynophore 1–1,7 × 1–1,5 cm; pericarp thinly coriaceous; seeds 3–10 per capsule, ovate-subcircular, 5–7 mm.

A local endemic species in South West Africa, in semi-desert on sand, marble, and quartz and granite rock; 200–1000 m. Flowers in Dec., fruits in Febr.

S.W.A.—2115 (Karibib): Omaruru, *Giess 9190;* Pforte, *Merxmüller 1758.* 2214 (Swakopmund): near Walvis Bay, *Jensen s.n.* 2215 (Trekkopje): near Husab, *Pechuël-Loesche s.n.* 2416 (Maltahohe): Blässkranz, *Volk 931.*

5. **Adenia natalensis** *De Wilde* in Meded. Landbouwhogeschool Wageningen 71–18: 161, fig. 25 (1971). Type: Natal, *Gerrard* 1820 (K, holo.!; BM).

Subherbaceous climber, c. 4 m; glabrous. *Leaves* grey-green, not punctate beneath, deeply 3–5-palmately lobed or -partite, orbicular to 5-angular in outline, base subcordate, 5–9 × 6–10 cm, 5-subplinerved; lobes or blade-parts ovate to (ob)lanceolate, entire or 2–4 (–6)-lobed, apex rounded to subacute, 2–6 cm; margin entire; petiole 1,5–6 cm. *Glands* at blade-base 2, separate or ± contiguous, on the 1–2 mm wide peltate blade-base; blade-glands 2–6, submarginal or ± scattered. *Stipules* subtriangular, c. 0,5 (–1) mm, withering. *Inflorescences* peduncled for 1–4,5 cm, (1–) 2–6-flowered in ♂; tendril 0 or 1, 2–5 cm, sterile tendrils 8–12 cm. *Male flowers* campanulate, including the 4–4,5 mm long stipe 20–22 mm; hypanthium saucer-shaped, ± 5-saccate, 2–3 × 10 (–13) mm; calyx-tube (1,5–) 2–3 mm; calyx-lobes elongate-triangular, (sub)acute, 10–12 mm, subentire; petals obovate-oblong or ± spathulate-unguiculate, apex rounded, 7,5–9 mm, 3–5-nerved, denticulate, inserted at the same level as the corona; filaments 4–5 mm, connate for 1,5–2 mm, inserted at the base of the hypanthium; anthers c. 9 mm, (sub)obtuse; septa 1,5–2 mm high; corona consisting of hairs simple or branched, or of a laciniate membrane, (0,5–) 1 mm; disc-glands 1,7–2 mm. *Female flowers* and fruit not known.

This species is known from only two collections made in Natal, in the previous century; apparently it is a local endemic and rare species, not unlikely extinct by now. Presumably it is a species of low coastal forest.

NATAL.—Zululand: without precise locality, *Gerrard 1200; 1820.*

6. **Adenia hastata** *(Harv.) Schinz* in Bot. Jahrb. 15, Beibl. 33, 1: 3 (1892); Burtt Davy in Ann. Transv. Mus. 3: 121 (1912); Engl., Pflanzenw. Afr. 3, 2: 603 (1921); Harms in Notizbl. Bot. Gart. Berl. 8: 295 (1923); in Pflanzenfam. ed. 2, 21: 491 (1925); Burtt Davy, Fl. Transv. 1: 221 (1926); Liebenberg in Bothalia 3: 536, fig. 4–5 (1939); A. & R. Fernandes in Garcia de Orta 6: 256 (1958); Watt & Breyer-Brandwijk, Med. Pois. Pl. ed. 2: 828 (1962); De Wilde in Meded. Landbouwhogeschool Wageningen 71–18: 168 (1971). Type: Natal, *Gerrard* 1199 (TCD, holo.; BM; K; W).

Modecca hastata Harv., Thes. Cap. 2: 43, tab. 167 (1863).

Adenia schlechteri Harms in Bot. Jahrb. 33, 1: 150 (1902). Type: Transvaal, *Schlechter* 11747 (BM; BR; HBG; K; P; Z).

Herbaceous climber to 4 m, growing from a tuberous rootstock; glabrous. *Leaves* pale- or greyish-green, often punctate beneath, simple, broadly ovate to ± hastate, base cordate to truncate, apex obtuse to acute, up to 2 mm subapical mucronate, 1,5–10 (–14) × 1,5–10 (–13) cm, 3–7-(sub)plinerved; margin entire; petiole (0,5–5 (–10) cm. *Gland(s)* at blade-base 1 or 2, contiguous, situated on the not- or shortly-peltate, sometimes 2-lobed blade-base; blade-glands 0–2 (–4), often conspicuous, (sub)marginal or near the top; leaves of juvenile specimens with peltate base, without glands. *Stipules* narrowly triangular, acute, 1,5–2 mm. *Inflorescences* peduncled for 0,5–4 (–11) cm, up to 12-flowered in ♂, 1–3-flowered in ♀; tendril (0 or) 1, 1–5 cm; sterile tendrils up to 12 cm. *Male flowers* tubiform to infundibuliform, including the 2–6 mm long stipe (13–) 15–30 cm; hypanthium cup-shaped, tapering, 2–3 mm; calyx-tube (5–) 7–15 × (3–) 4–5 mm, lobes ovate to elliptic, obtuse, 4–7 mm, densely fimbriate; petals linear-lanceolate, 5–7 (–9) mm, 1–3-nerved, entire to densely fimbriate, inserted (1–) 2–8 mm above the corona; filaments 3–7 mm, connate up to 2,5 mm, inserted at the base of the hypanthium or on a short androgynophore; anthers 4–7 mm, obtuse; septa 1–2,5 mm high; corona of fine hairs c. 1–2 mm; disc-glands 1–1,5 mm. *Female flowers* campanulate-infundibuliform, including the 0,5–2,5 mm long stipe (8–) 10–18 mm; hypanthium tapering, 1–2 mm; calyx-tube 3–6 mm, lobes ovate, 4–7 mm, crenulate-laciniate; petals linear, acute, 3–5 (–6) mm, 1-nerved, entire or ± fimbriate, inserted 1,5–3 mm above the corona; staminodes 2–2,5 mm; septa up to 0,2 mm high; corona hairs 1–1,5 mm; disc-glands c. 0,5 mm; gynophore 1–1,5 mm; ovary ellipsoid to subglobose, (2–) 2,5–4,5 mm; styles connate for 0,3–1,5 mm, free style-arms 1,5–2,5 mm; stigmas palmately branched, papillate, each 2–2,5 mm diam. *Fruit* 1–2 per inflorescence, subglobose, excluding the 1–3 mm long gynophore 2–3,5 (–6) × 2–3,5 (–5) cm; pericarp coriaceous, sometimes spongy inside; seeds 2–25 per capsule, ellipsoid, 6–7 (–8) mm.

A species of savanna, stony slopes, rocky hillsides, stream banks, and forest along water-courses, found in Natal, Transvaal and Swaziland; also in southern Mozambique; 0–1200 m. Flowers in Aug.–Jan., fruits Nov.–Jan.

Leaves without glands at the apex; glands at blade-base on a ± semi-orbicular appendage; petals in ♂ flowers entire to remotely serrate-fimbriate, or fimbriate only at the base ....(a) var. *hastata*

Leaves with glands at the apex; glands at blade-base on 2 separate auricles or on a bi-lobed appendage; petals in ♂ flowers (mostly densely) fimbriate ....(b) var. *glandulifera*

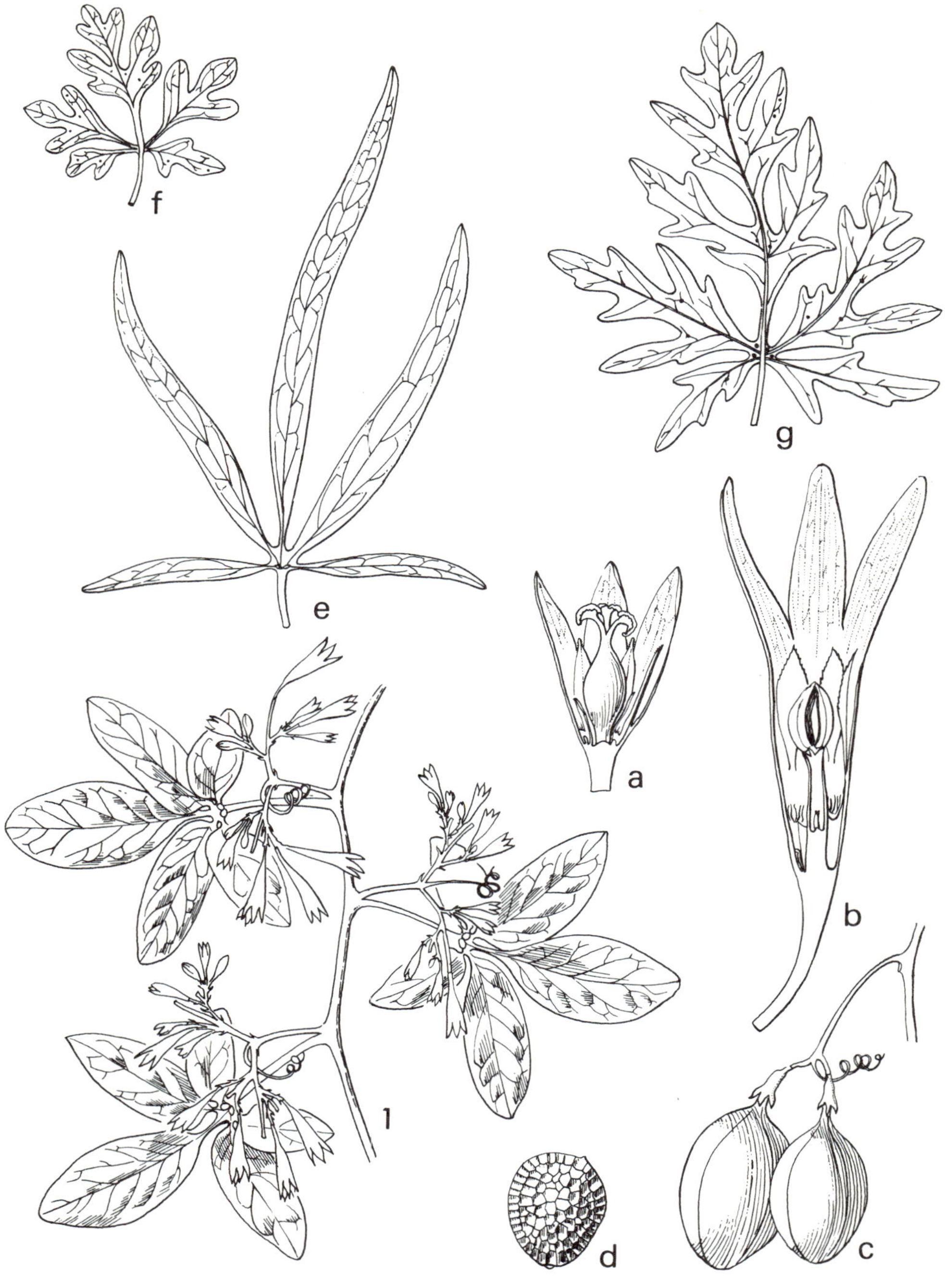

FIG. 36.—1, **Adenia digitata,** flowering stem, × ½ (*Fries, Norlindh & Weimark* 3415); a, female flower, longitudinal section, × 2½ (*Chase* 7887); b, male flower, longitudinal section, × 2½ (*Liebenberg* 3366); c, fruits, × ½ (*Marques* 3443); d, seed, × 2½ (*Munro* s.n.); e–g, variation in leaf shape, all × ½: e (*Young* 33291), f (*Galpin* 677), g (*Munro* s.n.) (after De Wilde, 1971).

(a) var. **hastata.**

De Wilde in Meded. Landbouwhogeschool Wageningen 71–18: 170 (1971).

Leaves without glands at the apex. *Gland(s)* at blade-base single, or 2 glands contiguous on a single, semi-orbicular, not lobed appendage. *Male flowers* 13–25 mm; petals entire or margin irregularly serrate-fimbriate, or fimbriate only at base; anthers in anthesis reaching to or nearly to the throat of the calyx-tube.

The typical variety occurs in Natal, the eastern Transvaal lowveld, and in southern Mozambique, at 0–700 m altitude.

TRANSVAAL.—2331 (Phalaborwa): Shingwedzi area, *Van der Schijff 3850*. 2431 (Acornhoek): Skukuza Camp, *Codd 5732*. 2531 (Komatipoort): Sigaas River, *Van der Schijff 1305;* Komatipoort, *Schlechter 11747*.

NATAL.—2632 (Bela Vista): Ndumu Game Reserve, *Ward 4520*. 2732 (Ubombo): Pongola Poort, *Strey 4643*. 2830 (Dundee): near Tugela Ferry, *Dyer 4389*. 2831 (Nkandla): Mahlabatini, *Gerstner 4234*.

Strey 4643, and probably also *Codd* 10281, both from northern Natal, possibly represent a separate form with large fruits up to 6 cm in diameter. These specimens have relatively large leaves, and were collected in forest along water-courses.

(b) var. **glandulifera** *De Wilde* in Meded. Landbouwhogeschool Wageningen 71–18: 170 (1971). Type: Transvaal, *Thorncroft* 2034 (PRE!).

Leaves with 2 (–4)-paired, sometimes bulging glands at, or just below the apex. *Glands* at blade-base 2, on 2 separate auricles, or glands contiguous on the bi-lobed ± peltate blade-base. *Male flowers* 15–30 mm; petals moderately to densely fimbriate; anthers in anthesis remaining well below the throat of the calyx-tube.

Found in Natal, Swaziland and Transvaal, at 500–1200 m altitude; it has a more western distribution than var. *hastata*.

TRANSVAAL.—2530 (Lydenburg): near Nelspruit, *Burtt Davy 1489; Leach 11540*. 2531 (Komatipoort): Barberton, *Galpin 563*.

SWAZILAND.—2631 (Mbabane): Bulunga Poort, *Compton 31759*.

NATAL.—2731 (Louwsburg): near Nongoma, *Acocks 13019; Gerstner 2345;* Pongola, *Edwards 3188; Nel 215*. 2732 (Ubombo): Mkuzi, *Venter 5204*.

The specimens *Gerstner* 2895, *Edwards* 3188, *Nel* 215, and *Compton* 31759 from Natal and Swaziland are more or less intermediate between the 2 varieties, but in all 4 the glands at the apex of the leaves are present.

7. **Adenia digitata** *(Harv.) Engl.* in Bot. Jahrb. 14: 375 (1891); Burtt Davy in Ann. Transv. Mus. 3: 121 (1912); Engl., Pflanzenw. Afr. 3, 2: 605 (1921); Burtt Davy, Fl. Transv. 1: 221 (1926); Steyn, Tox. Pl. S. Afr. 310, fig. 43A, 44, 45, 47A (1934); Henkel, Woody Pl. Natal 118 (1934); Liebenberg in Bothalia 3: 541, 527, 530–532, fig. 1–3, 14–17, pl. 6–36 (1939); A. & R. Fernandes in Garcia de Orta 6: 259 (1958); Letty, Wild Flow. Transv. 225 (1962); Watt & Breyer-Brandwijk, Med. Pois. Pl. ed. 2: 826, fig. 218 (1962); De Wilde in Meded. Landbouwhogeschool Wageningen 71–18: 188, fig. 30, 31 a–h (1971). Type: Zululand*, *Owen* s.n. (T.C.D., holo.; K, photo.!).

Modecca digitata Harv., Thes. Cap. 1: 8, tab. 12 (1859); in F.C. 2: 500 (1862). *M. senensis* (Klotzsch) Mast. in F.T.A. 2: 513 (1871); Hook.f. in Bot. Mag. t.7763 (1901).

Clemanthus senensis Klotzsch in Peters, Reise Mossamb. Bot. 143 (1862). Type: Mozambique, *Peters* s.n. (B†, holo.; PRE, photo.!).

Adenia senensis (Klotzsch) Engl. in Bot. Jahrb. 14: 375 (1891); Harms in Pflanzenfam. ed. 2, 21: 491 (1925); Bak. f. in J. Linn. Soc. (Bot.) 40: 73 (1910); Engl., Pflanzenw. Afr. 3, 2: 605 (1921); Burtt Davy, Fl. Transv. 1: 221 (1926); A. & R. Fernandes in Garcia de Orta 6: 258 (1958). *A. stenophylla* Harms in Bot. Jahrb. 26: 238 (1899); Engl., Pflanzenw. Afr. 3, 2: 605 (1921); Burtt Davy, Fl. Transv. 1: 222 (1926). Type: Transvaal, *Wilms* 941 (B†, holo.; PRE, photo.!). *A. multiflora* Pott in Ann. Transv. Mus. 5: 235 (1917); Harms in Notizbl. Bot. Gart. Berl. 8: 298 (1923); Burtt Davy, Fl. Transv. 1: 221 (1926). Type: Transvaal, *Fehrson* sub TRV 13786 (PRE, holo.!). *A. angustisecta* Burtt Davy in Kew Bull. 1921: 280 (1921); Fl. Transv. 1: 222 (1926). Type: Transvaal, *Mundy* 4700 (BOL, holo.; PRE, photo.!). *A. buchananii* Harms in Engl., Pflanzenw. Afr. 3, 2: 605 (1921). Lectotype: Malawi, *Buchanan* 244 (K, lecto.) (see De Wilde, l.c.).

(Sub)herbaceous climber 0,2–3 m growing from a variable tuber. *Leaves* beneath greyish to glaucous, densely punctate or not, deeply (3–) 5-parted or (pseudo-) foliolate, in outline suborbicular with cordate base, 4–18 × 3–17 cm; leaflets variable, entire to deeply (2–) 3–5 (–10)-lobed, ovate or obovate to linear, base long-attenuate or acute, apex rounded to acute, 1,5–15 × 0,7–4 (–7) cm, margin entire, petiolule up to 2 cm; petiole 1–9 cm. *Glands* at blade-base 2, on two separate ± upward directed auricles 1–2 mm diam. at the transition of petiole to blade; blade-glands 2–4, situated between the insertions of the leaflets, and each leaflet with 0–8 glands, scattered or submarginal. *Stipules* narrowly triangular to lanceolate, acute, 1–3 mm, withering. *Inflorescences*

*It must be remembered that Miss Owen also collected near Rustenburg in the Transvaal and these gatherings were also placed as "Zululand" by Harvey.

peduncled for up to 7 cm, (1–) 5–20 (–60)-flowered in ♂, 1–10-flowered in ♀; tendril 1, 2–10 cm, sterile tendrils up to 15 cm. *Male flowers* tubular-infundibuliform, including the 3–12 (–15) mm long stipe (14–) 20–38 mm; hypanthium cup-shaped, (1–) 2–3,5 × 2–4 (–5) mm; calyx-tube (5–) 8–12 mm; calyx-lobes ovate or oblong to lanceolate, (sub)obtuse, (4–) 7–11 mm, dentate-fimbriate; petals lanceolate, acute, dentate or fimbriate, 6–12 mm, (1–) 3-nerved, inserted at the same level as, or up to 5 mm above the corona; filaments 3,5–9 (–12) mm, connate for at least half-way, inserted on an androgynophore 1–3,5 mm; anthers 3–6 mm, ± inward curved and at the top clinging by the c. 0,2 mm long papillate apiculae; septa 0,5–3 mm high; corona hairs c. 0,5–2 mm, rarely absent; disc-glands 0,5–1,5 mm. *Hermaphrodite flowers* 20–25 mm. *Female flowers* tubular-infundibuliform, including the 2–7 mm long stipe 15–26 mm; hypanthium c. 2–4 × 2–4 mm; calyx-tube 4–8 mm, calyx-lobes ovate to oblong, obtusish, 5–7 mm, entire; petals lanceolate to linear, acute, (sub)entire, 2–7 mm, 1 (–3)-nerved, inserted at the same level as, or up to 4 mm above the corona; staminodes 3–5 mm, up to 1 mm connate; septa up to 0,5 mm high; corona hairs 0,3–1 mm; disc-glands c. 1 mm; gynophore 2–4 mm; ovary ovoid to oblong, (4–) 5–6 mm; styles connate for 1–1,5 mm, style-arms 1–1,5 mm; stigmas subreniform, woolly-papillate, each 2–3 mm diam. *Fruit* 1–3 per inflorescence, ovoid to ellipsoid (to oblong), excluding the (2–) 5–12 mm long gynophore (2,5–) 3–5,5 (–7,5) × (1,5–) 2–3,5 (–4) cm; pericarp coriaceous, brilliant yellow to red when fresh, sometimes spongy inside, smooth; seeds (10–) 20–60 per capsule, (ob)ovate to ellipsoid, 6–8 mm. FIG. 36.

Found in Natal, Swaziland and Transvaal, outside the Flora area widely distributed in Angola, Tanzania, Zambia, Rhodesia, Malawi, Mozambique and Botswana. Grows in savanna, rocky and grassy places, on termite mounds, forest fringes, etc., in various soil types; 0–1850 m. Never gregarious. Flowers and fruits mainly Oct.–March.

TRANSVAAL.—2229 (Waterpoort): near Ingwe Motel, *Jacobsen 1905*. 2230 (Messina): near Lake Funduzi, *Story 4856*. 2231 (Pafuri): Baiandbai, *Lang sub TRV 32154*. 2327 (Ellisras): Junction of Limpopo and Palala Rivers, *Mogg 24566*. 2329 (Pietersburg): *Breyer sub TRV 24215*. 2330 (Tzaneen): Westfalia Estate, *Scheepers 515*. 2428 (Nylstroom): Mosdene, *Galpin 13196*. 2429 (Zebediela): Potgietersrus, *Leendertz sub TRV 6007*. 2430 (Pilgrims Rest): The Downs, *Rogers sub TRV 18889*. 2526 (Zeerust): Groot Marico, *Liebenberg S.2*. 2527 (Rustenburg): Brits, *Mogg 14609*. 2528 (Pretoria): near Rust de Winter, *Codd 2236*. 2529 (Witbank): Loskop Dam, *Theron 1943*. 2530 (Lydenburg): near Nelspruit, *Schlieben 8449*. 2531 (Komatipoort): Barberton, *Galpin 677*. 2627 (Potchefstroom): Experimental Farm, *Theron 5*. 2630 (Carolina): Lothair, *Forester 28*.

SWAZILAND.—2631 (Mbabane): Hlatikulu, *Ben Dlamini s.n.*

NATAL.—2632 (Bela Vista): Ndumu Game Reserve, *Moll 4249*. 2730 (Vryheid): near Vryheid, *Gerstner 2330*. 2731 (Louwsburg): Nongoma, *Gerstner 4707*. 2732 (Ubombo): near Ingwavuma, *Gerstner s.n.* 2831 (Nkandla): 6 km N. of Nkwaleni, *Acocks 12954*.

This is an extremely variable taxon, but Liebenberg l.c. has convincingly shown that all forms belong to one single species.

The fruits are poisonous to humans.

8. **Adenia wilmsii** *Harms* in Bot. Jahrb. 26: 238 (1899); Engl., Pflanzenw. Afr. 3, 2: 603 (1921); Harms in Pflanzenfam. ed. 2, 21: 491 (1925); Burtt Davy, Fl. Transv. 1: 222 (1926); Liebenberg in Bothalia 3: 538, 526, 532, fig. 13 (1939); Letty, Wild Flow. Transv. 225 (1962); De Wilde in Meded. Landbouwhogeschool Wageningen 71–18: 198 (1971). Type: Transvaal, *Wilms* 961 (B†, holo.; BM; K; L; P; Z).

Erect herb 10–30 (–50) cm, without tendrils, glabrous, growing from a tuberous rootstock. *Leaves* ± glaucous-green, not punctate beneath, (5–) 7-pseudo-foliolate, broadly ovate to suborbicular in outline, base ± cordate, 2–8 (–12) × 2–6 (–10) cm, 5–7-plinerved; leaflets sessile, entire to pinnately 3–7-lobed, ovate-oblong, base attenuate, apex acute, 1–6 (–10) cm, the middle leaflet largest; margin entire; petiole 2–8 (–10) cm. *Glands* at blade-base 2, on two auricles at the transition to the petiole; blade-glands (0–) 4–6, situated at the very base in between the leaflets, sometimes a few submarginal. *Stipules* lanceolate, 1–3 mm. *Inflorescences* peduncled for 0,5–5 cm, 1–3-flowered in ♂, 1 (–3)-flowered in ♀; no tendrils. *Male flowers* tubiform-infundibuliform, including the 4–6 mm long stipe 20–25 mm; hypanthium cup-shaped, 2–3 mm; calyx-tube 7–8 × 2–5 mm, lobes elliptic to oblong, obtuse, 6–7,5 mm, entire; petals oblanceolate, obtusish, 7–10 mm, 3–5-nerved, subentire, inserted at the same level as, or up to 3 mm above the corona; filaments 4–5,5 mm, connate for 1,5–2,5 mm, inserted on a short androgynophore; anthers 3,5–4,5 mm; septa 2–3 mm high; corona hairs 0,5–1 mm; disc-glands c. 0,5 mm. *Female flowers* tubiform-campanulate, including the 2–3 mm long stipe c. 12 mm; hypanthium c. 1 mm; calyx-tube 5–6 × 3–4 mm, lobes elliptic-oblong, obtuse, c. 4 mm, entire; petals oblanceolate, subobtuse, c. 6

mm, 3-nerved, subentire, inserted at the same level as the corona; staminodes c. 2,5 mm; septa c. 1 mm high; corona hairs c. 0,5 mm; disc-glands c. 0,5 mm; gynophore 1,5 (–2) mm; ovary ovoid-ellipsoid, 4–4,5 mm; styles connate for 0,7 mm, style-arms c. 1 mm; stigmas subglobose, papillate, each c. 1,5 mm diam. *Fruit* 1 per inflorescence, ellipsoid, excluding the c. 5 mm long gynophore c. 4 × 2,5 cm; pericarp thinly coriaceous; seeds not known.

This species is a local endemic in Transvaal, not found elsewhere, growing in rocky places, dolomite outcrops, or on red loam soil; at c. 1500 m. Flowers and fruits Sept.–Dec.

TRANSVAAL.—2530 (Lydenburg): near Lydenburg, *Liebenberg 3488; Strey & Schlieben 3116; Wilms 961; Van Wyk 7.*

9. **Adenia repanda** *(Burch.) Engl.* in Bot. Jahrb. 14: 375 (1891); Harms in Pflanzenfam. ed. 2, 21: 490, fig. 224 (1925); Liebenberg in Bothalia 3: 534, 525, 532, fig. 12 (1939); A. & R. Fernandes in Garcia de Orta 6: 658 (1958); C.F.A. 4: 219 (1970); Watt & Breyer-Brandwijk, Med. Pois. Pl. ed. 2: 828 (1962); Schreiber in F.S.W.A. 89 (1968); De Wilde in Meded. Landbouwhogeschool Wageningen 71–18: 237, fig. 37 (1971). Type: Cape, Griqualand West, *Burchell* 2486/2 (K, holo.!; PRE).

Paschanthus repandus Burch., Trav. 1: 543 (1822); Schinz in Bot. Jahrb. 15, Beibl. 33, 1: 3 (1892); Marloth Fl. S. Afr. 2, 2: 197, fig. 130 (1925). *P. jäggii* Schinz in Mém. Herb. Boiss. 20: 23 (1900). Type: as for *Jäggia repanda* Schinz.

Modecca paschanthus Harv. in F.C. 2: 500 (1862), nom. illegit. *M. repanda* (Burch.) Druce in Rep. Bot. Exch. Cl. Brit. Is. 636 (1917).

Jäggia repanda Schinz in Verh. Bot. Ver. Prov. Brandenb. 30: 254 (1888); Harms in Bot. Jahrb. 24: 169 (1897). Type: S.W. Africa, *Schinz* s.n. (Z, holo.!).

Suberect herb or ± woody climber 0,2–2 m, glabrous, growing from a tuberous rootstock. *Leaves* grey-glaucous, sometimes punctate beneath, entire or irregularly repand to lobed, obovate to linear, base attenuate to subcordate, apex subacute to obtuse or retuse, sometimes curved, 2–15 × 0,2–2 (–6) cm, ± penninerved; lobes up to 5 at either side of blade, up to 1 cm; petiole 0,1–1 cm. *Glands* at blade-base 2, one at each side of the base of the midrib; blade-glands 0–10, submarginal, mostly corresponding with the lobes, and one apical or subapical gland in which the midrib ends. *Stipules* narrowly triangular, acute, 1–1,5 mm. *Inflorescences* sessile or peduncled for up to 2 cm, 1–5-flowered in ♂, 1–2 (–3)-flowered in ☿ and ♀; tendril (0 or) 1, up to 4 cm; sterile tendrils up to 5 cm; flowers polygamous or dioecious. *Male flowers* tubiform-infundibuliform, including the 2–3 (–4) mm long stipe 15–24 mm; hypanthium including calyx-tube 9–14 × 2–5 mm; calyx-lobes oblong-lanceolate, obtuse, 4–7,5 mm, entire; petals lanceolate, obtuse to subacute, 5–8 mm, (1–) 3-nerved, subentire, inserted near the throat of the calyx-tube, 8–12 mm above the base of the hypanthium; filaments 3,5–6 mm, free, inserted in the calyx-tube 4–5 mm above the base of the hypanthium; anthers 4–6 mm; septa 0; corona 0; disc-glands 0. *Hermaphrodite flowers* ± intermediate between male and female flowers; anthers smaller. *Female flowers* tubiform-campanulate, including the c. 1 mm long stipe 8–11 mm; hypanthium including calyx-tube 3,5–5 × 3–4 mm; calyx-lobes oblong to lanceolate, obtuse, 3–5 mm, entire; petals lanceolate, obtuse or acute, 1,5–2 mm, 1 (–3)-nerved, entire, inserted 2,5–3,5 mm above the base of the hypanthium; staminodes 2–3 mm, inserted near the base of the hypanthium; septa 0; corona 0; disc-glands 0; gynophore 2–3 mm; ovary ovoid to ellipsoid, 2,5–4 mm; styles connate for 0,5–1,5 mm, style-arms 0,5–0,7 mm; stigmas subglobose, papillate, each 1–1,5 mm diam. *Fruit* 1–2 per inflorescence, subglobose, excluding the 2–5 mm long gynophore 1,5–2,5 (–3) × 1,2–2,5 cm; pericarp coriaceous, sometimes spongy inside; seeds (2–) 5–12 per capsule, suborbicular to broadly ovate, 7–8 mm.

This species, which represents a monotypic section *Paschanthus* in *Adenia,* is distributed in rather restricted areas in Southern Africa, viz. the northern Cape Province, the Soutpansberg area in the Transvaal, and throughout South West Africa; it also occurs in Angola, Zambia, Rhodesia, and Botswana. The species is often recorded as ''rare'' or ''uncommon'', and grows in open or partly shaded places in woodland, among rocks, on sandy soils (''Kalahari sand''), red sand, and granite; 500–1 600 m. Flowering time Oct.–Febr., fruits Nov.–April.

S.W.A.—1715 (Ondangua): Ondangua, *De Winter & Giess 6853.* 1817 (Tsintsabis): near Tsintsabis, *Boss sub TRV 35549.* 1917 (Tsumeb): Kududamm, *Volk 413.* 1920 (Tsumkwe): Gautscha Pan, *Story 6284.* 2117 (Otjosondu): Quickborn, *Bradfield 75.* 2217 (Windhoek): Windhoek, *Rogers 29798.* 2718 (Grünau): Karasberg, *Pearson 8166.*

TRANSVAAL.—2229 (Waterpoort): between Soutpan and Waterpoort, *Obermeyer, Schweickerdt & Verdoorn 246.* 2230 (Messina): Tshipise, *Van der Schijff 5215.* 2329 (Pietersburg): Vivo, *Bremekamp & Schweickerdt 206.*

CAPE.—2624 (Vryburg): Armoedsvlakte, *Mogg 8769*. 2723 (Kuruman): near Kuruman, *Marloth 1350*. 2824 (Kimberley): Barkly West, *Acocks 1561*. 2922 (Prieska): near Prieska, *Bryant 345*.

According to *Bryant* 345 "greedily eaten by stock, which may partly account for its rarity", but also reported as "deadly poisonous to man".

10. **Adenia gummifera** *(Harv.) Harms* in Pflanzenfam. 3, 6a, Nachtr. 1: 255 (1897); ed. 2, 21: 490 (1925); Burtt Davy in Ann. Transv. Mus. 3: 121 (1912); Fl. Transv. 1: 222 (1926); Henkel, Woody Pl. Natal 110 (1934); Liebenberg in Bothalia 3: 535, 523, 532, fig. 10 & 11 (1939); A. & R. Fernandes in Garcia de Orta 6: 256 (1958); Watt & Breyer-Brandwijk, Med. Pois. Pl. ed. 2: 828 (1962); F. White, For. Fl. N. Rhod. 267 (1962); Letty, Wild Flow. Transv. 225 (1962); De Wilde in Acta Bot. Neerl. 17: 131, fig. 2h (1968); in Meded. Landbouwhogeschool Wageningen 71–18: 261 (1971). Type: South Africa, "inter Omsamculo et Omcomas", *Drège* 5211 (P, holo.!; S).

Modecca gummifera Harv. in F.C. 2: 500 (1862).

Ophiocaulon gummifer (Harv.) Mast. in F.T.A. 2: 518 (1871); Harms in Engl., Pflanzenw. Ost Afr. 2, C: 281 (1895) (as *gummiferum*); Fries, Wiss. Ergebn. Rhod. Kongo 157 (1914). *O. cissampeloides* sensu Bak.f., Fl. Maur. and Seychelles 106 (1877); in J. Linn. Soc. (Bot.) 40: 74 (1910).

Adenia rhodesica Suesseng. in Trans. Rhod. Sc. Ass. 43: 13 (1951). Type: Rhodesia, *Dehn* 696/52 (M, holo.!; BM; SRGH).

Subligneous climber up to 30 m, up to 10 cm thick at base; twigs green or grey-green, often pruinose. *Leaves* rarely punctate, entire to deeply 3 (–5)-lobed, orbicular to ovate or rhomboid, or ± 3 (–5)-angular in outline, base cordate to truncate, apex obtuse or retuse, rarely subacute, 2,5–11 × 2,5–11 cm, 3-plinerved and with 1 pair of straight nerves from the midrib ending in marginal glands; reticulation rather fine, distinct; margin entire; lobes obtuse, up to 4 cm; petiole 1,5–11 cm. *Gland* at blade-base single, on a median subcircular to spathulate appendage 1–3 mm; blade-glands 0–4, often 2 glands rather approximate to the axils of or contiguous with the upper side nerves; marginal glands minute, 3–7 on either side of the blade. *Stipules* broadly rounded to triangular, finely lacerate, 0,5 (–1) mm. *Inflorescences* peduncled for (0,5–) 1–12 (–16) cm, up to 35-flowered in ♂, 2–6-flowered in ♀; tendrils 0 or 1, 1–4 cm, sterile tendrils simple or 3-fid, 5–20 cm. *Male flowers* ± campanulate, including the 2–6 mm long stipe 11–17 mm; hypanthium cup-shaped, 1–2 (–2,5) × 2–4 mm; calyx-tube 0; sepals lanceolate, subobtuse, (7–) 8–10 × 2–3 mm, margin up to 0,2 mm laciniate, punctate; petals (ob)lanceolate, obtuse, (6–) 8–11 mm, 3-nerved, finely laciniate-serrulate, remotely punctate; filaments (1–) 2–3,5 mm, connate for (0–) 0,5–1,5 (–2) mm, inserted at the base of the hypanthium; anthers 3–6 mm, obtuse, up to 0,1 mm apiculate; septa 0–0,2 mm high; corona 0; disc-glands 0. *Female flowers* ± campanulate, including the c. 0,5 mm long stipe 5,5–8 mm; hypanthium flattish, c. 0,5 × 2–2,5 (–3) mm; calyx-tube 0; sepals oblong, c. 4–6,5 mm, entire, punctate; petals lanceolate-linear, 2–4,5 mm, 1–3-nerved, sub-entire, sparingly punctate or not; staminodes c. 0,5 mm; septa 0; corona 0; disc-glands 0; gynophore c. 0,5 mm; ovary ovoid, 3–4,5 mm; style 0–0,5 mm; stigmas (sub)sessile, subreniform, laciniate-papillate, each c. 1–1,5 mm diam. *Fruit* 1–4 per inflorescence, ovoid to ellipsoid, sometimes ± 3 (–6)-angular, excluding the c. 1 mm long gynophore 2,5–4 (–4,5) × 1,7–3 cm; pericarp woody-coriaceous, c. 0,2 mm thick, smooth or finely pitted or granulate; seeds 30–50 per capsule, subovate, 3,5–5,5 mm.

Widely distributed in eastern Africa, from Somalia in the north to the Kei River in the south; also in Seychelles. It is a locally common, vigorous trailer, found in forest and scrub at 0–1 800 m. Flowering time mainly Sept.–April.

TRANSVAAL.—2229 (Waterpoort): 30 km N. of Louis Trichardt, *Buitendag 1075*. 2230 (Messina): Pepiti Falls, *Smuts & Gillett 3221*. 2231 (Pafuri): Punda Milia, *Codd 5966*. 2329 (Pietersburg): 8 km W. of Louis Trichardt, *Obermeyer, Schweickerdt & Verdoorn 355*. 2330 (Tzaneen): Westfalia Estate, *Scheepers 744*. 2429 (Zebediela): Sekukuniland, *Barnard 128*. 2430 (Pilgrims Rest): Mariepskop, *Van der Schijff 4736*. 2431 (Acornhoek): Lothian, *Strey 3573*. 2530 (Lydenburg): Sabie, *Louw 2398*. 2531 (Komatipoort): Barberton, *Galpin 782*.

SWAZILAND.—2631 (Mbabane): near Mbabane, *Compton 30370*.

NATAL.—2632 (Bela Vista): Kosi Bay, *Strey & Moll 3829*. 2732 (Ubombo): Sordwana Bay, *Vahrmeijer & Tölken 314*. 2830 (Dundee): near Tugela Ferry, *Dyer 4389*. 2831 (Nkandla): Eshowe, *Gerstner 2641*. 2832 (Mtubatuba): Hluhluwe Game Reserve, *Ward 3962*. 2930 (Pietermaritzburg): Berea, Durban, *Medley Wood 5502*. 2931 (Stanger): Hawaan Forest, *Ross & Moll 2265*. 3030 (Port Shepstone): Dumisa, *Rudatis 523*.

CAPE.—3129 (Port St. Johns): Port St. Johns, *Galpin 3461*. 3228 (Butterworth): Kentani, *Pegler 869*.

In South Africa the typical variety, var. *gummifera* occurs; specimens from a restricted area in Zambia have been described as var. *cerifera* (De Wilde, l.c. p. 264).

5363a 3. SCHLECHTERINA

Schlechterina *Harms* in Bot. Jahrb. 33: 148 (1902); in Ber. Deutsch. Bot. Ges. 24: 177, t.12 (1906); in Engl., Pflanzenw. Afr. 3, 2: 596 (1921); Pflanzenfam. ed. 2, 21: 485, fig. 221 (1925); Marloth, Fl. S. Afr. 2, 2: 198 (1925); Ross, Fl. Natal 251 (1972); De Wilde in Blumea 22: 48 (1974). Type: *S. mitostemmatoides* Harms.

Crossostemma sensu Hutch., Gen. Fl. Pl. 2: 370 (1969), pro parte.

Low perennial climber or suberect shrublet, mostly provided with tendrils, glabrous, growing from a perennial rootstock. *Leaves* simple, elliptic to linear, entire to deeply pinnately lobed, shortly petiolate, margin (sub)entire to dentate. *Glands* on petiole 0 or 1 (or 2) pairs at the top. *Stipules* small, caducous. *Tendrils* axillary. *Inflorescences* axillary, (sub)sessile, few-flowered, often arranged in short shoots from the supra-axillary bud; bracts and bracteoles small. *Flowers* hermaphrodite, whitish; stipe articulate to short pedicel; hypanthium small, shallowly cup-shaped. *Sepals* 3 or 4, free, imbricate, elliptic to oblong. *Petals* 2–4, free, elliptic to oblong. *Corona* single, composed of threads connate at base into a low tube, inside set with additional hair-like appendages. *Disc* absent. *Androgynophore* short. *Stamens* 6–8, connate at base into a shallow cup, often with small lobes (staminodes) on its margin in between the filaments; anthers dorsifixed, versatile, ellipsoid-oblong, obtuse, 2-locular. *Ovary* ellipsoid-oblong, on a short gynophore; placentas (3 or) 4, each with 3–8 ovules; style single; stigma single, flattish, 3- or 4-lobed. *Fruit* a stipitate 3- or 4-valved capsule, ellipsoid-oblong, fusiform; valves coriaceous. *Seeds* flattened, ellipsoid, arillate; testa crustaceous, scrobiculate.

One species in tropical East Africa, just entering northern Natal.

Related to the West African genus *Crossostemma*, but the latter is distinguished by a conspicuous intra-staminal disc, and the absence of a gynophore.

Schlechterina mitostemmatoides *Harms* in Bot. Jahrb. 33: 148 (1902); in Ber. Deutsch. Bot. Ges. 24: 177 (text fig.), t.12 (1906); Bak. f. in J. Linn. Soc. (Bot.) 40: 73 (1911); Engl., Pflanzenw. Afr. 3, 2: 596 (1921); Harms in Pflanzenfam. ed. 2, 21: 485, fig. 221 (1925); A. & R. Fernandes in Garcia de Orta 6: 249 (1958); Ross, Fl. Natal 251 (1972). Type: Mozambique, Lourenço Marques, *Schlechter* 11681 (B, holo.†).

S. mitostemmatoides var. *holzii* Harms in Ber. Deutsch. Bot. Ges. 24: 184 (1906); Engl., Pflanzenw. Afr. 3, 2: 596 (1921); Harms in Pflanzenfam. ed. 2, 21: 486 (1925). Syntypes: Tanzania, near Dar-es-Salaam, *Holz* 1070, 1070a; 1086 (B†).

Small liana or subscandent shrublet to 3 m, glabrous, with perennial rootstock; older stem corky, shoots often lenticellate; tendrils 3–10 (–15) cm. *Leaves* simple, variable (heterophyllous), elliptic to lanceolate (to linear), apex acute to (long) acuminate, base acute to attenuate, margin entire or regularly to irregularly dentate, or leaves pinnately lobed to various depth, the lobes acute to rounded, 2,5–13 × 0,7–4,5 cm; leaves in sapling-, or sterile- or juvenile shoots often lanceolate to linear, (deeply) pinnately lobed, the lobes broad or narrow, 10–30 × 0,2–2 cm; petioles 4–12 mm, in juvenile leaf-forms 0–4 mm; glands on petiole absent, or 1 (or 2) pair(s) at the apex; glands on blade-margin several, minute, mostly at the tips of the teeth; stipules subtriangular-linear, c. 0,5 mm. *Inflorescences* 1–3-flowered, often arranged in short shoots from the supra-axillary bud; bracts subtriangular, 0,5–1 mm. *Flowers* glabrous, stipe 6–25 mm; hypanthium small, shallowly cup-shaped, c. 3–4,5 mm wide; sepals 3 or 4, elliptic to oblong, obtuse, 6–11 × 3–6 mm; petals 2–4, elliptic to oblong, obtuse, 5–10 mm long; corona single 5–8 mm high, composed of threads ± connate at base into a tube 0,5–2 mm, free parts of threads 4–6 mm, inside tube and at base of free threads a zone of short hair-like appendages 0,5–1 mm. *Disc* absent. *Androgynophore* c. 1 mm. *Stamens* 6–8 (often 7), 7–10 mm, connate at base into a cup c. 1 mm, often with small lobes on its margin in between the filaments; anthers ellipsoid-oblong, obtuse, 2–3 mm. *Gynophore* 2–2,5 mm; ovary ellipsoid-oblong, 2–2,5 mm, ± (3- or) 4-angled; style single, 1,5–2 mm;

FIG. 37.—**Schlechterina mitostemmatoides,** flowering stem, × 2/3; a, flower, × 3 (*Faulkner* 1773); b–e, variation in leaf shape, all × 2/3: b (*Strey* 10439), c (*Pedro* 328), d (*Sim* 20660), e (*Marques* 2242).

stigma single, 3- or 4-lobed, 2–2,5 mm diam. *Fruit* a (3- or) 4-valved capsule, ellipsoid-oblong, acute at both ends, excluding the c. 1 cm long gynophore 4,5–5 × 2,5–3 cm; seeds rather few, ellipsoid, c. 8 mm. FIG. 37.

This species enters the Flora area in the very north-east of Natal, growing in coastal scrub and forest. Its main area of distribution is Kenya, Tanzania, Zanzibar and Mozambique, where it is recorded from sandy soils and old coral reefs, at an altitude from 0–700 m.

NATAL.—2632 (Bela Vista): 8 km N.E. of Makanes Pont, *Ross 2370; Strey 10439.* 2732 (Ubombo): Makanes Pont, *Venter 5122.*

The species displays a remarkable variability in leaf shape (see description).

5369 4. BASANANTHE

Basananthe *Peyr.* in Bot. Ztg. 17: 101 (1859); (in Wawra & Peyr., Sertum Benguelense in) Sitz. Ber. Acad. Wien 38: 569 (1860); Hook.f. in Benth. & Hook.f., Gen. Pl. 1: 812 (1867); Welw. in Trans. Linn. Soc. Lond. 27: 27 (1871); Mast. in F.T.A. 2: 508 (1871); Trans. Linn. Soc. Lond. 27: 639 (1871); Hiern, Cat. Afr. Pl. Welw. 1, 2: 382 (1898); De Wilde in Blumea 21, 2: 327 (1974). Type: *B. littoralis* Peyr.

Tryphostemma Harv., Thes. Cap. 1: 32, t.51 (1859); in F.C. 2: 499 (1862); Hook.f. in Benth. & Hook.f., Gen. Pl. 1: 811 (1867); Mast. in F.T.A. 2: 507 (1871); Engl. in Bot. Jahrb. 14: 387 (1891) (including sect. *Eutryphostemma*, sect. *Neotryphostemma*, and sect. *Basananthe*); Pflanzenw. Afr. 3, 2: 598 (1921); Harms in Notizbl. Bot. Gart. Berl. 8: 291 (1923) (including 'Reihen' *Appendiculatae* and *Exappendiculatae*); in Pflanzenfam. ed. 2, 21: 487 (1925); Hutch. & Pearce in Kew Bull. 1921: 257 (1921) (including series *Lobatifoliae* and *Integrifoliae*); Phill., Gen. ed. 2: 518 (1951); A. & R. Fernandes in Garcia de Orta 6: 250, 662 (1958); Hutch., Gen. Fl. Pl. 2: 371 (1967); Ross, Fl. Natal 251 (1972).

Carania Chiov., Fl. Somala 175 (1929).

Annual or perennial herbs or small climbers, rarely shrubs, glabrous or hairy, with or without tendrils. *Leaves* simple (not lobed) or lobed, sessile or petiolate; margin entire or mostly dentate with small glandular teeth; apex mostly mucronate. *Stipules* small, linear; false stipules in some species, developed from the supra-axillary bud. *Tendrils* axillary, replacing central flower of cyme, or absent. *Inflorescences* axillary, cymose, 1–3-flowered, sessile or peduncled; bracts and bracteoles small, linear, often forming an involucre. *Flowers* bisexual (sometimes functionally unisexual), campanulate, greenish; stipe indistinctly articulate at base to the short pedicel; hypanthium rather narrow, flattish, rarely shallowly cup-shaped. *Sepals* 5 (or 6), oblong to lanceolate, free. *Petals* absent, or 1–2, or (4 or) 5 (or 6), oblong to lanceolate, (sub)obtuse, free, usually smaller than the sepals. *Outer corona* consisting of a ± barrel-shaped tube bearing a ring of filiform processes (threads), bluish, and mostly with a ring of small inward curved teeth. *Disc* low, annular, rarely absent. *Inner corona* membranous, cup-shaped, margin entire or lobulate, in *B. berberoides* from Somalia forming 5 small cups around the bases of the filaments. *Stamens* 5 (or 6–9); filaments inserted in the upper half inside the corona, free; anthers basifixed, ellipsoid to lanceolate, (sub)sagittate, 2-locular. *Ovary* ellipsoid, superior, mostly sessile, 1-celled, with 3 (or 4) placentas; styles 3 (or 4), free or partially united; stigmas globose, small. *Fruit* a sessile or shortly stipitate 3- (or 4)-valved capsule, ellipsoid; valves coriaceous. *Seeds* 1 or a few, arillate, ellipsoid to reniform, ± compressed; testa coriaceous, mostly rugose, blackish.

A genus of 25 species in central, east and southern Africa; 5 species occur in the Flora area.

As I have explained in my revision (l.c.), I have refrained from subdividing the genus into sections.

False stipules, as used in key and descriptions, are stipule-like appendages much larger than the minute true stipules, developed from the supra-axillary serial bud or (short) shoot.

False stipules large, foliaceous; petals present:
  Leaves 3–9-lobed:
    Sepals 3–4,5 mm, with keeled nerves; leaves 3–7 (–9)-lobed .... 1. *B. pedata*
    Sepals 5–7,5 mm, not keeled; leaves 3-lobed .... 2. *B. triloba*
  Leaves simple, rarely partly faintly lobed .... 3. *B. polygaloides*
False stipules absent, or the first leaves (cataphylls) of supra-axillary shoot not distinctly appearing as false stipules; petals present or absent:
  Petals absent, or 1 or 2; leaves not lobed .... 4. *B. sandersonii*
  Petals present, 4 or 5; leaves, at least in part, deeply 3-lobed .... 5. *B. heterophylla*

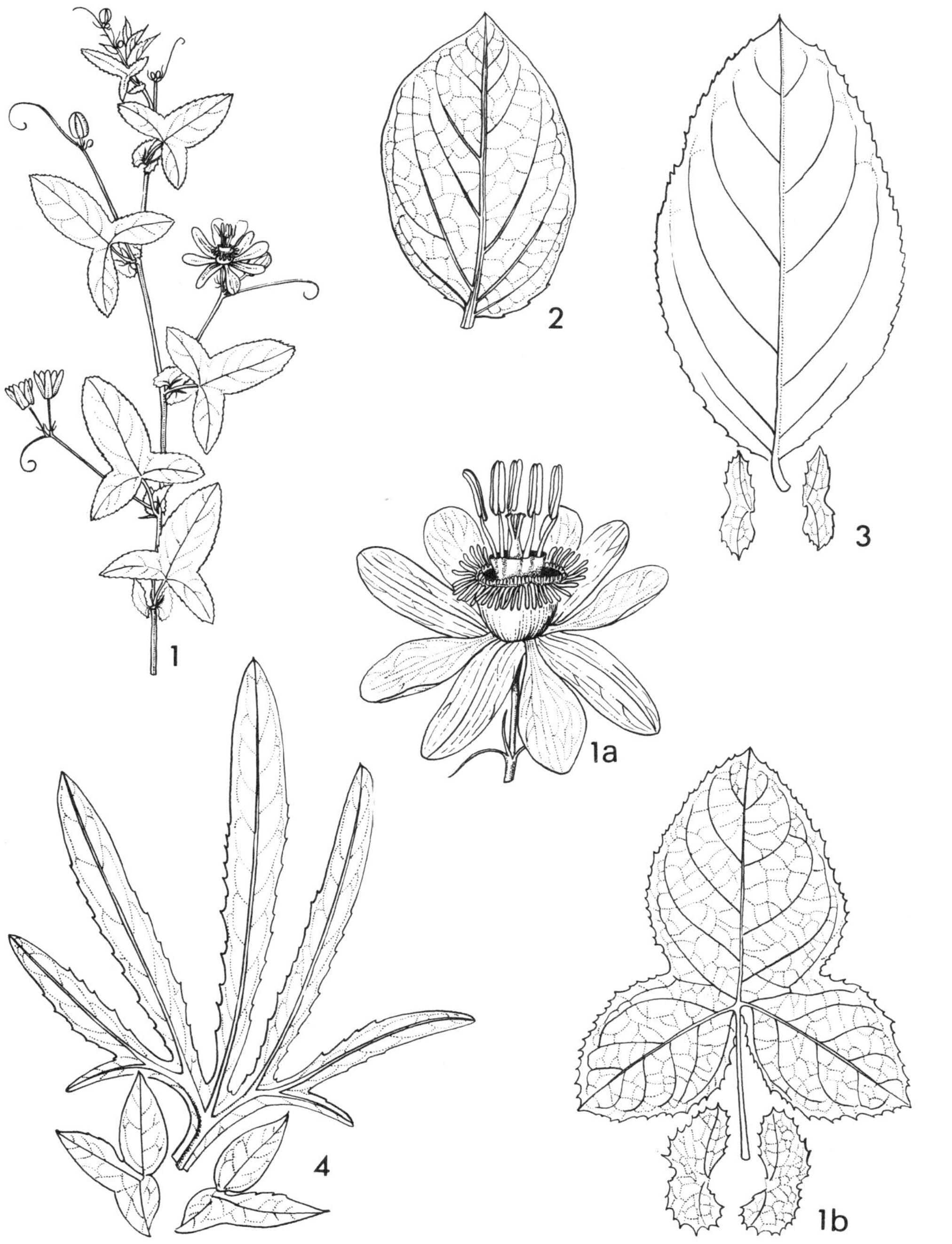

FIG. 38.—1, **Basananthe triloba,** flowering stem, × 1; 1a, flower, × 4 (*Pott* 5454); 1b, leaf, × 1 (*Compton* 31222); 2. **B. sandersonii,** leaf, false stipules absent, × 1 (*Acocks* 11795); 3, **B. polygaloides,** leaf and false stipules, × 1 (*Harrison* 296); 4, **B. pedata,** leaf and false stipules, × 1 (*Wild & Drummond* 7001) (after De Wilde, 1973).

1. **Basananthe pedata** *(Bak.f.) De Wilde* in Blumea 21: 333, fig. 1a, 5 (1974). Type: Rhodesia, *Rand* 67 (BM).

Tryphostemma pedatum Bak.f. in J. Bot. Lond. 37: 436 (1899); Hutch. & Pearce in Kew Bull. 1921: 263 (1921). *T. schlechteri* Schinz in Vjschr. Naturf. Ges. Zürich 55: 243 (1911); Schreiber in F.S.W.A. 89: 2 (1968). Type: Transvaal, *Schlechter* 4596 (B, holo.!; HBG). *T. arenophilum* Pott in Ann. Transv. Mus. 5: 234 (1915). Type: Transvaal, *Leendertz* 2062 (PRE, holo.!; K). *T. harmsianum* Dinter in Feddes Rep. 24: 304 (1928). Type: S.W. Africa, *Dinter* 872 (B†, holo.; M).

Erect annual or biennial herb up to 50 cm, glabrous or slightly scabrous, often with some spreading-erect branches from the base; tendrils absent. *Leaves* deeply pedately 3–7 (–9)-lobed, the lobes elliptic to linear, (0,5–) 1–8 cm, obtuse to acute-acuminate, margin remote- to densely dentate-mucronate (serrate) up to 1,5 mm; leaf-base subcordate to long-cuneate, decurrent in the alate petiole 0,2–1,5 cm. *Stipules* 2–6 mm; false stipules foliaceous, asymmetrical, 0,5–2 cm, acute-acuminate, mucronate. *Inflorescences* 1- or 2-flowered; peduncle up to 2 cm; bracts 2–4 mm. *Flowers* glabrous; stipe 1,5–3,5 mm; hypanthium 1–2 mm wide. *Sepals* 2,5–4,5 mm, with (2–) 3-keeled or winged green nerves. *Petals* 2–4,5 mm. *Outer corona-tube* (0,5–) 0,7–1,3 mm, threads (0,5–) 0,7–1 mm. *Disc* 0,1–0,2 mm. *Inner corona* cup-shaped, 0,3–0,8 mm. *Stamens* 5; filaments 1–1,5 mm; anthers 0,5–1 mm. *Ovary* 0,5–1,5 mm; styles free, 0,6–2,3 mm. *Fruit* subsessile, (0,7–) 1 cm, containing 1 seed c. 7 mm. FIG. 38: 4.

Growing mostly in sandy soil of riverbeds or dunes, or in gravelly, or sometimes rocky soil; 400–1 200 m; occurs locally in Transvaal and in north-eastern South West Africa; also in Zambia, Rhodesia and Botswana.

S.W.A.—1716 (Enana): near Oshandi, *De Winter & Giess 7027*. 1718 (Kuring-kuru): Omuramba Mpungu, *De Winter 3902*. 1721 (Mbambi): Shamvura, *De Winter & Marais 4600*. 1819 (Karakuwisa): Karakuwisa, *Dinter 7269*. 1820 (Tarikora): 16 km N. of Tamso, *De Winter & Marais 4727*.

TRANSVAAL.—2329 (Pietersburg): Kalkbank, *Acocks 8848*. 2428 (Nylstroom): near Warmbaths, *Leendertz 2062*. 2528 (Pretoria): 14 km E. of Hammanskraal, *Meeuse 9573*.

2. **Basananthe triloba** *(H. Bol.) De Wilde* in Blumea 21: 335, fig. 1b, 5 (1974). Type: Mozambique, Delagoa Bay, *Bolus* 7606 (K, holo.!; BM).

Tryphostemma trilobum H. Bol. in Hook. Icon. Pl. 19: t.1838 (1889); Schinz in Bot. Jahrb. 15, Beibl. 33: 3 (1892); Hutch. & Pearce in Kew Bull. 1921: 261 (1921); Harms in Pflanzenfam. ed. 2, 21: 488 (1925); A. & R. Fernandes in Garcia de Orta 6: 252 (1958). *T. schinzianum* Harms in Pflanzenfam. 3, 6a: 72, 75, fig. 25a, 26c (1893); ed. 2, 21: 488, fig. 217c (1925); in Engl., Pflanzenw. Ost Afr., C: 281 (1895); Bak.f. in J. Linn. Soc. (Bot.) 40: 73 (1911); Hutch. & Pearce in Kew Bull. 1921: 263 (1921); A. & R. Fernandes in Garcia de Orta 6: 253 (1958). Type: Mozambique, *Stuhlmann* 835 (B†, holo.; BM, drawing; HBG). *T. sagittatum* Hutch. & Pearce in Kew Bull. 1921: 262 (1921); Harms in Pflanzenfam. ed. 2, 21: 488 (1925). Syntypes: Transvaal, *Galpin* 505 (PRE!); *Kirk* 74 (K!).

Perennial herb, several shoots creeping or climbing to 70 (–250) cm, branched or not from a woody rootstock, glabrous; tendrils up to 10 cm. *Leaves* ± sagittate, 3-lobed, 1,5–8 × 1–7,5 cm, the lobes suborbicular to oblong, up to 5,5 × 3 cm, apex obtuse to acute, the middle lobe sometimes largest; margin simple or double (densely) serrate-dentate 1–2 mm; petiole 0,1–4 cm. *Stipules* 3–8 mm; false stipules ± asymmetrical, 2-lobed, serrate-mucronate, 0,5–3,5 × 0,2–1 cm. *Inflorescences* (1- or) 2-flowered; peduncle 1–6 cm, bracts 2–3 mm. *Flowers* glabrous; stipe 4–7 mm; hypanthium 2,5–3 mm wide. *Sepals* 5–7,5 mm. *Petals* 4–6,5 mm. *Outer corona-tube* 1,7–2,5 mm, threads 0,5–1,3 mm. *Disc* c. 0,2 mm. *Inner corona* cup-shaped, 1,2–2 mm. *Stamens* 5; filaments 2–2,5 mm (sometimes alternating with small tooth-like appendages); anthers 1,2–1,8 mm. *Ovary* up to 1 mm stipitate, 1–1,5 mm; styles (1–) 2,5–4 mm, free or up to 1 mm connate. *Fruit* (excluding the 2–4 mm long gynophore) c. 1,2–1,5 cm, containing 2 or 3 seeds c. 7 mm. FIG. 38: 1.

Distributed in Natal, Swaziland and Transvaal, extending to the north in Mozambique and Rhodesia; found on sandy or stony soil, in grassland, scrub, and open forest; 0–1 200 m. Flowers and fruit mainly Sept.–Febr.

TRANSVAAL.—2431 (Acornhoek): Klaserie, *Killick & Strey 2515*. 2530 (Lydenburg): Lowveld Botanic Garden, *Buitendag 214*. 2531 (Komatipoort): near Pretorius Kop, *Codd 5668;* Barberton, *Galpin 505*.

SWAZILAND.—2631 (Mbabane): near Stegi, *Compton 28394; 31222*.

NATAL.—2732 (Ubombo): near Jozini, *Strey 10250*. 2831 (Nkandla): near Hlabisa, *Gerstner 5022*.

3. **Basananthe polygaloides** *(Hutch. & Pearce) De Wilde* in Blumea 21: 338, fig. 1d, 5 (1974). Type: Natal, *Medley Wood* 10339 (K, holo.!).

Tryphostemma polygaloides Hutch. & Pearce in Kew Bull. 1921: 263 (1921); Harms in Pflanzenfam. ed. 2, 21: 488 (1925); Ross, Fl. Natal 251 (1972).

Perennial herb, up to 100 cm, with woody rootstock, glabrous; tendrils 2–8 (–10) cm, or none. *Leaves* simple or rarely faintly 3-lobed, ovate to elliptic, 2–7,5 × 1–5 cm, base subcordate to acute, apex subacute to acute-acuminate, margin single- or double serrate-dentate 1–3 mm deep; petiole 0,1–0,5 cm. *Stipules* 2,5–8 mm; false stipules asymmetrical, sharply dentate, 0,4–1,5 cm. *Inflorescences* (1- or) 2-flowered; peduncle 1–5,5 cm, bracts 2–6 mm. *Flowers* glabrous; stipe 5–15 mm; hypanthium 3–4,5 mm wide. *Sepals* 7–10 mm. *Petals* 6,5–9 mm. *Outer corona-tube* 2–2,5 mm, with a plicate, inward-folded edge; threads 1–2 mm. *Disc* c. 0,2 mm. *Inner corona* cup- or funnel-shaped, 1,5–2 mm. *Stamens* 5; filaments 2,5–4,5 mm (sometimes alternating with small tooth-like appendages); anthers c. 2 mm. *Ovary* 1,5–2mm; styles 3,5–4,5 mm, connate for 0,5–2,5 mm. *Fruit* (excluding the 2–4 mm long gynophore) 1,5–2,2 cm, containing 2–4 seeds 8–9 mm. FIG. 38: 3.

A local endemic of South Africa, in Zululand (Natal), not occurring elsewhere. Grows in grassland, bushes, and coastal forest, always reported from sandy soil; 0–200 m. Flowers and fruits Jan.–March, and Sept.–Nov.

NATAL.—2732 (Ubombo): Mpangazi, *Strey 5055.* 2832 (Mtubatuba): 26 km N. of Mtubatuba, *Acocks 13082;* near Cape Vidal, *Codd 10166;* Richards Bay, *Venter 4867.*

4. **Basananthe sandersonii** *(Harv.) De Wilde* in Blumea 21: 339, fig. 2a–c, 5 (1974). Type: Natal, *Sanderson* 1864 (TCD, holo.;.K).

Tryphostemma sandersonii Harv., Thes. Cap. 1: 33, t.51 (1859); in F.C. 2: 499 (1861); Hutch. & Pearce in Kew Bull. 1921: 265 (1921); Ross, Fl. Natal 251 (1972). *T. natalense* Mast. in Trans. Linn. Soc. Lond. 27: 639 (1871), nom. illegit. Type as for *T. sandersonii*. *T. longifolium* Harms in Bot. Jahrb. 33: 149 (1902); Hutch. & Pearce in Kew Bull. 1921: 264 (1921); A. & R. Fernandes in Garcia de Orta 6: 252 (1958). Type: Tanzania, *Busse* 673 (B†, holo.; EA). *T. viride* Hutch. & Pearce in Kew Bull. 1921: 265 (1921); Ross, Fl. Natal 251 (1972). Syntypes: Transvaal, *Bolus* 7602 (K!); *Galpin* 931 (PRE!; K!); *Thorncroft* 4366 (K!). *T. friesii* Norlindh in Bot. Notiser 107, fig. 9, 10 (1934); A. & R. Fernandes in Garcia de Orta 6: 251, 252 (1958). Type: Rhodesia, *Fries, Norlindh & Weimarck* 3112 (S, holo.!; BM!; BR!; PRE!).

Perennial herb 2–60 cm, glabrous, 1 to several shoots erect or ± prostrate at base from a rootstock; tendrils sometimes present, 0,5–2 (–5) cm. *Leaves* simple, 2–16 × 0,7–4 (–5,5) cm, suborbicular or broadly ovate (or obovate) or elliptic to lanceolate; base subcordate or rounded to acute (to attenuate), apex broadly obtuse (to retuse) to (mostly) acute, margin remotely serrate-dentate 0,5–1 mm, especially towards base; petiole 0–0,3 (–0,5) cm. *Stipules* 1,5–5 mm; false stipules absent. *Inflorescences* 1–3-flowered; peduncle up to 4,5 cm; bracts 1–4 mm. *Flowers* glabrous; stipe 3–17 mm; hypanthium 2–4 mm wide. *Sepals* 5–7, (the inner ones petaloid), obtuse, 4,5–8,5 mm. *Petals* 0 (–2). *Outer corona-tube* 1–2 (–3) mm, threads 0,7–1,5 (–2) mm, sometimes ± branched. *Disc* 0,1–0,5 mm. *Inner corona* cup-shaped 1–1,5 (–2,5) mm. *Stamens* 5; filaments 1,5–3,5 mm, anthers 0,7–1,3 mm. *Ovary* up to c. 1 mm stipitate, 1–1,5 mm; styles 3 or 4, free, 3–4 mm. *Fruit* (excluding the 1–3 mm long gynophore) 1,2–2 cm, containing 1–4 seeds 6–8 mm. FIG. 38: 2.

Distributed in eastern South Africa, in the north-eastern Cape Province, Natal, Swaziland, and eastern Transvaal; extending into Rhodesia, Mozambique and southern Tanzania; found in (regularly burnt) grassland, in open scrub, forest edges, on sandy and rocky soil; 0–1 700 m; flowers and fruits can be found throughout the year, but mostly in the period Sept.–Oct.

TRANSVAAL.—2329 (Pietersburg): Haenertsburg, *Pott sub TRV 13379.* 2330 (Tzaneen): Westfalia Estate, *Scheepers 760.* 2530 (Lydenburg): 37 km N.E. of Machadodorp, *Acocks 16617.* 2531 (Komatipoort): Barberton, *Galpin 931.*

SWAZILAND.—2531 (Komatipoort): Havelock Mine, *Miller 3075.* 2631 (Mbabane): near Mbabane, *Compton 28059.*

NATAL.—2730 (Vryheid): near Utrecht, *Thode A378.* 2831 (Nkandla): near Katazo, *Acocks 11795.* 2930 (Pietermaritzburg): near Pinetown, *Hilliard & Burtt 6887.* 2931 (Stanger): near Stanger, *Moll 2201.* 3030 (Port Shepstone): near Highflats, *Dyer 4134.*

CAPE.—3029 (Kokstad): Clydesdale, *Tyson 2106.* 3129 (Port St. Johns): Mkambati, *Codd 9727.*

This is a highly polymorphic species, especially in habit, leaf shape, as well as in flower size. The narrow-leaved forms described as *T. viride* are linked up with broad-leaved forms by numerous transitional specimens.

5. **Basananthe heterophylla** *Schinz* in Verh. Bot. Ver. Prov. Brandenb. 30: 252 (1888); De Wilde in Blumea 21: 353, fig. 4k, 6 (1974). Type: S.W. Africa, Amboland, *Schinz* 435 (K, iso.!).

Tryphostemma heterophyllum (Schinz) Engl. in Bot. Jahrb. 14: 388 (1891) in obs.; 15: 577 (1893); Pflanzenw. Afr. 3, 2: 599 (1921); Hutch. & Pearce in Kew Bull. 1921: 261 (1921); Harms in Pflanzenfam. ed. 2, 21: 488 (1925); Schreiber in F.S.W.A. 89: 2 (1968).

Annual or biennial to 50 cm, erect, sometimes ± branched at base, glabrous; tendrils absent. *Leaves* up to 8 × 8 cm, mostly deeply 3-lobed, sometimes partly simple, the lobes oblong-lanceolate, 2–7 × 0,7–2 cm, apex obtuse to acute, margin up to 1 (–2) mm deep serrate-setaceous dentate; leaf-base subcordate to subacute; petiole 1–5 cm, shortly winged, with gland-teeth. *Stipules* 3–6 mm; false stipules absent. *Inflorescences* 1- or 2-flowered; peduncles 0,2–1,5 cm; bracts 3–7 mm. *Flowers* glabrous; stipe 2–4 mm; hypanthium 2,5–4 mm wide. *Sepals* 7–15 mm, long-acute, the outer with 2 submarginal prominent nerves. *Petals* 5–7 mm, (sub)acute. *Outer corona-tube* 1–1,5 mm, threads 2,5–3,5 mm. *Disc* 0,2–0,4 mm. *Inner corona* cup-shaped, 0,5–0,8 mm. *Stamens* 5; filaments (inserted near the base of the cup) 4–5 mm, anthers narrow, 2–2,5 mm. *Ovary* 1,5–2 mm; styles 2,5–6 mm, connate for 1–3,5 mm. *Fruit* (sub)sessile, 1,5–1,8 cm, containing 1–4 seeds c. 7–8 mm.

Restricted to, but rather widely distributed in the northern half of South West Africa, with one collection from south-western Botswana. It grows in red or white sand, and on sand dunes; 1 000–1 500 m. Flowering and fruiting time Dec.–April.

S.W.A.—1920 (Tsumkwe): near Tsumkwe, *Giess, Watt & Snyman 11051*. 2115 (Karibib): Karibib, *Dinter 6834*. 2116 (Okahandja): Okahandja, *Bradfield 323*.

Specimens usually have (deeply) 3-lobed leaves, but occasional plants with partly simple leaves have been found.

The persistent cotyledons measure 2–2,5 × 1–1,5 cm, with a petiole of 1–1,3 cm.

5372 5. PASSIFLORA

Passiflora *L.*, Sp. Pl. 2: 955 (1753); Gen. Pl., ed. 5: 410 (1754); DC., Prodr. 3: 322 (1828); Benth. & Hook.f., Gen. Pl. 1: 810 (1867); Mast. in Trans. Linn. Soc. Lond. 27: 593 (1871); in Fl. Bras. 13: 531 (1872); Harms in Pflanzenfam. ed. 2, 21: 495 (1925); Killip in Field Mus. Publ. Bot. 19, 1: 1–613 (1938); Hutch., Gen. Fl. Pl. 2: 370 (1967); De Wilde, Fl. Malesiana 1, 7: 407 (1972). Type: *P. incarnata* L.

For generic synonyms see Harms, l.c. and Killip, l.c.

Mostly perennial climbing herbs to large lianas, rarely (not in Africa) shrubs or trees, glabrous or hairy, provided with tendrils. *Leaves* (mostly) alternate, (deeply) lobed or simple, palminerved or penninerved, petiolate; margin mostly dentate, often with small gland-teeth; petiole with or without glands; blade-glands present or not. *Stipules* minute to large. *Inflorescences* sessile or peduncled, 1- to many-flowered, with or without a simple tendril; or flowers rarely collected into pseudo-racemes; bracts and bracteoles small to large, forming a conspicuous involucre or not. *Flowers* hermaphrodite, 5-merous; hypanthium saucer-shaped to cylindrical. *Sepals* and petals free, often brightly coloured; petals mostly resembling sepals, membranous, sometimes absent. *Corona* extra-staminal, variously shaped, simple or mostly composed of a usually complicated outer corona consisting of threads, and flat or plicate inner coronas, with in addition a nectary ring or annulus or not. *Androgynophore* mostly distinct, 3 mm or more. *Stamens* 5 (–8), free (or in some Asian spp. partly connate), in older flowers mostly reflexed; anthers dorsifixed, versatile, elliptic to linear. *Gynophore* absent, or sometimes up to 7 mm. *Ovary* globose to fusiform; styles 3 (or 4), free or connate at base; stigmas capitate. *Fruit* mostly indehiscent, ± baccate, often with coriaceous exocarp, globose or ellipsoid, or rarely fusiform, containing many seeds.

About 370 species, of which c. 350 occur in the Americas, and 20 species in S.E. continental Asia, Indo-Australia and the West Pacific. The genus is not indigenous in Africa. Species described from Madagascar and the Mascarene Is. pertain to early introductions from America.

Several species are introduced in South Africa as ornamentals, e.g. *P. coerulea* L., or for the edible fruits with delicate flavour, e.g. *P. edulis* Sims, *P. laurifolia* L. and *P. quadrangularis* L.

Three species, *P. foetida* L., *P. suberosa* L., and *P. subpeltata* Ortega are locally established weeds in many tropical countries; *P. edulis* Sims is often cultivated and sometimes escapes. These species are marked in the key with an asterisk, and treated fully.

The genus was subdivided by Harms (1925) into 21 sections; Killip (1938) accepts 22 subgenera, and many sections and series for the American species.

FIG. 39.—1, **Passiflora coerulea,** flowering stem, × 1; a, fruit, × 1 (*Robertson* s.n.).

Bracts and bracteoles inconspicuous, filiform or linear, not forming an involucre; flowers small, c. 1,5 cm diam., greenish-yellow, without petals .... 1. *P. suberosa**

Bracts and bracteoles conspicuous, foliaceous, forming an involucre; flowers much larger, variously coloured, provided with petals:

- Involucre bracts finely and deeply divided .... 2. *P. foetida**
- Involucre bracts not divided:
  - Leaves not lobed, (sub)pinnately nerved; stem 4-angular, distinctly winged .... *P. quadrangularis*
  - Leaves usually lobed or partite, palmately nerved; stem not angular or winged:
    - Stipules lanceolate or filiform; involucre bracts serrate-denticulate .... 3. *P. edulis**
    - Stipules foliaceous; involucre bracts (sub)entire:
      - Stipules 1–2 cm long, falcate, remotely dentate; leaves (3–) 5 (–9)-lobed, incisions nearly to the base; corona threads with bluish or purplish tinge .... *P. coerulea*
      - Stipules 1,5–4 cm long, straight, entire; leaves 3-lobed to about the middle; corona threads white .... 4. *P. subpeltata**

1. **Passiflora suberosa** *L.*, Sp. Pl. 2: 958 (1753); Killip in Field Mus. Nat. Hist. Chicago Bot. 19, publ. 407: 88 (1938). Type: 'Dominica' (Hispaniola) (LINN).

Climber or creeper to 6 m, perennial, glabrous or glabrescent or pubescent to various degree; stem ± angular, corky when older. *Leaves* entire or usually 3-lobed up to $^{4}/_{5}$ the blade, subcircular to ovate or oblong in outline, 3–5-plinerved, membranous or subcoriaceous, base rounded or cordate, 4–10 × 4–14 cm, lobes triangular to lanceolate, apex acute to acuminate; margin entire; petiole 0,5–4 cm, with 2 small obconical or wart-like glands at about the middle; blade-glands usually absent. *Stipules* linear, 5–8 mm. *Inflorescences* sessile, 1- or 2-flowered, with a central tendril, simple, 3–12 cm; pedicels 1–2 cm, jointed about half-way; bracts setaceous, caducous, c. 1 mm. *Flowers* 1–2 cm diam., pale greenish-yellow; hypanthium saucer-shaped, 3–5 mm wide; sepals ovate to lanceolate, subobtuse, 5–10 mm; petals absent; corona threads in 2 series, 2–6 mm; operculum (inner corona) plicate, minutely fimbriate; disc annular. *Androgynophore* 2–4 mm; filaments ± subulate, 2–3 mm; anthers 1–2 mm; ovary subglobose-ellipsoid, 1–2 mm, glabrous; styles 2–3 mm. *Fruit* a berry, (sub)globose, glabrous, 0,8–1,5 cm diam., purple-blackish; seeds several to many, subovoid, 3–4 mm long.

Introduced in many parts of the Old World tropics, up to 2 500 m alt.; it is sometimes cultivated as an ornamental in gardens, while escaped plants can be found on roadsides, in disturbed shady places, sandy places near the coast, etc.

NATAL.—2930 (Pietermaritzburg): near Pietermaritzburg, *Acocks 20017*. 3030 (Port Shepstone): Ifafa River heights, *Strey 8312*.

2. **Passiflora foetida** *L.*, Sp. Pl. 2: 959 (1753); Killip in Field Mus. Nat. Hist. Chicago Bot. 19: 474 (1938). Type: South America, (perhaps Lesser Antilles; fide Killip) (LINN).

Climber or creeper to 4 m, annual or biennial, subglabrous or whitish or yellowish-brown hairy to various degree; ill-odoured; stem terete. *Leaves* entire, or usually 3 (–5)-lobed up to half-way, suborbicular to ovate in outline, 3–5-subplinerved, usually membranous, base cordate, 3–10 × 3–10 cm; lobes up to 4 cm, usually acute-acuminate, margin (sub)entire with coarser gland-tipped hairs; petiole 1–6 cm; stipules subreniform, 0,5–1 cm, deeply cleft into filiform gland-tipped processes; glands absent, except gland-tipped stronger hairs or processes on petiole, stipules, bracts, etc. *Inflorescences* sessile, 1 (–2)-flowered, the straight peduncle 2–6 cm inserted beside a simple tendril 5–15 cm; bracts and bracteoles (1–) 2–4 cm, deeply 2–4 times pinnatisect, with filiform gland-tipped segments, forming an involucre just below, and enveloping the flower. *Flowers* c. 3–5 cm diam., pale pinkish or lilac, rarely white; hypanthium short, saucer-shaped; sepals ovate-oblong to lanceolate, c. 1,5–2 cm, awned 2–4 mm dorsally just below the apex; petals oblong to lanceolate or ± spathulate, slightly shorter than the sepals; corona consisting of 2 outer series of threads, c. 1 cm long, several inner series of capillary threads 1–2 mm; operculum membranous, ± erect, denticulate; disc conspicuous, annular. *Androgynophore* 4–6 mm; filaments flattened, c. 5–6 mm; anthers 3–5 mm; ovary globose to ellipsoid, c. 2–3,5 mm, usually glabrous; styles 4–5 mm. *Fruit* a rather dry berry, (sub)globose, usually glabrous, 1,5–3 cm diam., yellowish to orange, ± enveloped by persistent involucre; seeds many, subovoid to ± cuneiform, c. 4–5 mm long, obscurely 3-dentate at apex.

A common alien in parts of the subtropics and all tropical areas, up to 2 500 m altitude. It is found on coastal sands, in waste places, in ruderal and disturbed vegetation, and roadsides. It is sometimes cultivated and often escapes. Killip (l.c.) recognized many varieties, but most African material can be referred to var. *hispida* (DC.) Killip ex Gleason, a variety with glabrous ovary, and usually 3-lobed leaves, the whole plant with a rather hispid tomentum.

NATAL.—2930 (Pietermaritzburg): Merebank, *Baijnath 180;* Isipingo Flats, *Ward 6244.*

3. **Passiflora edulis** *Sims* in Bot. Mag. 45: t.1989 (1818); Killip in Field Mus. Nat. Hist. Chicago Bot. 19: 393 (1938). Type: cult. in Europe, probably originally from Brazil (fide Killip, l.c.).

Climber to c. 15 m, perennial, glabrous throughout (except ovary); stem sometimes ± angular. *Leaves* 3-lobed up to ¾ the blade, rarely without lobes, suborbicular to broadly ovate in outline, 5–11 × 6–12 cm, 3-plinerved, subcoriaceous, base acute to cordate, lobes elliptic to oblong, up to 8 cm, apex acute, shortly acuminate; margin serrate; petiole 1–4 cm; glands on petiole 2, wart-like, situated at transition of or up to 0,5 cm below the blade; blade-glands absent. *Stipules* lanceolate-linear, c. 1 cm. *Inflorescences* 1-flowered, the straight peduncle (pedicel) 3–6 cm, inserted beside a simple tendril 5–20 cm; bracts and bracteoles ovate, acute, 1,5–2 cm, glandular-serrate, forming an involucre. *Flowers* 4–7 cm diam., white, corona threads purplish towards base; hypanthium cup-shaped, c. 1 × 1–1,5 cm; sepals oblong, 2–3 cm, corniculate; petals oblong, 1–2,5 cm, obtuse; corona composed of several series of threads, the outer 2 series 0,5–2,5 cm, the inner ones much shorter; operculum membranous, incurved, crenulate-fimbriate; disc (limen) cupuliform, entire or crenulate. *Androgynophore* 6–8 mm, thickened towards base; filaments sub-subulate, 6–8 mm; anthers 8–10 mm; ovary subglobose to ellipsoid, 3–5 mm, glabrous or shortly pubescent; styles 10–12 mm. *Fruit* berry-like, with coriaceous-leathery pericarp, globose to ellipsoid, excluding the 0,5–1,5 cm long gynophore, 4–5 cm diam., glabrous, purplish or yellow; seeds many, ellipsoid, 5–6 mm.

This species is cultivated for its edible, well-flavoured fruit. Sometimes escaped in waste gardens, disturbed places and forest fringes; in the tropics up to 2 500 m alt.

TRANSVAAL.—2230 (Messina): near Lake Funduzi, *Story 4847*. 2527 (Rustenburg): Jacksonstuin, *Van Vuuren 196*. 2530 (Lydenburg): Vertroosting Nature Reserve, *Muller 2492*, 2531 (Komatipoort): Ida Doyer Nature Reserve, *Muller 2119*.

NATAL.—2831 (Nkandla): Port Durnford, *Venter 1882*.

4. **Passiflora subpeltata** *Ortega,* Nov. Rar. Pl. Hort. Madrit. 6: 78 (1798); Killip in Field Mus. Nat. Hist. Chicago Bot. 19: 436 (1938). Type: Mexico (specimen in Herb. Madrid, fide Killip).

Herbaceous creeper or climber to 5 m, perennial, essentially glabrous throughout; stem terete. *Leaves* 3-lobed to about half-way, suborbicular in outline, 4–10 × 4–11 cm, (3–) 5-plinerved, herbaceous, base rounded, truncate or cordate, subpeltate; lobes elliptic to oblong, up to 5 cm, apex obtuse or acutish, c. 1 mm mucronate; margin entire except for a few gland-teeth in or near the lobe-sinuses; petiole 3–6 cm; glands on petiole 2–5, scattered or ± paired at about the middle, slender, up to 1 mm long; blade-glands absent. *Stipules* ovate-oblong, straight, 1,5–4 cm long, entire or with a few minute gland-teeth, apex mucronulate. *Inflorescences* 1-flowered, the straight peduncle 3–6 cm, inserted besides a simple tendril 4–12 cm; bracts and bracteoles (broadly) ovate 1–1,5 cm, acute, entire or with a few minute gland-teeth at base, forming an involucre. *Flowers* 4–5,5 cm diam., white; hypanthium broadly cup-shaped, 7–10 mm wide; sepals oblong, 2–2,5 cm, obtuse, with a subapical horn 0,5–1 cm; petals oblong, 1,5–2 cm, acutish; corona composed of 4 or 5 series of threads, those of the outer 2 series 1 (–1,5) cm long, those of inner series 2–6 mm; operculum subplicate, fimbriate-laciniate for about half or less, and with a fringe of inward curved dentiform processes; disc annular; limen with lobulate edge, erect or ± reflexed at the top, closely surrounding androgynophore. *Androgynophore* 10–12 mm; filaments 5–6 mm, dilated; anthers 5–7 mm; ovary ellipsoid, c. 4 mm, glabrous; styles 8–10 mm. *Fruit* ± leathery, ellipsoid or subglobose, excluding the 1,5–2 cm long gynophore 3,5–4,5 cm long, greenish turning yellow; seeds many, ellipsoid, 4–5 mm.

Originating from tropical America and cultivated as an ornamental; sometimes profusely escaped, e.g. in abandoned plantations, on forest edges, etc.; up to 1 000 m, in the tropics up to 2 500 m.

TRANSVAAL.—2530 (Lydenburg): Lowveld Botanic Garden, *Buitendag 615*.

NATAL.—2930 (Pietermaritzburg): Nagle Dam, *Wells 1354*. 3030 (Port Shepstone): Umgai, *Strey 10583*.

Killip, in 1947, named the identical material in the East African Herbarium as the related and similar species *P. eichlerana* Mast. (1872). This species, according to Killip (1938), differs from true *P. subpeltata* mainly by the operculum, which in *P. eichlerana* is laciniate for about half or more, and by the presence of inward curved dentiform processes; in *P. subpeltata* the operculum is fringed only at the margin, with the entire portion much longer than the teeth or fringes. All the abundant material now available from East Africa, as well as similar material from India, Malesia, Australia, Hawaii, etc., is doubtlessly conspecific. The operculum in all these alien specimens is dissected up to about half-way or less, thus rendering the distinction between *P. eichlerana* and *P. subpeltata* for these specimens difficult. As both species are related in their original country, the introduced plants are possibly a deviating form of one or the other, or a hybrid. As *P. subpeltata* is the older name, I have chosen to accept it over *P. eichlerana*.

ACHARIACEAE

by D. J. B. Killick

Herbaceous climbers, subherbaceous shrublets or acaulescent herbs; leaves alternate, simple, lobed or not lobed, sometimes all radical, crenate or serrate; stipules present or 0. *Plants* monoecious. *Inflorescence* a raceme or flowers 1–several, axillary. *Sepals* 3–5, usually linear, sometimes adnate to corolla-tube. *Corolla-tube* campanulate; lobes 3–4, about as long as the tube or shorter. *Stamens* 3–5, inserted at base of corolla-tube or in throat; anthers 2-thecous, sometimes cohering. *Glands* inserted at base of corolla-tube. *Ovary* superior, 1-locular, with few to many ovules on 3–5 parietal placentas; style 3–5-lobed with the lobes sometimes 2-fid. *Fruit* a capsule, enclosed in a perianth corolla. *Seed* subglobose, pitted or tubercled; embryo straight; cotyledons flat; endosperm present.

Genera 3, species 3, occurring in South Africa and Swaziland. Achariaceae was included by Bentham & Hooker f. (1867) in Passifloraceae, but differs in the sympetalous campanulate corolla and, in two genera, the stamens which are adnate to the corolla-tube. According to Hutchinson (1967) the habit of *Ceratiosicyos* recalls that of some Cucurbitaceae, hence Achariaceae represents something of a link between Passifloraceae and Cucurbitaceae.

Acaulescent herbs with radical leaves ....3. **Guthriea**
Plants not acaulescent:
  Erect shrublets; male flowers axillary, 1–several; sepals present in female flowers; fruit ellipsoid, up to 1 cm long ....2. **Acharia**
  Herbaceous climbers; male flowers racemose; sepals absent in female flowers; fruit siliquiform, up to 8 cm long ....1. **Ceratiosicyos**

5374 1. CERATIOSICYOS

Ceratiosicyos *Nees* in Eckl. & Zeyh., Enum. 281 (1836); Harv., Gen. S. Afr. Pl. 107 (1838); in F.C. 2: 501 (1862); Benth. & Hook. f., Gen. Pl. 1, 3: 814 (1867); Phill., Gen. ed. 2: 519 (1951). Type species: *C. ecklonii* Nees.

Herbaceous climber. *Leaves* alternate, simple, petiolate, exstipulate. *Male flowers* racemose. *Sepals* 4 or 5. *Corolla-tube* campanulate; lobes 4 or 5. *Stamens* 5, inserted at base of corolla-tube, alternating with an equal number of glands adnate to corolla-tube; filaments free. *Ovary* 0. *Female flowers* solitary. *Calyx* absent. *Corolla* as in male flowers. *Glands* 5. *Ovary* shortly stalked, with several ovules on 4 or 5 parietal placentas; styles 4 or 5, 2-lobed at apex. *Fruit* an elongated ribbed capsule, several seeded. *Seeds* with tuberculate testa; embryo straight; cotyledons ovate, distinctly veined; endosperm copious, fleshy.

One species found in the Transvaal, Swaziland, Natal and the Cape. The generic name is compounded from the Greek for pod or siliqua and cucumber in reference to the nature of the fruits.

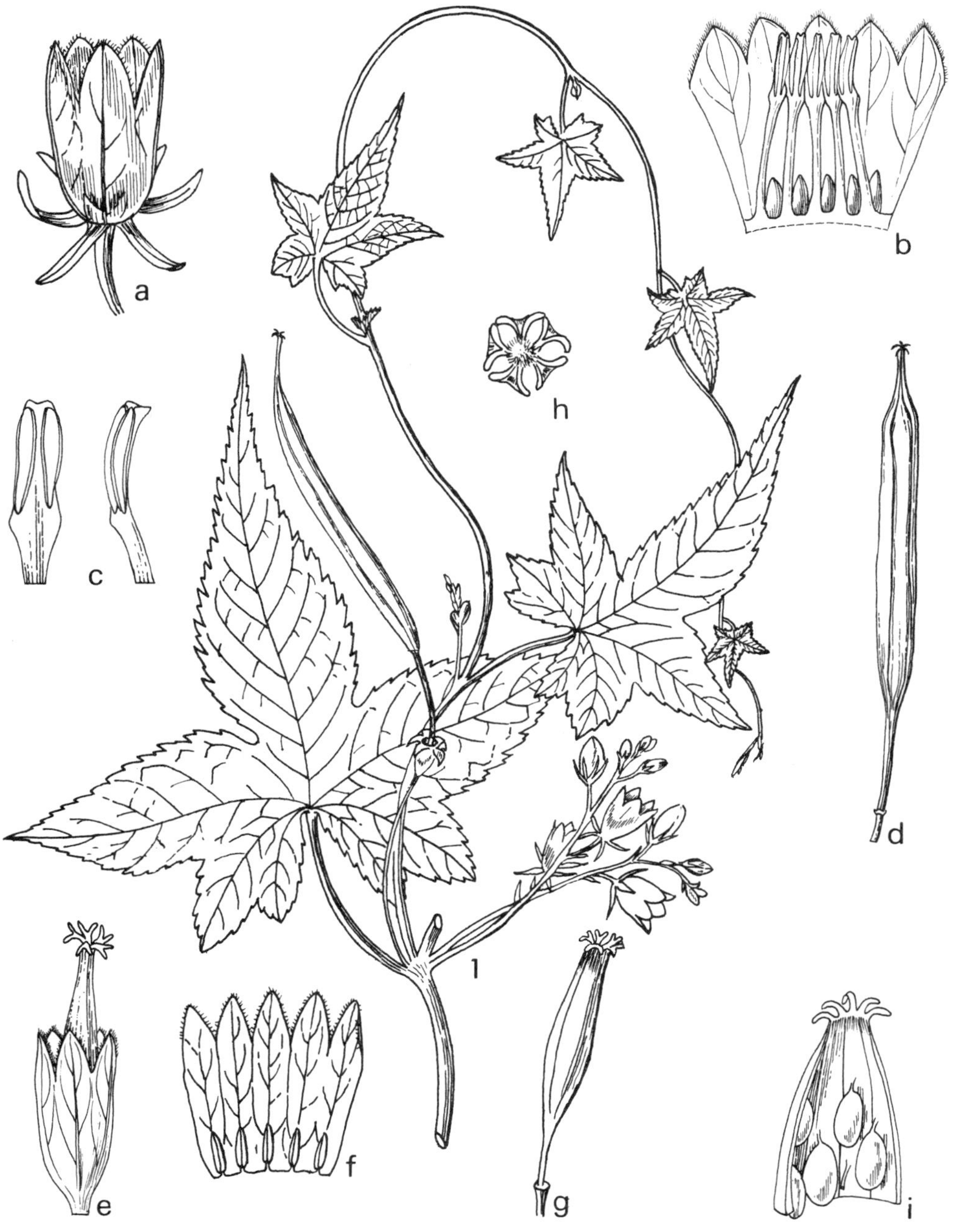

FIG. 40.—1, **Ceratiosicyos laevis,** flowering and fruiting stem, × 1; a, male flower, × 5; b, corolla of male flower opened out, × 6; c, upper part of stamens, front and side views, × 5; d, fruit, × 1; e, female flower, × 8; f, corolla of female flower opened out; g, pistil, × 8; h, stigma, × 10; i, portion of ovary opened out to show attachment of ovules, × 20 (mostly after Harvey in Ann. Nat. Hist. ser. 1, 3: t.10, 1839).

Ceratiosicyos laevis *(Thunb.) A. Meeuse* in Bothalia 8: 20 (1962). Type: Cape, *Thunberg* s.n. (UPS, holo.!; PRE, photo.).

Bryonia laevis Thunb., Prodr. 1: 13 (1794).

Ceratiosicyos ecklonii Nees in Eckl. & Zeyh., Enum. 281 (1836); Harv., Gen. S. Afr. Pl. 107 (1838); Arn. & Harv. in Ann. Nat. Hist. ser. 1, 3: 421, t.10 (1839); Harv. in F.C. 2: 501 (1862); Marloth, Fl. S. Afr., 2,2: 200, t.131 (1925). Type: Cape, "in nemoribus ad Krakakamma, Adow & Olifantshoek (Uitenhage), tum in Albany & Kafferland", *Ecklon* 1797 (BOL!; G!; K!; L!; S!).

Herbaceous climber, glabrous. *Leaf-blade* palmately 5–7-lobed with lobes acuminate and margins serrate, 3–14 cm long, 3–12 cm wide; petiole 2–14 cm long. *Male flowers* yellowish-green. *Sepals* filiform-linear, 2,5–3 mm long, 0,2 mm wide, slightly swollen at apex, spreading. *Corolla-tube* 3,5–5 mm long; lobes slightly shorter than the tube, ovate-oblong, ciliate, cohering together by the cilia giving the corolla an urceolate appearance. *Glands* linear or obovate, 2 mm long, fleshy. *Stamens* with linear filaments, 5–6 mm long, dilated upwards; anthers oblong, c. 3 mm long, cohering, reflexed at apex. *Female flowers* yellowish-green. *Corolla-tube* 2–3,5 mm long; lobes 2–3 mm long. *Glands* narrowly ovate, up to 1,5 mm long, fleshy. *Ovary* stipitate, linear-elongate, 3–4 mm long; styles short, 2-lobed with recurved lobes. *Fruit* elongated, cylindrical up to 8 cm long, 0,8 cm wide, tapered at both ends. *Seeds* subglobose to oblong, 3–4 mm diam., tuberculate. FIG. 40.

A forest climber distributed from Wilderness in the south-eastern Cape, into Natal and Swaziland, and northwards as far as Duiwelskloof in the northern Transvaal.

TRANSVAAL.—2330 (Tzaneen): Westfalia Estate, Duiwelskloof, *Scheepers 397*. 2430 (Pilgrims Rest): 1,6 km from Graskop to Kowyns Pass, *Jordaan 100;* The Downs, *Junod 4197*.

SWAZILAND.—2631 (Mbabane): Duiker Bush, *Compton 25949*.

NATAL.—2731 (Louwsburg): Wendelane Spruit, Nongoma, *Gerstner 4417*. 2829 (Harrismith): Draycott Hill, *Acocks 11436*. 2830 (Dundee): 20 km from York on Rietvlei Road, *Marais 350*. 2831 (Nkandla): Ngoye, Ubisana Valley, *Venter 1313*. 2832 (Mtubatuba): Hluhluwe Game Reserve, *Hitchins & Ward 2*. 2929 (Underberg): Deepdale, near Umkomaas River, *Strey 4815*. 2930 (Pietermaritzburg): "Ehlatini", Karkloof, *Moll 3468*. 2931 (Stanger): Umgeni River Mouth, *Schlechter 2845* (K). 3030 (Port Shepstone): Uvongo, *Strey 9598;* Dumisa, *Rudatis 470* (K).

CAPE.—3029 (Kokstad): Mount Currie, *Tyson 472*. 3126 (Queenstown): N'Zebanya Mountain, near Queenstown, *Galpin 1932*. 3128 (Umtata): 23 km from Umtata on road to Ngqeleni, *Story 1327*. 3226 (Fort Beaufort): Katberg, *Hutton s.n.* (K). 3227 (Stutterheim): Hogsback, *Acocks 11018*. 3228 (Butterworth): Kentani, *Pegler 903*. 3322 (Albert): Wilderness, *Van Niekerk 213*. 3323 (Willowmore): Stinkhoutkloof, Knysna, *Forest Officer s.n.* 3325 (Port Elizabeth): "Vista", ± 5 km from Paterson on road to Mimosa, *Retief 18*. 3326 (Grahamstown): Alexandria Forest, *Johnson 1059*.

A readily distinguishable herbaceous climber with 5–7-lobed leaves and elongated, ribbed capsular fruit up to 8 cm long.

5375 2. ACHARIA

Acharia *Thunb.*, Prodr. 1: 14, t. s.n. (1794); Fl. Cap. ed. Schult. 37 (1823); Harv., Gen. S. Afr. Pl. 409 (1838); Arn. & Harv. in Ann. Nat. Hist. ser. 1,3: 420, t.9 (1839); Harv. in F.C. 2: 501 (1762); Benth. & Hook.f., Gen. Pl. 1,3: 814 (1867); Phill., Gen. ed. 2: 520 (1951). Type species: *A. tragodes* Thunb.

Shrubby herb. *Leaves* alternate, simple petiolate, stipulate. *Flowers* axillary, 1–several, cernuous. *Male flowers: Sepals* 3 or 4. *Corolla* pubescent; tube campanulate; lobes 3 or 4, slightly winged or keeled. *Glands* 3 or 4, at base of corolla-tube, ciliate. *Stamens* 3 or 4, free; anthers oblong with cells widely separated. *Ovary* 0. *Female flowers: Sepals* 3 or 4, larger than in male, pubescent. *Corolla* larger than in male, pubescent; tube campanulate, keeled; lobes ovate. *Glands* 3 or 4, at the base of the corolla-tube. *Ovary* subsessile or stalked, with 3–5 ovules on 3–5 parietal placentas, densely pubescent; style 3–5-lobed. *Fruit* a capsule, enclosed in persistent corolla, 3–5-valved, terminated by the persistent style. *Seed* 1, narrowly winged (arillate) on one side.

An endemic, monotypic genus extending from the Uitenhage District in the Cape into Natal. The genus was named in honour of Erik Acharius (1757–1819), pupil of Linnaeus, lichenologist and physician of Vadstena in Sweden.

Acharia tragodes *Thunb.*, Prodr. 1: 14, t. s.n. (1794); Fl. Cap. ed. Schult. 37 (1823); Harv., Gen. S. Afr. Pl. 409 (1838); Arn. & Harv. in Ann. Nat. Hist. ser. 1,3: 420, t.9 (1839); Harv. in F.C. 2: 502 (1862); Martin & Noel, Fl. Albany & Bathurst 77 (1960). Type: *Thunberg* s.n. (UPS, holo!).

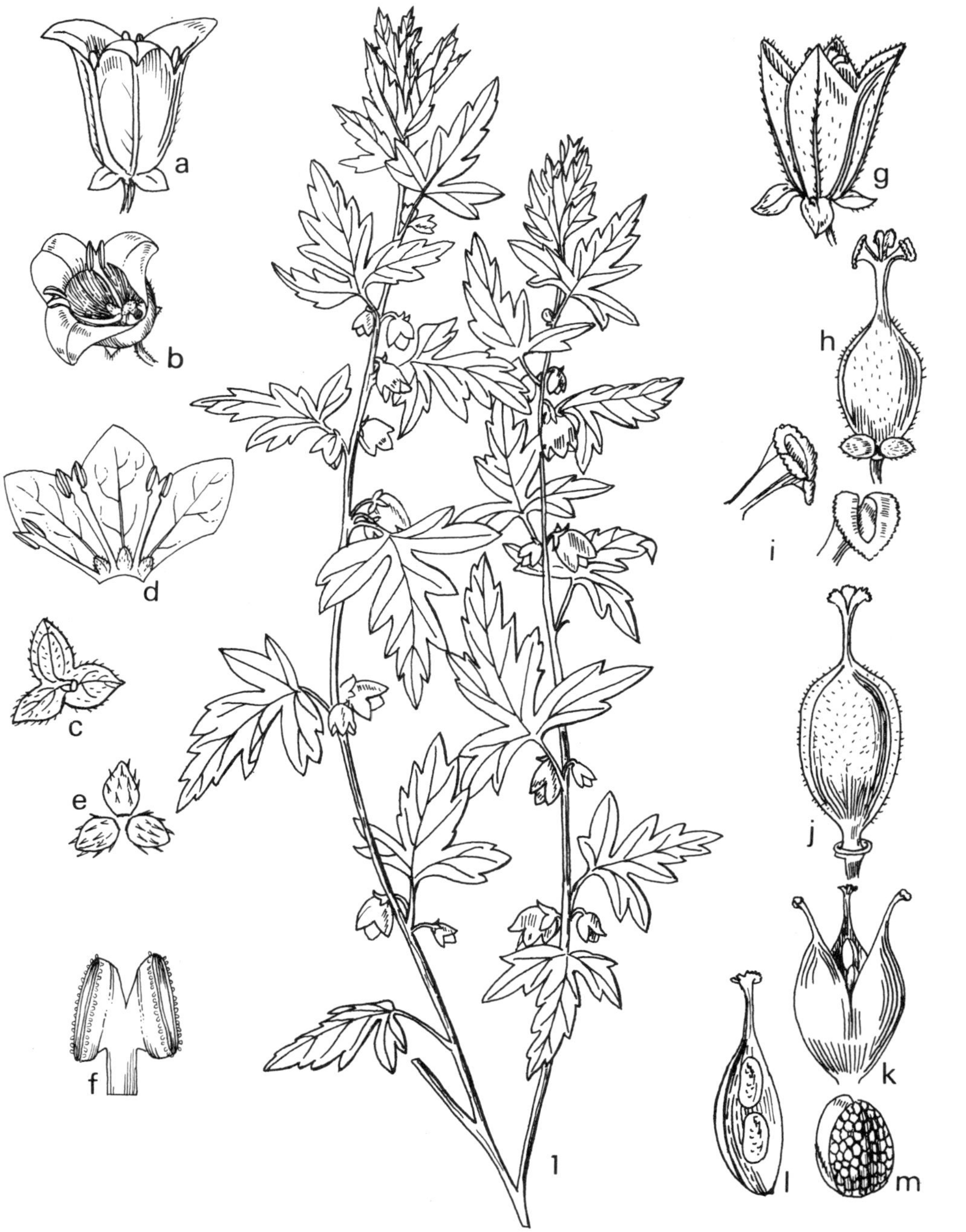

FIG. 41.—1, **Acharia tragodes,** flowering stems, × 1; a, male flower, × 5; b, view of male flower from above, × 4; c, sepals, × 3; d, corolla tube opened out, × 4; e, basal glands, × 6; f, glandular anther, × 10; g, female flower, × 4; h, pistil with basal glands, × 10; i, stigmas, × 10; j, stalked fruit, × 3; k, dehiscing fruit, × 3; l, fruit valve with seeds, × 3; m, seed, × 4 (mostly after Harvey in Ann. Nat. Hist. ser. 1, 3: t.9, 1839).

Perennial shrubby herb up to 40 cm high. *Leaf-blade* deeply palmately-lobed, 1,5–6,3 cm long, 1,5–6 cm wide, subchartaceous, pubescent; petiole 1,5–3 cm long, pubescent. *Flowers* 1 or more (of one or both sexes) together, shortly pedicellate. *Male flowers: Sepals* linear to narrowly elliptic, 1,8–3 mm long, 0,5–0,75 mm wide, somewhat recurved, pubescent. *Corolla-tube* 3–4 mm long; lobes triangular, 2–2,2 mm long. *Glands* ovate, 1 mm long, 0,75 mm wide. *Stamens* with filaments 0,5–1,2 mm long; anthers 1–1,5 mm long, glandular. *Female flowers: Sepals* elliptic, 2–3,5 mm long, 1,5 mm wide. *Corolla-tube* 4–5,5 mm long, lobes ovate, 3 mm long, 3–4 mm wide. *Glands* ovate, c. 1 mm long, pubescent. *Ovary* subsessile or with stalk 1 mm long, ellipsoid, 2–2,5 mm long, 1,5–2 mm wide; style terete, 2 mm long, lobes 1 mm long, with flabelliform stigmas. *Fruit* ellipsoid, 8–10 mm long, 5–8 mm wide, pubescent. *Seed* subglobose, 3 mm diam., reticulate. FIG. 41.

A rather rare species occurring in scrub and woods from the Uitenhage District in the Cape coastwise to Durban in Natal.

NATAL.—2931 (Stanger): Port Natal, *Drège s.n.* (G).

CAPE.—3228 (Butterworth): Kentani, *Pegler 757;* near Kei River Mouth, *Flanagan 50;* Collywobbles, Idutywa, *Van Breda 876.* 3325 (Port Elizabeth): Van Stadensberg, *Zeyher 4654* (P;S); Elandsrivier, *Ecklon & Zeyher 105,2; Wall 17* (S). 3326 (Grahamstown): Hopewell, *Acocks 16145.*

A perennial, shrubby herb growing up to 40 cm high with deeply palmately-lobed leaves and ellipsoid fruit up to 1 cm long. The specific epithet has been variously spelled as *tragodes, tragoides* and *tragioides,* but there is no reason why there should be any departure from the original spelling, *tragodes.*

5376 3. GUTHRIEA

Guthriea *H. Bol.* in Hook. Icon. Pl. t.1161 (1876); Marloth, Fl. S. Afr. 2,2: 200 (1925); Phill. Gen. ed. 2: 520 (1951). Type species: *G. capensis* H. Bol.

An acaulescent herb. *Leaves* rosulate, alternate, simple, petiolate. *Flowers* axillary, solitary, pedicelled. *Male flowers: Sepals* 5, adnate to corolla for almost whole of their length, shortly exceeding corolla. *Corolla* campanulate; lobes 5, reflexed. *Glands* 5, fleshy, inserted at base of corolla-tube. *Stamens* 5, inserted in throat of corolla; filaments slightly longer than anthers, flattened; anthers with cells attached to a broad connective. *Female flowers: Sepals* 4 or 5, adnate to corolla for half their length, about half as long as corolla. *Corolla-tube* campanulate; lobes 4 or 5. *Glands* 4 or 5, fleshy, inserted at base of corolla-tube. *Ovary* sessile, with 10–15 ovules on 4 or 5 parietal placentas; style 4- or 5-lobed. *Fruit* a 4- or 5-valved capsule enclosed in the corolla. *Seeds* several, arillate; embryo straight.

A genus of one species occurring in the eastern Cape, Natal and Lesotho. The genus was named in honour of Francis Guthrie, Professor of Mathematics, South African College, Cape Town.

Guthriea capensis *H. Bol.* in Hook. Icon. Pl. t.1161 (1876); Marloth, Fl. S. Afr. 2,2: 200 (1925). Type: Cape, "in saxosis in monte Gnadouw–Sneeuwbergen in ditione Graaff Reinet", et "in monte Oudeberg prope Graaff Reinet in locis graminosis", *Bolus* 818 (BOL, holo.!; K!; PRE!; SAM!; Z!).

An acaulescent herb with a sub-horizontal rhizome and thick, fleshy roots. *Leaf-blade* broadly ovate, occasionally broadly elliptic, 4–7 cm long, 3,5–6,5 cm wide, apex obtuse to round, base cordate, margins crenate, digitately nerved, discolorous, dark green and shiny above, paler below, bullate, glabrous; petiole 3–12 cm long. *Flowers* with pedicels 2–8 cm long. *Male flowers: Sepals* linear, 1,5 cm long, projecting beyond corolla, c. 2,5 mm. *Corolla* 1,3–1,5 cm long, 1,2 cm wide at mouth, pubescent inside in lower half; lobes ovate, 4 mm long, 3 mm wide. *Glands* ovate, 2 mm long, 1,5 mm wide, thick, fleshy, cream. *Stamens* with filaments 3,5 mm long, anthers 2–2,5 mm long, glandular. *Female flowers: Sepals* linear, c. 6 mm long, free for 2–3 mm. *Corolla-tube* 1,3–1,5 cm long, 1,2–1,4 cm wide at mouth; lobes transversely ovate, 1–3 mm long, 6 mm wide, erect. *Glands* ovate to obovate, 2,5 mm long, 2 mm wide, cream. *Ovary* ellipsoid, 8 mm long, 4 mm wide, brown gland-dotted; style subterete, 4–5 mm long; lobes simple, 1 mm long, spreading. *Fruit* triangular-ellipsoid, 1,2 cm long, 1 cm wide. *Seeds* ellipsoid, 3,5 mm long, 2 mm wide, minutely tuberculate, pubescent, black. FIG. 42.

FIG. 42.—1, **Guthriea capensis,** flowering plant, × 1; a, male flower opened out, × 1½; b, female flower opened out, × 1½; c, transverse section of ovary, × 2; d, fruit with persistent style and corolla, × 1½; e, seed, × 5 (Mostly after Fitch in Hook. Icon. Pl. t.1161, 1876).

Herb occurring in grassland on mountains in the eastern Cape and Natal from the Sneeuwbergen to Mont aux Sources and in Lesotho.

NATAL.—2828 (Bethlehem): Sentinel area, Mont aux Sources, *Killick & Marais 2199; Schweickerdt 707;* summit of Mont aux Sources, *Flanagan 2103.* 2929 (Underberg): Giants Castle, *Symons 345.*

LESOTHO.—2928 (Marakabies): Mamalope, *Jacot Guillarmod 766.*

CAPE.—3027 (Lady Grey): Ben McDhui, Wittebergen, *Galpin 6630.* 3028 (Matatiele): Naude's Nek, *Hilliard 5212* (NBG). 3124 (Hanover): Compassberg, *Acocks 16424.* 3224 (Graaff Reinet): Oudeberg, *Bolus 818; 1044* (K); "Gnadouwsberg", *Bolus 818.*

An acaulescent herb with radical, rosulate, bullate leaves dark green and shiny above and paler below, with flowers mostly hidden under the leaves. Bolus lumped his type material of *G. capensis* from Gnadouw Mountain (probably Nadouwsberg, 40 km E. of Graaff Reinet) and Oudeberg (probably Ouberg 20 km N.W. of Graaff Reinet) under one number, *Bolus* 818. The specimen in Bolus Herbarium from Gnadouw Mountain with notes by Bolus in his own handwriting is presumably the holotype rather than the one from Oudeberg marked 'typus auctoris', but not in Bolus's hand.

LOASACEAE

by L. E. CODD

Herbs, rarely woody, mostly clothed with rough hairs. *Leaves* alternate or opposite, entire or variously divided, exstipulate. *Inflorescence* a raceme or cyme, rarely a head or flowers solitary; bracteoles present. *Flowers* bisexual, regular. *Calyx-tube* adnate to the ovary; lobes 4–5, imbricate or contorted, persistent, often enlarged in fruit. *Petals* 4–5, free, inserted in the throat of the calyx-tube, induplicate-valvate. *Nectary scales* petaloid, cucullate, alternating with the petals, appendaged or scales absent. *Stamens* many, rarely few, inserted with the petals, often collected in bundles opposite the petals; filaments sometimes beaded; anthers often didymous, 2-thecous, opening by longitudinal slits; staminodes present or absent. *Ovary* inferior or nearly completely so, 1–3-chambered, with 1–many pendulous ovules from the apex of the ovary chambers or on parietal placentas; style entire or 2–3-fid; stigma simple or capitate. *Fruit* a capsule, 1–3-locular, 2–5-valved. *Seed* often minute; embryo straight; endosperm present or absent.

Genera 15, species about 230; 14 of the genera and all but two of the species are natives of tropical or South America, with the remaining genus, *Kissenia,* in Africa.

5388 KISSENIA

Kissenia *R. Br. ex Endl.,* Gen. Pl. Suppl. 2: 76 (1842) (err. *Fissenia*); T. Anders. in J. Linn. Soc. (Bot.) 5, Suppl. 1: 43 (1861); Harv. in F.C. 2: 503 (1862); Gen. ed. 2: 119 (1868); Mast. in F.T.A. 2: 501 (1871); Gilg in Pflanzenfam. 3, 6a: 114 (1894); Phill., Gen. ed. 2: 521; (1951); Dandy in Kew Bull. 20: 451 (1966).

Subshrub, scabrous, covered with brittle hairs armed with minute recurved barbs. *Leaves* alternate, petiolate, slightly or deeply lobed. *Inflorescence* a scorpioid, bracteate cyme. *Calyx-tube* obconic, 10-ribbed, densely villous; lobes 5, equal, longer than the tube, oblong-spathulate, enlarged in fruit. *Petals* shorter than the calyx-tube, concave, hairy without. *Nectary scales* 5, petaloid. *Stamens* many, some of the outer stamens with a hairy scale at the base and with or without anthers; filaments linear, all or mostly beaded. *Ovary* inferior, 2–3-chambered, with 1 or rarely 2 apical pendulous ovules in each chamber; the portion of the ovary projecting above the calyx-tube flattened and densely villous; style simple, somewhat angled; stigma simple. *Fruit* woody, ribbed, 2–3-locular, 1–3-seeded, indehiscent, crowned with the enlarged calyx-lobes, densely villous. *Seed* compressed; cotyledons flat, thick, fleshy, oblong; endosperm 0.

FIG. 43.—1, **Kissenia capensis,** flowering and fruiting twig, × 1; a, fruit, × 1½; b, stamen, × 6; c, d, nectary scale, × 6; e, flower, × 1½ (*Marloth* 1226).

A genus of 2 closely related species with a disjunct distribution, one in southern Arabia, Somalia and Ethiopia and the other in the western parts of Southern Africa.

The name is derived from a place in southern Arabia now known as Kishin or Qishn (Dandy, 1966). *Fissenia* is considered to be an unintentional orthographic error on the part of Endlicher for *Kissenia,* the ms. name provided for it by Robert Brown.

Kissenia capensis *Endl.,* Gen. Pl. Suppl. 2: 76 (1842) (err. *Fissenia*); Harv., Thes. Cap. 1: 61 (1859) (err. *Fissenia*); Dandy in Kew Bull. 1926: 177 (1926); in Kew Bull. 20: 451 (1966); Roessler in F.S.W.A. 93: 1 (1968). Type: Namaqualand, between Verleptpram and Orange River Mouth, *Drège.*

Cnidone mentzelioides E. Mey. ex Drège, Cat. Pl. Exsicc. Afr. Austr. 27 (1838), nom. nud.; Zwei Pfl. Doc. 93 (1843), nom. nud.

Kissenia mentzelioides R. Br. ex Harv., Thes. Cap. 1: sub. t.98 (1859) (err. *Fissenia*), nom. syn. *K. spathulata* R. Br. ex T. Anders. in J. Linn. Soc. (Bot.) 5, Suppl.: 43 (1861); Harv. in F.C. 2: 503 (1862); Mast. in F.T.A. 2: 501 (1871); Marloth, Fl. S. Afr. 2,2: 200, t.68, f.132 (1925); nom. illegit.

Pallid, scabrid subshrub 25–120 cm tall; bark whitish. *Leaves* ovate to ovate-oblong, 6–9 cm long, 4–7 cm broad, markedly scabrid, usually 3–5-lobed; base truncate to subcordate, apex obtuse; margin irregularly dentate; petiole 2–5 cm long. *Flowers* in terminal, scorpioid cymes; bracts ovate-lanceolate, 5–9 mm long, persistent. *Calyx-tube* densely hirsute; lobes pale green, turning yellow, oblong, 17–20 mm long, 4 mm broad, finely hispid on both surfaces. *Petals* white, spathulate, concave, 9–12 mm long, 5 mm broad, pubescent without. *Nectary scales* usually 5, sometimes fewer or with one smaller than the rest, strap-shaped, cucullate, 4–5 mm long, 1 mm broad, recurved at the apex, hairy within. *Stamens, ovary, fruit* and *seed,* as for generic description. FIG. 43.

Often locally common in sandy soil between boulders, on river banks and dry water courses in north-western Cape Province and the western half of S.W. Africa.

S.W.A.—1812 (Sanitatas): Anabib, *De Winter & Leistner 5742.* 2014 (Welwitschia): Welwitschia, *Galpin & Pearson 7631.* 2015 (Otjihorongo): 65,6 km W. of Outjo, *De Winter 3062 A.* 2114 (Uis:) near Brandberg, *Rodin 2747.* 2115 (Karibib): Usakos, *Marloth 1226.* 2215 (Trekkopje): Namib Desert Park, *Jensen 90.* 2315 (Rostock): near Gobabeb, *Koch A 21.* 2316 (Nauchas): Farm Ubib, *Merxmüller & Giess 28116.* 2416 (Maltahöhe): Buellsport, *Strey 2150.* 2518 (Tses): Karub, *Wiss 972.* 2617 (Bethanie): Seeheim, *De Winter & Giess 6421a.* 2618 (Keetmanshoop): Keetmanshoop, *Dinter 3556.* 2718 (Grunau): Klein Karas, *Oertendahl 90a.* 2818 (Warmbad); near Warmbad, *Verdoorn & Dyer 1776.* 2819 (Ariamsvlei): Ham River, *Galpin 14147.*

CAPE.—2817 (Vioolsdrift): Richtersveld, *Hardy 1665.* 2818 (Warmbad): Wolveton, *Schlechter 11439.* 2820 (Kakamas): Keimoes, *Marloth 14029.* 2822 (Glen Lyon): Langeberg, *Acocks 495.* 2917 (Springbok): 48 km N. of Steinkopf, *Hartmann 1587.* 2919 (Pofadder): near Pella, *Pearson 3544.* 2922 (Prieska): 21 km E. of Draghoender, *Codd 1228.* 2923 (Douglas): Kameelfontein, *Bryant J114.*

BEGONIACEAE

by O. M. HILLIARD*

Herbs or undershrubs, mostly fleshy, generally erect, sometimes creeping or climbing by means of adventitious roots, or acaulescent with a rhizome or tuber. *Leaves* alternate, rarely subradical, petiolate, generally asymmetrical, digitately nerved, margins entire, toothed, lobed or dissected. *Stipules* 2, conspicuous, persistent or caducous. *Inflorescences* axillary, cymose, bracteate. *Flowers* monoecious, actinomorphic or zygomorphic, showy. *Male flowers*: tepals usually 2 or 4, rarely many or 0, petaloid. *Stamens* numerous, filaments free or connate; anthers 2-thecous, dehiscing laterally. *Ovary* rudimentary or 0. *Female flowers*: tepals 2–5, rarely many, petaloid. *Staminodes* 0 or rarely represented by glands. *Ovary* inferior or rarely half-inferior, 2–4-, rarely 1-, celled, usually 3-celled and 3-winged or angled, ovules numerous on projecting simple or lobed axile placentae. *Styles* 2–5, free or connate at the base, usually 2-fid, the stigmas often

*Department of Botany, University of Natal, Pietermaritzburg. All type specimens cited were seen by the author unless otherwise stated.

twisted, papillose. *Fruit* a capsule, rarely a berry. *Seeds* numerous, minute, testa reticulate, endosperm scanty or 0, embryo straight.

A family of 5 genera and about 900 species, most of them in the genus *Begonia*. Mainly natives of the warm, wet parts of America and Asia, relatively few in Africa and represented here only by the genus *Begonia*.

5397 BEGONIA

Begonia *L.*, Sp. Pl. 1: 1056 (1753); Gen. Pl. ed. 5: 475 (1754); Klotzsch in Abh. Ak. Wiss. Berlin 1854: 125 (1855), reimp. Beg. Gattung Art. (1855); Sond. in F.C. 2: 480 (1862); A.DC. in DC., Prodr. 15: 266 (1864); Benth. & Hook.f., Gen. Pl. 1: 841 (1867); Hook.f. in F.T.A. 2: 569 (1871); Warb. in Pflanzenfam. 3,6a: 121 (1894); Engl., Pflanzenw. Afr. 3,3: 612 (1921); in Pflanzenfam. ed. 2, 21: 548 (1925); Phill., Gen. ed. 2: 522 (1951); Irmscher in Bot. Jahrb. 81: 106 (1961).

Caulescent or acaulescent herbs or undershrubs with watery stems and leaves, generally erect, sometimes creeping or climbing, sometimes with rhizomes or tubers. *Leaves* alternate, rarely subradical, petiolate, usually asymmetrical, entire, lobed or partite, irregularly toothed, green or sometimes richly coloured or spotted. *Male flowers*: tepals 2 or 4, rarely many or 0, petaloid. *Stamens* numerous, filaments free or connate at the base. *Female flowers*: tepals often 5 or 6, sometimes 4, petaloid. *Ovary* usually 3-, rarely 2-, 4- or 5- celled; ovules numerous, placentae axile, projecting, simple or lobed. *Styles* usually as many as the loculi, free or connate at the base, 2-fid, the stigmas entire or branched, often twisted, papillose. *Fruit* usually a capsule, 3-winged or 3-angled, rarely terete or 4-angled, or a berry. *Seeds* numerous, minute, endosperm 0.

About 900 species described, mainly from tropical America and Asia. Perhaps only 5 native in South Africa.

Placentae bilamellate (they appear bifurcate in transverse section):
- Stems and leaves glabrous, leaf margin lobed as well as toothed .... 1. *B. sonderana*
- Stems and leaves hairy, leaf margin toothed but not, or scarcely, lobed .... 2. *B. hirtella*

Placentae entire:
- Leaves symmetrical or nearly so, suborbicular, all, or nearly all, radical .... 3. *B. geranioides*
- Leaves mostly asymmetrical, cauline:
  - Flowers orange (but often drying pink), male tepals generally 4 .... 4. *B. sutherlandii*
  - Flowers white or pink, male tepals usually 2:
    - Primary leaves up to c. 13 × 7 cm, seldom less than 7 × 3 cm, ratio of length of vein adjacent to main vein in broader half of leaf to length of vein running to leaf apex (main vein) 1: (1,1–) 1,3 (–1,5) . 5. *B. homonyma*
    - Primary leaves up to 8 × 3,5 cm, more usually c. 5 × 2 cm or less; vein ratio 1: (1,3–) 1,7 (–2,5); if in the range 1: 1,3–1,4, then leaves smaller than the lower limits for *B. homonyma* .... 6. *B. dregei*

1. **Begonia sonderana** *Irmscher* in Bot. Jahrb. 81: 156 (1961) (as *B. sonderiana)*. Type: Transvaal, Barberton Distr., summit of Saddleback Mt., *Galpin* 821 (PRE, holo.; BOL; GRA; K; NBG).

B. caffra sensu Burtt Davy, Fl. Transv. 1: 232 (1926). *B. dregei* sensu Burtt Davy, l.c. (1926). *B. sonderana* var. *transgrediens* Irmscher in Bot. Jahrb. 81: 160 (1961). Type: Transvaal, Sabie, *Van der Merwe* 311 (PRE, holo.). *B. sp.*, Letty, Wild Flow. Transv. 204, pl. 101, 2 (1962).

A glabrous, herbaceous perennial herb, often tuberous or the stem decumbent at the base, the nodes swollen and rooting. *Stems* up to 1 m tall, stout, fleshy, often pinkish or reddish, simple or branched, leafy throughout. *Stipules* lanceolate, membranous, brown. *Leaves* obliquely ovate-acuminate, up to 18 × 14 cm, but often only about half that size, base cordate, margin 5–7-lobed, lobes more or less triangular, coarsely and irregularly toothed or occasionally crenate, light green and waxy above, paler below, veins often reddish, petioles up to 15 cm long, often reddish. *Inflorescences* in the upper leaf axils. *Flowers* up to 3 cm across, usually pink, sometimes white or white flushed pink, tepals of ♂ flowers usually 4, occasionally 2 or 3, of ♀ flowers usually 5. *Fruits* 3-winged, more or less rounded-oblong or cuneiform in outline, 2× 2–3 cm across the wings, placentae split, appearing bifurcate in transverse section. FIGS. 44: 3; 45: 3.

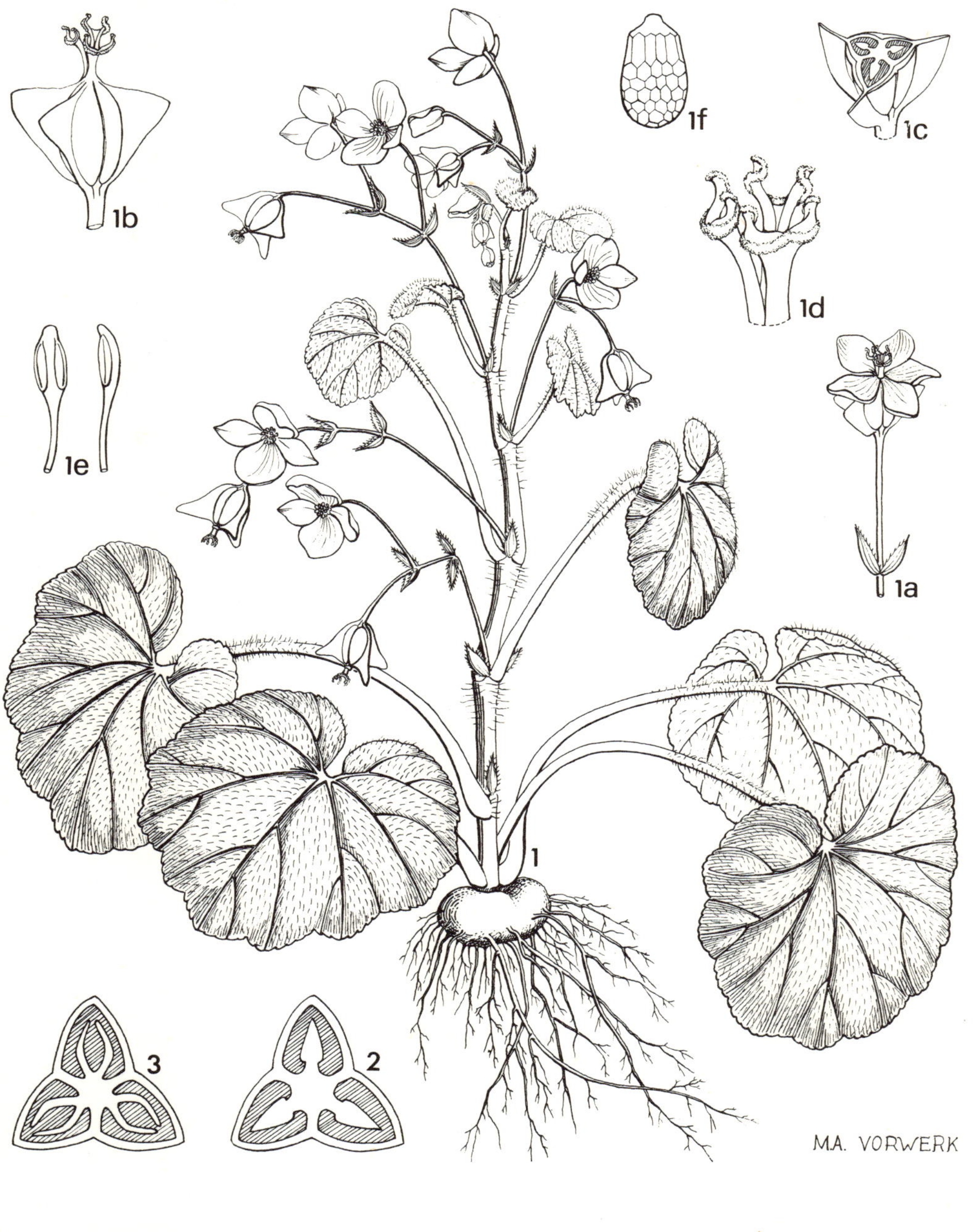

FIG. 44.—1, **Begonia geranioides,** whole plant, × 4/5; 1a, female flower, × 4/5; 1b, ovary, style and stigma, × 2; 1c, ovary cut transversely, × 2; 1d, stigma, × 5; 1e, anther, front and side view, × 6; 1f, seed, × 18 (*Stewart* 1640). 2, **B. sutherlandii,** transverse section of ovary with entire placentae (*Hilliard & Burtt* 5647). 3, **B. sonderana,** transverse section of ovary with split placentae (*Hilliard & Burtt* 5962).

Ranges from the eastern highlands of Rhodesia and Mozambique to the mountains of the eastern and north-eastern Transvaal and Swaziland. Favours broken cliff faces and rock falls in forest or on forest margins, or shady rock outcrops on steep grass slopes between about 1 000 and 1 850 m above sea level. Flowers mainly from January to March.

TRANSVAAL.—2230 (Messina): Sibasa, *Junod sub TRV 21218*. 2330 (Letaba): Woodbush, *Codd 9419*. 2430 (Pilgrims Rest): The Downs, *Junod 4194*; Mariepskop, *Hilliard & Burtt 5962*; Graskop, *Galpin 14443*. 2530 (Lydenburg): Sabie, *Taylor 1905*. 2531 (Komatipoort): 16 km W. of Havelock Mine, *Codd 6423*.

SWAZILAND.—2531 (Komatipoort): near Piggs Peak, *Compton 32244*. 2631 (Mbabane): Sheba's Veil, *Compton 27278*; Hlatikulu, *Stewart 111*.

This species was for long confused with *B. dregei* and *B. homonyma*, but is easily distinguished by its split, not entire, placentae.

2. **Begonia hirtella** *Link*, Enum. Hort. Berol. 2: 396 (1822). A.DC. in Mart., Fl. Bras. 4: 344 (1861). Type: cult. Berlin, material from Brazil (not seen).

B. villosa Lindl. in Bot. Reg. 15: t.1252 (1829). Type: cult. hort. R.H.S., seed from Brazil.

A robust, branched herb. *Stems* several from the base and up to 2 cm diam. there, branched, up to 75 cm tall, green, fleshy, thickly clothed in long, coarse, white hairs, glabrescent. *Stipules* ovate-lanceolate, about 1 × 0,5 cm, membranous, whitish, apex acuminate, margin fimbriate-ciliate. *Leaves* obliquely ovate, up to 10 × 8 cm, apex acute, base truncate or nearly so, margins irregularly crenate-serrate, not or scarcely lobed, green above, thinly villous, paler below, hairs nearly confined to the veins, petioles up to 9 cm long, fleshy, villous. *Inflorescences* in the upper leaf axils, peduncles up to about 5 cm long, usually shorter than the petiole. *Flowers* up to c. 8 mm across, pure white, tepals of ♂ flowers 4, of ♀ flowers 5. *Fruits* 3-winged, rounded-oblong to cuneiform in outline, c. 2 cm across the broadest part of the wings, placentae split, appearing bifurcate in transverse section. FIG. 45: 2.

A native of Brazil, long cultivated in Europe, *B. hirtella* appears to have established itself in certain Zululand forests, having twice been collected without indication that it was not growing naturally.

NATAL.—2732 (Ubombo): Gwalaweni Forest, *Vahrmeijer & Drijfhout 2000*. 2831 (Nkandla): Eshowe, *Lawn 1073*.

It is an annual, though it will persist for several seasons in cultivation, and seeds itself freely. Male and female flowers open at the same time, and it is clearly self-fertile, so that there is massive seed production.

Only *B. sonderana* among native species has split placentae like *B. hirtella*, but is easily distinguished by its much larger and generally pink flowers and glabrous, distinctly lobed leaves.

3. **Begonia geranioides** *Hook.f.* in Bot. Mag. 92: t.5583 (1866); Irmscher in Bot. Jahrb. 81: 124 (1961). Type: "imported from Port Natal by Messrs. Backhouse of York"(K).

A tuberous, often stemless, perennial herb. *Leaves* several from the crown, petioles up to 20 cm long, blade suborbicular, up to 8 × 11 cm, base deeply cordate, apex very obtuse, margin often shallowly and irregularly lobed, obscurely crenate, both surfaces and generally the petiole thinly clad with long white hairs, light green, juveniles sometimes white-spotted, veins often pinkish or reddish. *Flowering stems* up to 30 cm high, often much shorter, usually leafless, occasionally with a few leaves similar to but smaller than the radical ones, glabrous or sparsely hairy, often pinkish or reddish, bracts in pairs subtending each di- or trichotomy, ovate, margins sometimes ciliate-fimbriate. *Flowers* 2–3 cm across, pure white, tepals of ♂ flowers generally 4, of ♀ generally 5. *Fruits* 3-winged, cuneiform in outline, about 2 × 2 cm across the broadest part of the wings, placentae undivided. FIG. 44: 1; 45: 4.

Confined to forests between about 600 and 1 375 m above sea level from Karkloof in south-central Natal to the Zuurberg and Ngeli slopes on the Cape-Natal border near Weza. Grows on damp earth banks and rock faces, often forming large colonies. Flowers mainly from December to March.

NATAL.—2929 (Underberg): Polela, Hlabeni Forest, *Fernando sub NH 9616*. 2930 (Pietermaritzburg): Karkloof, *Moll 3541*; Town Bush Valley, *Canham 12*; Inanda, Groenberg, *Johnson 1323*; near Richmond, *Barker 5230*; near Ixopo, *Crewe 71*; Mid Illovo, farm Ismont, *Strey 8354*. 3030 (Port Shepstone): near Dumisa, *Hilliard & Burtt 3400*. 3130 (Port Edward): Beacon Hill, *Strey 6072*.

CAPE.—3029 (Kokstad): Zuurberg, *Tyson 1716*.

Easily recognized by its tuft of suborbicular hairy leaves. The plant can become dwarfed under dry conditions.

4. **Begonia sutherlandii** *Hook.f.* in Bot. Mag. 94: t.5689 (1868); Burtt Davy, Fl. Transv. 1: 232 (1926); Pole Evans in Flow. Pl. Afr. 8: t.283 (1928); Irmscher in Bot. Jahrb. 81: 162, t.8 fig. 1 (1961); Letty, Wild Flow. Transv. 204, pl. 101, 3 (1962); Batten & Bokelmann, Wild Flow. E. Cape 106, pl. 87, 1 (1966); Wilczek in F.C.B. Begoniaceae: 50 (1969). Type: Natal, without precise locality, Dec. 1861, *Sutherland* (K).

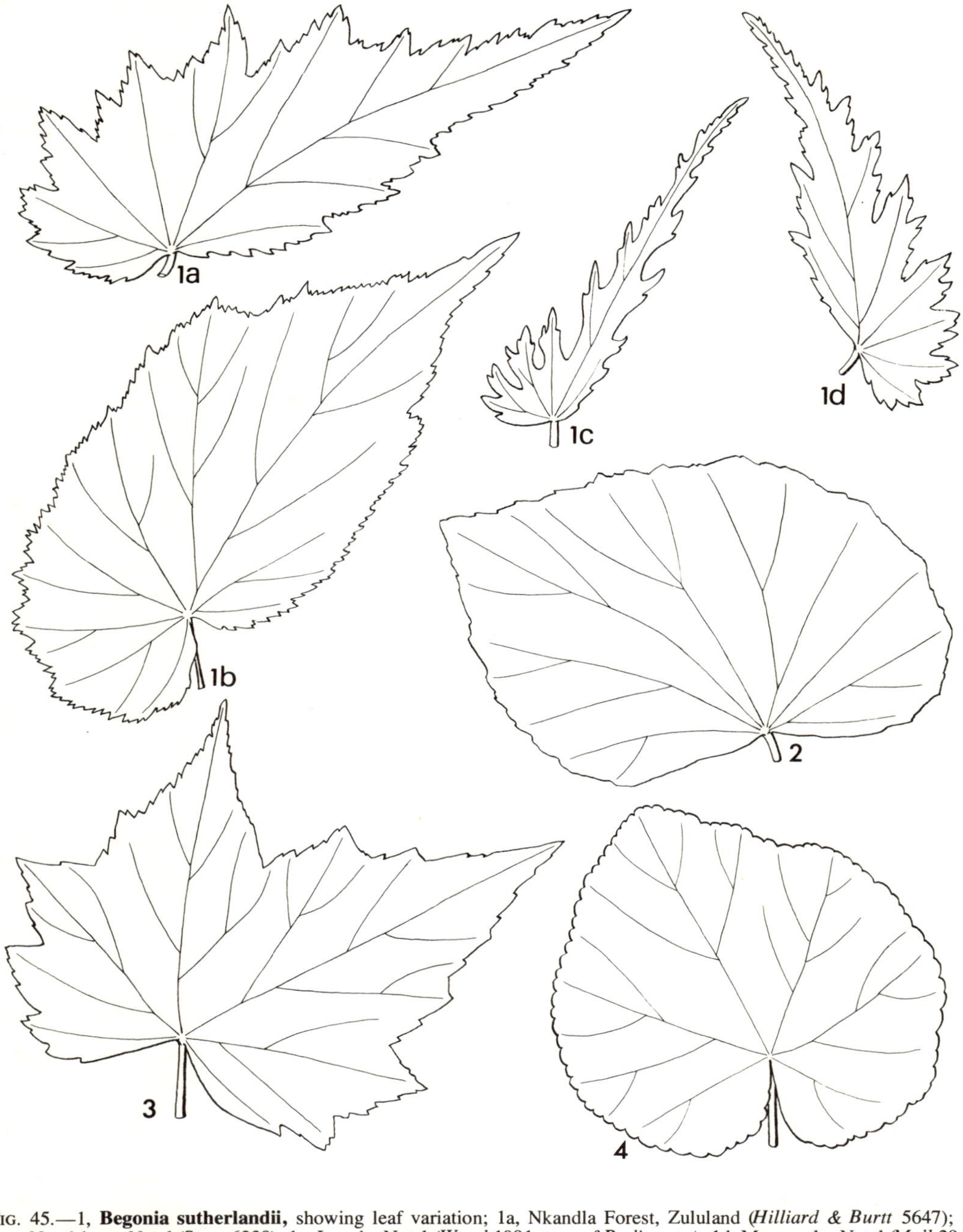

FIG. 45.—1, **Begonia sutherlandii,** showing leaf variation; 1a, Nkandla Forest, Zululand (*Hilliard & Burtt* 5647); 1b, Noodsberg, Natal (*Strey* 6239); 1c, Inanda, Natal (*Wood* 1081, type of *B. dissecta);* 1d, Mapumulo, Natal (*Moll* 2945). 2, **B. hirtella,** leaf (*Vahrmeijer & Drijfhout* 2000). 3, **B. sonderana** (*Hilliard & Burtt* 5962). 4, **B. geranioides** (*Wood* 749). All × 3/5.

B. sutherlandii Hook.f. var. *latior* Irmscher in Bot. Jahrb. 81: 165, t.9 figs. 1 & 2 (1961), forma *latior* Irmscher, l.c. Type: Zambia, above Mukoma, Inono Falls, 900 m, *Richards* 3699 (B, holo., not seen: K). *B. dissecta* Irmscher, l. c. 81: 178, t.11, fig. 2 (1961). Type: Natal, Inanda, *Wood* 1081 (K, holo.; NH; NBG). *B. buttonii* Irmscher, l.c. 81: 178 (1961). Type: Natal, Durban, (Port Natal) *Button* 8 (K). *B. suffruticosa* Meisn. var. *gueinziana* A.DC. in DC., Prodr. 15: 385 (1864). Type: Natal, Durban (Port Natal), *Gueinzius* 248 (W, not seen). *B. gueinziana* (A.DC.) Irmscher, l.c. 81: 177 (1961).

An herbaceous perennial with one or several stems from a tuber. *Stems* up to 1 m tall, simple or branched, slender, fleshy, leafy throughout, they and the leaves glabrous to thinly hairy, green or partially or wholly tinged red as are the stipules, petioles, veins and sometimes the leaf margins as well. *Stipules* lanceolate to ovate, margins often fimbriate-ciliate. *Leaves* very oblique, elliptic-lanceolate to ovate, up to 25 × 15 cm, but often only two-thirds or less that size, base cordate to truncate, apex acuminate to long-acuminate, margin acutely and irregularly serrate, frequently lobed as well, or occasionally more deeply cut, thin-textured, light bright green above with a satiny sheen, paler below, petioles slender, up to c. 15 cm long. *Inflorescences* axillary and terminal, many-flowered. *Flowers* up to 3,5 cm across, very showy, varying from pale to deep orange or brick red (often drying pink), tepals of ♂ flowers 4, of ♀ flowers usually 5. *Fruits* 3-winged, cuneiform in outline, up to 2 × 3 cm across the broadest part of the wings, placentae entire. FIG. 44: 2; 45: 1.

One of the most widespread of the African Begonias, ranging from Katanga, Tanzania, Zambia and the eastern highlands of Rhodesia through the mountains of the eastern Transvaal and Swaziland to Natal, where it is common in forest patches up to about 1 900 m above sea level, and into the eastern Cape (Transkei) as far south as Kei Mouth. Grows in forest, on the steeply sloping forest floor or on earth banks, rock faces or mossy boulders, sometimes epiphytic, often forming large colonies. Flowers mainly in December and January.

TRANSVAAL.—2430 (Pilgrims Rest): The Downs, *Junod 4183;* Mariepskop, *Hilliard & Burtt 5969.* 2530 (Lydenburg): Dullstroom, *Galpin 13731;* Lunsklip Falls, *Codd 10013.* 2531 (Komatipoort): Barberton, *Galpin 727.* 2630 (Carolina): Ermelo, *Collins sub TRV 6531.*

SWAZILAND.—2531 (Komatipoort): near Havelock, *Compton 29845.*

NATAL.—2730 (Vryheid): Nyati Mt., *Hilliard & Burtt 5897.* 2829 (Harrismith): Oliviershoek Pass, *Hilliard 4910;* Tintwa Mts, *Doidge sub PRE 31569.* 2831 (Nkandla): Nkandla Forest, *Hilliard & Burtt 5647;* Eshowe district, *Gerstner 3204.* 2929 (Underberg): Bulwer, Marwaqa Mt., *Marais 1448;* Cathkin Park, *Galpin 11764.* 2930 (Pietermaritzburg): Umgeni Poort, *Moll 1403;* Little Noodsberg, *Strey 6250;* Table Mt., *McClean 152.* 2931 (Stanger): near Mapumulo, *Moll 2945.* 3029 (Kokstad): Ngeli Mt., *Hilliard & Burtt 5753.* 3030 (Port Shepstone): farm Ellesmere, *Strey 7054;* Paddock, *Strey 6193.* 3130 (Port Edward): Beacon Hill, *Strey 6070.*

CAPE.—3028 (Matatiele): near Mt. Frere, *Story 947.* 3029 (Kokstad): Tabankulu Mt., *Hilliard & Burtt 6526;* Fort Donald, *Tyson 1655.* 3128 (Umtata): Baziya Mt., *Baur 391.* 3129 (Port St. Johns): Port St. Johns, *Flanagan 2516.* 3228 (Butterworth): Kei Mouth, *Flanagan 419.*

There is considerable variation in stature, leaf size and leaf cutting, in indumentum, and in the presence or absence of red tints in the vegetative parts of this species, the most handsome and, by virtue of its beautiful orange flowers, the most easily recognized of the South African Begonias. Favouring as it does banks, cliffs and boulders, it is very susceptible to the effects of drought, and there is little doubt that environmental factors profoundly influence stature and leaf size. There seems to be some correlation too between light intensity and the development of the red coloration in stems, petioles, veins and so on. The degree of lobing and cutting of the leaf margin is more probably under genetical control. The margin is always sharply serrate and may be shallowly to deeply lobed, even deeply dissected, but the variation is continuous. Nearly all specimens from the Transvaal and Swaziland are noticeably hairy with both leaf surfaces, and often the petiole as well, clothed in long (up to 3 mm) hairs, while in Natal, the Transkei and eastern Cape, leaves are generally glabrous or have a sparse covering of very short hairs. However, around Sabie and Barberton in the Transvaal, plants may be hairy or nearly glabrous, while specimens with long hairs on the leaf surfaces but scarcely extending onto the petiole occur in widely separated areas in Natal and the Cape, for example, in Ndwedwe district *(Strey* 6131; 6064; 6250), Umzinto district *(Rudatis* 811; *Strey* 7054) and at Port St. Johns in the Transkei *(Flanagan* 2516 and *Flanagan* sub herb. Bolus 8717).

5. **Begonia homonyma** Steud., Nom. Bot. ed. 2: 194 (Aug. 1840); B.L. Burtt in Notes Roy. Bot. Gard. Edinb. 32: 274 (1973). Type: cult. Edinb. ex Berlin.

B. sinuata E. Mey. ex Otto & Dietr. in Allg. Gartenztg. 4: 357 (5 Nov. 1836), nom. illegit., non Wall. ex Meisn. (Aug. 1836). Type: cult. Berlin (not seen). *B. sinuata* Grah. in Edinb. New Phil. Journ. 24: 192 (1838); in Bot. Mag. t.3731 (1839), nom. illegit., non Meisn. (1836). Type as for *B. homonyma*. *B. caffra* Meisn. in Linnaea 14: 501 (1841); Sond. in F.C. 2: 481 (1862); Irmscher in Bot. Jahrb. 81: 136 (1961). Type as for *B. sinuata* Otto & Dietr. *B. dregei* Otto & Dietr. var. *sinuata* A.DC. and var. *caffra* A.DC. in DC., Prodr. 15: 384 (1864). *B. favargeri* Rechinger in Ann. Naturh. Hofmus. Wien 20: 33 tab. 1 (1905); Thonner, Fl. Pl. Afr., t.107 (1915). Type: Natal, at the Umkomaas River, *Krook* (Pl. Penther. no. 2976, W, not seen). *B. rudatisii* Irmscher in Bot. Jahrb. 81: 129 (1961). Type: Natal, Umzinto distr., Dumisa, Friedenau, *Rudatis* 602 (B, not seen; E; K).

Augustia caffra (Meisn.) Klotzsch, Beg. Gatt. Art.: 81 (1855).

A glabrous, herbaceous perennial herb. *Stems* erect, 1 or several from a large tuber, up to 1 m tall, simple or branched, stout, fleshy, sometimes tinged red, leafy throughout.

Stipules oblong-lanceolate, membranous, brown. *Leaves* obliquely ovate, up to c. 13 × 7 cm, base shallowly cordate, apex broadly acute, margins entire, sinuate or broadly and shallowly lobed, the lobes rarely subserrate, thin-textured, crisp, light green and waxy above, paler below, the young ones often white-spotted, veins (in dried material) strongly raised below particularly near the basal sinus, often reddish, petioles up to 13 cm long, but often of the order 3–5 cm, decreasing in length upwards, sometimes reddish. *Inflorescences* in the upper leaf axils. *Flowers* up to 3 cm across, white, pink or white tinged pink, tepals of ♂ flowers usually 2, of ♀ flowers usually 5. *Fruits* 3-winged, about 2 × 2 cm across the expanded wings, placentae entire. FIG. 46: 1.

Recorded from the more coastal districts of Natal and the Transkei, up to about 900 m above sea level, from the southern end of the Lebombo Mts. to the Great Kei, on rock falls and cliffs in forest. Flowers from December to March.

NATAL.—2732 (Ubombo): Gwalaweni Forest, *Vahrmeijer & Drijfhout 2002*. 2930 (Pietermaritzburg): Bothas Hill, *McClean 116*. 2931 (Stanger): Umvoti Valley, near Glenmill, *Hilliard 5084*. 3029 (Kokstad): near Gundrift, *Acocks 12237*. 3130 (Port Edward): Umtamvuna Bridge, *Strey 5982*.

CAPE.—3029 (Kokstad): Clydesdale, *Tyson 2689*. 3129 (Port St. Johns): near Lusikisiki, *Strey 8930*. 3228 (Butterworth): Collywobbles, *Van Breda 878*. Grid uncertain: Gcalekaland, N. of Great Kei River, *Dinnie 665*.

B. homonyma together with *B. dregei* and the forms placed in synonymy thereunder represents a highly variable and complex group. *B. homonyma* can, however, be distinguished by its large obliquely ovate leaves with entire or shallowly lobed margins and veins strongly raised below particularly near the basal sinus. There is an almost clearcut discontinuity in the absolute size of the mature primary leaves of *B. homonyma* and *B. dregei:* in *B. homonyma* it scarcely falls below 7 × 3 cm and is more usually about 11 × 6 cm; in *B. dregei* the largest leaf measured was 8 × 3,5 cm, but more usually they are about 5 × 2 cm or less.

A more reliable measurement (one that must largely circumvent the effects of environmental factors and is an indicator of leaf shape rather than size) is the ratio obtained by dividing leaf breadth into leaf length (length of the vein running out to the leaf tip in this case = length, and the length of the adjacent vein in the broader half of the leaf = breadth). In *B. homonyma* this is 1: (1,1–) 1,3 (–1,5); in *B. dregei* 1: (1,3–) 1,7 (–2,5). The lower ratios in *B. dregei* (1: 1,3–1,4) are from small, deeply serrate or partly dissected leaves that cannot be confused with those of *B. homonyma*.

6. **Begonia dregei** *Otto & Dietr.* in Allg. Gartenztg. 4: 357 (1836); Meisn. in Linnaea 14: 502 (1840); Sond. in F.C. 2: 481 (1862); A.DC. in DC., Prodr. 15: 384 (1864), excl. var. *sinuata* and var. *caffra;* Irmscher in Bot. Jahrb. 81: 130 (1861). Type: Cape, between the Umtata and Umgaziana rivers, *Drège* (B, not seen).

B. parvifolia sensu Grah. in Edinb. New Phil. Journ. xxiv: 91 (1838); in Bot. Mag. t.3720 (1839). *B. suffruticosa* Meisn. in Linnaea 14: 502 (1840); Sond. in F.C. 2: 481 (1862); A.DC. in DC., Prodr. 15: 385 (1864) excl. var. *gueinziana* A.DC.; Irmscher in Bot. Jahrb. 81: 130 (1961), incl. forma *bolusii* Irmscher and forma *worsdellii* Irmscher. Type: Cape, between the Umgaziana and Umzimvubu rivers, *Drège* 8032 (B, not seen). *B. natalensis* Hook. in Bot. Mag. t.4841 (1855); Sond. in F.C. 2: 481 (1862); A.DC. in DC., Prodr. 15: 385 (1865); Irmscher in Bot. Jahrb. 81: 138 (1961). Type: the plate in Bot. Mag., from specimen collected by Capt. Garden in Natal, not preserved. *B. richardsiana* T. Moore in Gard. Chron. 1065 fig. 243 (1871); Regel in Gartenflora 21: 293 tab. 739 fig. b (1872); Houllet in Rev. Hort. 44: 333 fig. 35 (1872), sphalm. *richardsoniana;* R. A. Dyer in Flow. Pl. Afr. t.673 (1937). Type: spec. cult. R.H.S. (K). *B. partita* Irmscher in Bot. Jahrb. 81: 143, tab. 10, fig. 2 (1961). Type: Natal, Umzinto district, farms Campbellton and Glenrosa, 550 m, *Rudatis* 1876 (B, holo, not seen; NBG).

Augustia dregei (Otto & Dietr.) Klotzsch, Beg. Gatt. Art. 80 (1855). *A. natalensis* (Hook.) Klotzsch, l.c. 81 (1855). *A. suffruticosa* (Meisn.) Klotzsch, l.c. 81 (1855).

Glabrous (or occasionally a few fleshy setae on the upper leaf surface) perennial herbs. *Stems* up to 30 cm tall, one or several from a tuber, simple or branched, erect or sometimes decumbent at the base and there rooting, thick, fleshy, sometimes reddish, leafy throughout. *Stipules* oblong-lanceolate, membranous, brown. *Leaves* obliquely-ovate to obliquely ovate-lanceolate, up to 8 × 3,5 cm but more usually about 5 × 2 cm, base shallowly cordate to subtruncate, apex narrowly triangular-acute to long-attenuate and acuminate, margins rarely subentire to sinuate, more usually more or less regularly lobed or with 1 or 2 spreading lobes in the lower part of the broader half of the leaf, the narrower half lobed or not, the lobes in turn coarsely serrate, sometimes the leaf more deeply or even relatively finely dissected, somewhat fleshy, light green above, paler below, the young leaves often, the mature occasionally, white-spotted, veins green or reddish, not, or very rarely, raised below, petioles up to 9 cm long, often 4–5 cm, always decreasing in length upwards, green or reddish. *Inflorescences* in the upper leaf axils. *Flowers* up to 3 cm across, white, pink, or white tinged pink, tepals of ♂ flowers usually 2, of ♀ flowers usually 5. *Fruit* 3-winged, 1–2 × 1–2 cm across the expanded wings, placentae entire. FIG. 46: 2.

B. dregei ranges from Qudeni, Ngotshe and Ngoye forests just north of the Tugela in Zululand, through coastal Natal as far south as East London in the eastern Cape, usually below 600 m altitude. It therefore occupies virtually the same area as *B. homonyma*. Like that species, and other South African Begonias, it grows on rocky cliffs, steep

FIG. 46.—To show leaf variation in 1, **Begonia homonyma** and 2, **B. dregei;** 1a and 1b both from the Umvoti river, Natal (*Hilliard* 5084). 2a, Port St. Johns, Transkei (*Admiraal* 2286); 2b, Port St. Johns (*Strey* 6655); 2c and 2d, both on one specimen from the Bashee River mouth, Transkei (*Gordon-Gray* 1564); 2e, Mqanduli, Transkei (*Theron* 1504); 2f, Kei Mouth, Transkei (*Flanagan* 419); 2g, Cobe, Transkei (*Pegler* 813); 2h, Port St. Johns (*Sim* 20231); 2i, Kentani, Transkei (*Pegler* s.n.); 2j, hort. Kew Nov. 1860; 2k, Inanda, Natal (*Wood* s.n.); 2l, Eshowe, Zululand (*Lawn* 1447); 2m, Lupatana, Transkei (*Strey* 10223); 2n, Inanda (*Wood* 1197); 2o, Ngotshe Forest; Eshowe (*Hilliard & Burtt* 5657); 2p, Ngoye Forest, Zululand (*Wood* 10349); 2q, Ngoye Forest (*Hilliard & Burtt* 5640); 2r and s, both on a single specimen from Coffee Bay, Transkei (*Wells* 3539); 2t, Qora Mouth, Transkei (*Wells* 3595); 2u, Port St. Johns (*Bolus* 8914); 2v, Port St. Johns (*Mogg* s.n.); 2w, Manubi Forest, Transkei (*Worsdell* s.n.). 2a is the form described as **B. dregei,** 2j as **B. natalensis,** 2p as **B. partita,** 2u and 2v as **B. suffruticosa.** All × $^3/_5$.

earth banks and among rock falls in forest, and flowers mainly from December to March. Although the material may be classified into several groups, it is preferred not to adopt a formal infraspecific nomenclature for them at this stage.

GROUP 1 (*dregei* sens. strict.)

NATAL.—2831 (Nkandla): near Eshowe, *Lawn 517*.

CAPE.—3129 (Port St. Johns): Port St. Johns, *Admiraal 2286;* Mqanduli district, banks of Umnenga River, *Theron 1504*. 3228 (Butterworth): Kei Mouth, *Flanagan 419*.

GROUP 2 (*dregei* → *natalensis*)

NATAL.—3030 (Port Shepstone): Paddock, *Strey 5886;* St. Michaels-on-Sea, *Ross 1885*.

CAPE.—3129 (Port St. Johns): Port St. Johns, *Sim 20231*. 3228 (Butterworth): Kentani, *Pegler 813;* Bashee River Mouth, The Haven, *Gordon-Gray 2006*. 3327 (Peddie): East London, *Batten s.n.*

GROUP 3 (*natalensis* sens. strict.)

NATAL.—2830 (Dundee): Qudeni, *Gerstner 3233*. 2831 (Nkandla): near Eshowe, *Hilliard & Burtt 5657* (leaves here occasionally bilobed and thus approach *B. partita;* a similar specimen, *Lawn 1447* from near Eshowe, was designated *partita* × *natalensis* by Irmscher). 2930 (Pietermaritzburg): Inanda, *Wood 1197*. 3030 (Port Shepstone): farm Excelsior, *Hilliard & Burtt 3390*.

CAPE.—3129 (Port St. Johns): near Lusikisiki, *Strey 10198;* Lupatana, *Strey 10223*. 3227 (Stutterheim): Horseshoe, *Galpin 8212*.

GROUP 4 (*partita* sens. strict.)

NATAL.—2831 (Nkandla): Ngoye, *Wood 10349; Hilliard & Burtt 5640*. 3030 (Port Shepstone): Port Shepstone, *Dimock Brown 489*.

GROUP 5 (*suffruticosa* sens. strict.)

CAPE.—3129 (Port St. Johns): Port St. Johns, *Bolus 8914; Flanagan 2510; Galpin 3465; Howlett 21; Mogg 20157; Moss 4620; Schelpe 285; Schonland 4185*.

The extensive synonymy of *B. dregei* reflects the high degree of variability in leaf lobing and cutting and, to a lesser extent, in leaf shape. The specimens cited have been grouped to indicate this variation and the associated names. The variation is, however, continuous and is complicated by the fact that juvenile and secondary leaves are often entire or less deeply cut than mature primary leaves. For instance, in *Gordon-Gray* 2006 (cited in Group 2 above) some of the leaves, particularly the lower primaries, fit the criteria for *B. dregei,* while upper and secondary leaves fit those of *B. natalensis*.

Graham illustrated as *B. parvifolia* (Bot. Mag. t.3720) a plant received from the Berlin garden in 1836, the year in which *B. dregei* was described. The plants were undoubtedly the same and probably of the same stock. The Bot. Mag. plate may therefore be taken to depict "typical" *B. dregei,* having small obliquely ovate leaves, apex broadly acute, margin more or less regularly lobed, the lobes toothed. This leaf form grades imperceptibly into that of typical *B. natalensis:* leaf more elliptic-ovate in outline, with two broad lobes in the basal half and a long-attenuated upper half, margin typically serrate (see Bot. Mag. t.4841) but grading to sinuate or entire. Irmscher designated some intermediate plants as hybrids between *B. dregei* and *B. natalensis,* but they seem to be merely part of the continuous variation pattern (see fig. 46).

The name *partita* is attached to a specimen with deeply bilobed leaves, these two main lobes further lobed, margins entire or serrate. *B. suffruticosa* forma *worsdellii* has the deep bilobing of *B. partita* but in the marginal toothing begins to approach *B. suffruticosa*. *B. suffruticosa* is represented in herbaria by eight collections, all from Port St. Johns. The type locality "between the Umgaziana and Umzimvubu rivers" is also Port St. Johns. There seems little doubt that *B. suffruticosa* is merely a local variant with deeply dissected leaves, linked through *B. suffruticosa* forma *worsdellii* and other variants to *B. partita* and so to *B. natalensis* and *B. dregei*.

B. dregei is very closely related to *B. homonyma,* and the whole complex demands a thorough field and laboratory study. For that reason, no infraspecific categories are recognized here.

Both *B. dregei* and *B. homonyma* have been much confused with *B. sonderana,* a more northern species easily distinguished by its split placentae.

CACTACEAE

by A. A. OBERMEYER

Perennial, herbaceous, suffruticose or arborescent; predominantly succulents, rarely epiphytic. *Stems* woody or mostly succulent and variously modified, cylindrical, flattened or globose, bearing axillary, rounded, usually raised, woolly areoles which bring forth leaves, flowers, spine-clusters and often barbed bristles. *Leaves* present, usually early deciduous, often reduced or 0. *Flowers* solitary or rarely clustered, sessile (except *Pereskia*), bisexual, mostly regular. *Hypanthium* (receptacle-tube) long or short, naked or with leaf-like bracts and areoles containing reduced spines, wool and bristles and giving rise to a series of imbricate, sepaloid green segments which gradually change into a series of often very showy petaloid segments. *Stamens* numerous, inserted on the receptacle, free or adnate to the petaloid segments, anthers bilocular, opening lengthwise. *Ovary* inferior, 1-locular with 3 or more, many-ovuled, parietal placentas; style simple; stigmas 3–many. *Fruit* usually a berry, often edible, rarely a capsule; seeds few to numerous.

Genera 84, species approximately 2 000; widespread in the Americas with many naturalized and often troublesome elsewhere. A great many species representing a number of genera are grown by gardeners and hobbyists but only those which have become naturalized are dealt with here; most of these have been proclaimed as noxious weeds. One genus, *Rhipsalis*, appears to be indigenous.

Plants with succulent stems variously shaped; leaves 0 or few and then usually reduced and early deciduous; flowers solitary in each areole (rarely 2 or more in *Rhipsalis* but then small):

- Glochidia (barbed bristles) absent; stems terete, succulent, not articulated, supple when young; leaves absent; seeds black or brown, not encased in a long aril:
  - Terrestrial, large clambering plants without aerial roots; stems c. 2 cm in diam.; flowers c. 20 cm long, funnel-shaped ....1. **Harrisia**
  - Epiphytic or saxicolous plants with aerial roots; stems 3–6 mm in diam.; flowers c. 5 mm long ....2. **Rhipsalis**
- Glochidia present inside areoles; stems firm, terete or flattened, often articulated; leaves rarely present, usually small and early deciduous; seeds encased in a bony, white aril ....3. **Opuntia**

Plants with normal stems and leaves; flowers large, several from each areole ......4. **Pereskia**

5401a 1. HARRISIA

Harrisia *Britton* in Bull. Torr. Club 35: 561 (1908); Britton & Rose, The Cactaceae 2: 147 (1920); Hunt in Hutch., Gen. Flow. Pl. 2: 445 (1967). Type species: *H. (Cereus) gracilis* (Mill.) Britton.

Cereus, group 18, *Eriophori, A, Euharrisia*, Vaupel in Pflanzenfam. ed. 2, 21: 633 (1925).

Arborescent or clambering shrubs with long supple 3–12-ribbed or -angled stems, bearing several usually stout spines in each areole. *Flowers* nocturnal, solitary, large, funnel-shaped; outer segments linear to lanceolate, greenish or pink, inner white or pink; the cylindric hypanthium as long as the limb or longer, bearing pubescent areoles and scales; stamens numerous; ovary and young fruits tubercled. *Fruit* globose to obovoid-globose, with or without spines, scales deciduous; seeds numerous, small, dull black, somewhat verrucose.

About 20 closely related species recorded from tropical America to Argentina. One species recently naturalized in Natal.

Harrisia martinii *(Labouret) Britton* in Addisonia 2: 55 (1917); Britton & Rose, The Cactaceae 2: 155, t.19, f.3, t.20, f.2 (1920). Type: probably not preserved; origin unknown. Recorded from Argentina.

Cereus martinii Labouret, Ann. Soc. Hort. Haute Garonne (1854).

Plants with tuberous roots and long, fairly supple, clambering, much branched stems, 1–2 cm in diam., 4–5-angled; areoles numerous, placed fairly close together, raised and shortly pubescent bearing a stout central, patent or reflexed spine up to 3 cm long and several very short radiating spines. *Flower* about 20 cm long; outer segments linear, greenish; inner broader, white or pinkish. *Fruit* globose, 3,5 cm in diam., red, with small scales, the withered perianth long persistent; seeds small, black, scattered in a white pulp. FIG. 47.

Recorded from Natal in the neighbourhood of Pietermaritzburg where it has become troublesome in some areas, the birds transporting the seeds. Declared a weed throughout the Republic in 1968.

NATAL.—2930 (Pietermaritzburg): Ashburton, *Regional Officer sub PRE 32235; Paterson sub PRE 32308.*

FIG. 47.—1, **Harrisia martinii,** portion of flowering branch, × 2/3; a, fruit, × 2/3 (After Britton & Rose).

5416

2. RHIPSALIS

Rhipsalis *Gaertn.*, Fruct. Sem. 1: 137, t.28 (1788) nom. conserv.; Britton & Rose, The Cactaceae 4: 219 (1923); Vaupel in Pflanzenfam. ed. 2, 21: 617 (1925); Phill., Gen. ed. 2: 522 (1950); Hunt in Hutch., Gen. Flow. Pl. 2: 427 (1967), in F.T.E.A. Cactaceae: 5 (1968). Type species: *R. baccifera* (J. Mill.) Stearn (=*R. cassutha* Gaertn.).

Unarmed succulents, often epiphytic. *Stems* pendulous, branching freely, terete, angled and/or jointed or flattened and leaf-like, areoles small, woolly, bristly, rarely spinous. *Leaves* reduced to minute scales or 0. *Flowers* usually solitary, lateral, sessile, small. *Perianth segments* 5–12, with the outer sepaloid series short and free or nearly so, and the inner petaloid, fused below, mostly white. *Stamens* few to numerous, inserted at the base of the perianth or attached to the base of the petals. *Ovary* usually inferior with 3–several parietal placentas. *Berry* globular, translucent, smooth or with some scattered, reduced areoles containing bristles or spines, juice viscid; seeds oblong, irregularly angled, brown to black.

About 50 species in South America, Mexico and West Indies. A few of these naturalized elsewhere; one species, probably *R. cereusculus* Haw., is often cultivated in South Africa. The following species, doubtfully native, found wild in moist warm regions of Africa, Madagascar, Mascarenes and Sri Lanka.

Rhipsalis baccifera *(J. Mill.) Stearn* in adnot. Cactus J. 7: 107 (1939); Hunt in F.T.E.A., Cactaceae: 5, fig. 2 (1968); Troupin, Syll. Fl. Rwanda, Spermatophytes: VI, 47 with figure (1971). Type: description and plate, Class IX, Ord. 1 (1771) in J. Mill., Illustr. Sex. Syst. Linn. (1771–77).

Cassyta baccifera J. Mill., Illustr. Sex. Syst. Linn. Class IX, Ord. 1 (1771–77). Type as above.

R. cassutha Gaertn., Fruct. Sem. 1: 137, t.28 (1788); Hook. in Bot. Mag. 58: t.3080 (1831), as *R. cassytha*, the spelling subsequently adopted by most authors; Oliv. in F.T.A. 2: 581 (1871); Britton & Rose, The Cactaceae 4: 225 (1923); Vaupel in Pflanzenfam. ed. 2, 21: 619 (1925); Burtt Davy, Fl. Transv. 1: 237 (1926); Bally in J. E. Afr. Nat. Hist. 16: 44 (1942); Roland-Gosselin in Desert Plant Life 1947: 121 (1947). Type: ex hort. Kew, probably not preserved; origin unknown.

Epiphytic or saxicolous, small succulents with much branched, pendulous, succulent, terete, soft, glaucous-green stems about 3–6 mm in diam., bearing adventitious aerial roots covered with a thin velamen; branches clustered, articulated; areoles scattered, woolly, bearing few to many soft bristles up to 4 mm long which usually disappear with age but the juvenile form may persist for many years. *Flowers* 1–2 per areole, sessile, small, green, self-fertile; hypanthium bulbous, bearing a few reduced areoles with 1–2 soft short bristles; sepaloid segments few, minute, triangular; petaloid segments 4–6, oblong, irregular, c. 3 mm long, somewhat fleshy, greenish-white. *Stamens* 5–10, placed on receptacle, short, with short opposing locules, apiculate. *Ovary* embedded in bulbous hypanthium; style short. *Berry* globose or oblong-globose, 5–10 mm in diam., translucent, greenish-white, white, pink or red, the petaloid segments semi-persistent, sap viscid; seeds oblong, irregularly and obtusely angled, 1 mm long, black or dark brown, shiny, reticulate. FIG. 48.

Widespread in South and Central America; in moist warm regions of tropical west Africa to south eastern Africa, Madagascar, Mascarenes and Sri Lanka. Plentiful in Zululand, southern Natal and north eastern Cape, in coastal evergreen forests on trees or on rocks. Distributed by birds which are attracted to the berries; these contain a viscid juice which aids dispersal.

SWAZILAND.—2631 (Mbabane): Isateki Beacon south of Stegi, *Compton 27900; 27316*.

NATAL.—2731 (Louwsburg): Lebombo Mts., Border Cave, *Wells 4456;* Pongola Poort, *Repton 5982; Strey 4645;* Josini Dam, *Codd 10313*. 2831 (Nkandla): Umfolozi Game Reserve, *Bourquin 449a;* Eshowe–Entumeni Road, *Gerstner sub NH 22449; Lawn 435;* Ngoye Forest, *Wells & Edwards 51*. 2832 (Mtubatuba): Hluhluwe Game Reserve, *Ward 2057;* False Bay, *Ward 1635*. 2930 (Pietermaritzburg): Inchanga Valley, *Van der Merwe 2843*. 2931 (Stanger): Thrings Post, *Moll 2305*. 3030 (Port Shepstone): Paddock, *McClean 324; Strey 6268;* Kenterton, along Ifafa River Valley, *Bridgman sub PRE 32181;* Mgongongo, *Strey 8097*. 3130 (Port Edward): Beacon Hill East, *Strey 7230*.

CAPE.—3129 (Port St. Johns): Waterfall Bluff, Lusikisiki, *Van der Merwe 2515;* Luputana, *Strey 10235;* Port St. Johns, *Flanagan 2598;* West Gate, *Galpin 3197*. 3228 (Butterworth): 11 km from coast on Kei Mouth road in Kei River Valley towards Komga, *Dyer 4506*.

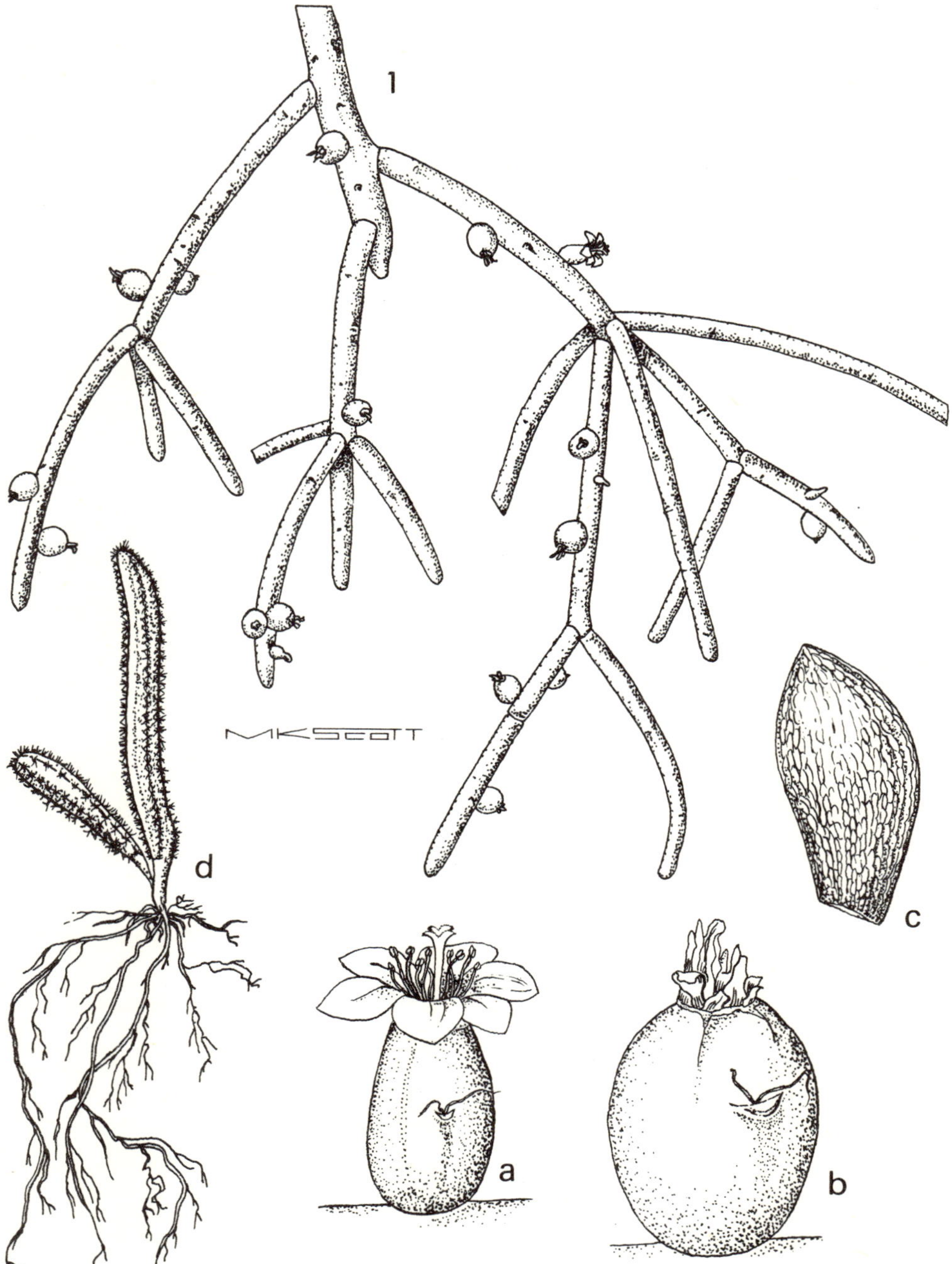

FIG. 48.—1, **Rhipsalis baccifera,** habit, × 1 (*Vahrmeijer & Hardy* 1692); a, flower, × 6; b, berry, × 6; c, seed, × 40; d, seedling, showing juvenile pubescence.

5417 OPUNTIA

Opuntia *Mill.*, Gard. Dict. Abr. ed. 4 (1754); Britton & Rose, Cactaceae 1: 42 (1919); Vaupel in Pflanzenfam. ed. 2,21: 647 (1925); Hunt in Hutch., Gen. Flow. Pl. 2: 435 (1967); F.T.E.A. Cactaceae: 1 (1968). Lectotype species: *O. compressa* (Salisb.) Macbr. (*Cactus opuntia* L., Sp. Pl.: 468, 1753, partly). For synonyms see Hunt in Hutch., Gen. Flow. Pl. 2: 435 (1967).

Arborescent, suffruticose or stoloniferous, gregarious, perennial succulents usually spreading through breakage of brittle joints (i.e. jointed succulent green branch-segments). *Roots* fibrous and/or tuberous. *Stems* in arborescent plants forming woody, cylindrical trunks; the joints cylindrical or flattened, linear to rounded, usually contracted below at the node, glabrous or pubescent, beset with few to many scattered areoles (metamorphosed axillary buds) which are variously pubescent, bear barbed bristles (glochidia) and usually one to many radiating, unequally long spines, besides secondary joints, leaves and flowers which are usually placed apically. *Leaves* when present mostly small, terete or subulate, usually early deciduous. *Flowers* solitary in each areole, fairly large with a green hypanthium (modified joint) beset with few to many areoles; sepaloid segments usually small, green, bract-like, grading into many yellow, orange or red, often showy, erect or spreading petaloid segments; stamens numerous; ovary embedded in the swollen hypanthium; style terete, stout, stigma-lobes 6–8. *Fruit* a berry, the epicarp fleshy, ovoid to globose, often with a concave apex; areoles few to many, the spines absent, few, or reduced to bristles; seed discoid or reniform, covered by a hard funicular aril, embedded in a soft juicy mesocarp.

About 240 variable species in North and South America. A small number have been in cultivation in the Old World for centuries and were used for food, forage and protective hedges. In recent times a number of garden escapes have become a terrible scourge in many countries where they have invaded large tracts of land. Their eradication is proving very costly. At present the most serious *Opuntia* infection is found in the eastern Cape, where *O. aurantiaca* and *O. ficus-indica* are invading more and more land and presently cannot be controlled.

A number of spineless hybrids from the United States have been introduced as stock feed in regions often plagued by drought.

Joints cylindrical, tuberculate:
- Leaves long-persisting, arcuate, 5–12 cm long; spines not sheathed, 1–2 per areole; flowers orange or greenish yellow .... 1. *O. exaltata*
- Leaves early deciduous or absent, up to 2,5 cm long; spines covered with papery sheaths, 10–20 per areole; flowers red:
  - Spines close together almost hiding the joint, whitish; small compact bushes up to 60 cm tall .... 2. *O. rosea*
  - Spines more widely spaced, light brown; open branching shrubs, 1–3 m tall .... 3. *O. imbricata*

Joints flattened or semi-terete, not tuberculate:
- Low spreading plants with semi-terete, densely spiny joints, 1,5–2,5 cm wide .... 4. *O. aurantiaca*
- Arborescent or shrubby with flat, rounded to narrowly obovate joints more than 6 cm wide:
  - Joints orbicular, about as long as broad:
    - Joints with close-set short dark thin spines c. 15 mm long, which are mostly deflexed: hypanthium cup-shaped, c. 2 cm long .... 5. *O. spinulifera*
    - Joints with some scattered atrophied areoles, the upper bearing 1–2 hard, flattened, slightly curved spines 3–5 cm long; hypanthium cylindrical, c. 4 cm long .... 6. *O. aff. O. lindheimeri*
  - Joints obovate, attenuate towards the base:
    - Areoles large, over 1 cm in diam., bearing 4–7 coarse hard brown spines up to 5 cm long, emerging from a bristly cushion; hypanthium and fruit curved .... 7. *O. dillenii*
    - Areoles smaller, about 5 mm in diam., with fewer, thinner greyish spines or unarmed; hypanthium and fruit straight:
      - Low spreading bushes 1–2 m tall; fruit obovoid, smooth, purple .... 8. *O. stricta*
      - Taller plants, often arborescent, forming trunks; fruit ovoid to obovoid with many scattered bristly areoles, red, yellow or purple:
        - Spines c. 1 or 2 per areole, 3–7 cm long; joints usually thin, bright green and shiny at least when young, the margin receding between the areoles in dried specimens; fruit obovoid, purplish red . 9. *O. vulgaris*
        - Spines 3–10 per areole, up to 3 cm long or absent in old plants; joints fairly thick, glaucous with straight margins; fruit ovoid to somewhat barrel-shaped, reddish or yellowish .... 10. *O. ficus-indica*

1. **Opuntia exaltata** *A. Berger,* Hort. Mortol. 410 (1912); Britton & Rose, The Cactaceae 1: 76 (1919); Hunt in F.T.E.A. Cactaceae: 2 (1968). Type: Described from living plants in Sir Thomas Hanbury's garden at La Mortola, Italy.

Much-branched spreading shrubs forming thickets, developing a trunk with age. *Joints* arcuate, cylindrical, tuberculate, c. 3 cm in diam., glaucous-green. *Leaves* linear to terete, 3–6 (–12) cm long, arcuate, fleshy, persistent. *Areoles* with a protruding white woolly pubescence; with 1 or 2 acicular, straight, strong yellow-brown spines up to 5 cm long. *Flowers* with the hypanthium forming a reduced apical spineless joint which in turn may proliferate; petaloid segments 2–3 cm long, orange or greenish yellow. *Fruit* not seen in South African material. ("Green, pear-shaped, 9 cm long, usually sterile; seeds large, irregular, 10 mm broad" fide Britton & Rose).

Recorded as a pest from the Orange Free State, Cape and Natal. Found both wild and cultivated in Ecuador, Peru, Bolivia and probably northern Chile, according to Britton & Rose.

TRANSVAAL.—2528 (Pretoria): Pretoria, *Gericke sub PRE 31403*. 2627 (Potchefstroom): On farm Leeuwfontein, Potchefstroom district, *Du Toit sub PRE 31402*.

O.F.S.—2827 (Senekal): Ficksburg district, *Dreyer sub PRE 24037*. 2926 (Bloemfontein): Bloemfontein district, *Dreyer sub PRE 24036*.

NATAL.—3030 (Port Shepstone): Umzimkulu, *Du Toit sub PRE 32673*.

CAPE.—3318 (Cape Town): Koelenhof, on a farm, *Carstens sub PRE 32672*.

Specimens at PRE were referred to *O. subulata* (Mühlenpfordt) Engelm. by Mr. E. J. Alexander of the New York Botanic Garden in 1949. Britton & Rose separate the two taxa, *O. subulata* and *O. exaltata,* although admitting they are very close. Mr. D. R. Hunt placed tropical east African material under *O. exaltata* where the South African material also appears to belong if the differences observed by Berger and Britton & Rose are of importance.

2. **Opuntia rosea** *DC.* in Mém. Mus. Hist. Paris 17: t.15 (1828); Prodr. 3: 471 (1828). Type: from Mexico.

O. pallida Rose, Smithson. Misc. Coll. 50: 507 (1908). Type: from Mexico.

Cylindropuntia rosea (DC.) Backebg., Die Cactaceae 1: 197, t.142 (1958).

Plants c. 60–80 cm tall, forming much-branched compact bushes; the many short curved branches massed in upper part of plant, fairly equal in size. *Joints* cylindrical, tuberculate, c. 4–5 cm in diam., densely covered with radiating spines (up to 20) from each raised areole, almost hiding the stems; spines c. 25–40 mm long, covered with whitish sheaths. *Flowers* c. 4 cm long, rarely developed; hypanthium narrowly ovoid, the tubercles tipped with some minute acicular spines and claw-like leaf rudiments. *Perianth segments* with the inner petaloid, c. 10 (–15) mm long, oblong, obtuse, wine-red, patent to recurved with age. *Stamens* and style shortly exserted, wine red. *Fruits* often proliferous, usually sterile, obtusely obovoid to globose, c. 3 cm long, swollen, nearly smooth.

Recorded from the Douglas district in the northern Cape Province, where there is a heavy infestation in dry thornveld.

CAPE.—2923 (Douglas): farm Kleinplaas, *Annecke & Zimmermann sub PRE 42223; 39302*.

3. **Opuntia imbricata** *(Haw.) DC.* Prodr. 3: 471 (1828); Britton & Rose, The Cactaceae 1: 63 (1919); Phill. in Fmg. S. Afr. 15: 121, t.8 (1940); Henderson & Anderson, Mem. Bot. Surv. S. Afr. 37: 223, f. 110 (1966); Lyman Benson in Fl. Texas 2: 227 (1969). Type: Introduced into cultivation by Loddiges in 1820; origin unknown.

Cereus imbricatus Haw., Rev. Pl. Succ. 70 (1821).

Erect, branched, spreading shrubs often arborescent with age, up to 3 m high. *Joints* cylindrical, up to 40 cm long, 2–3 cm in diam., dark green with the raised prominent tubercles capped by areoles bearing 8–30 stellately arranged spines, these variable in length, up to 3 cm long, brown, minutely barbed and at first covered with a white papery sheath. *Leaves* subulate, 1–2 cm, early deciduous. *Flowers* 4–6 cm long with the hypanthium forming an abbreviated apical joint; petaloid segments red, rotate; stamens and style purple. *Fruit* obovoid, up to 5 cm long, yellow, tuberculate, inedible, long persistent; seeds pale yellow, c. 3 mm in diam.

Recorded as a pest from the eastern Cape, Natal, south-western Transvaal and north-western Cape. Distributed in the south-western States of North America and Mexico.

TRANSVAAL.—2725 (Bloemhof): Schweizer-Reneke, *Du Toit sub PRE 24539*.

CAPE.—2824 (Kimberley): Kimberley Mining area, *Leistner 3507*. 3227 (Stutterheim): Cathcart, Riverside, 8 km N.N.W. of village, *Forester s.n.*

Common names: Kabelturksvy; Imbricate cactus.

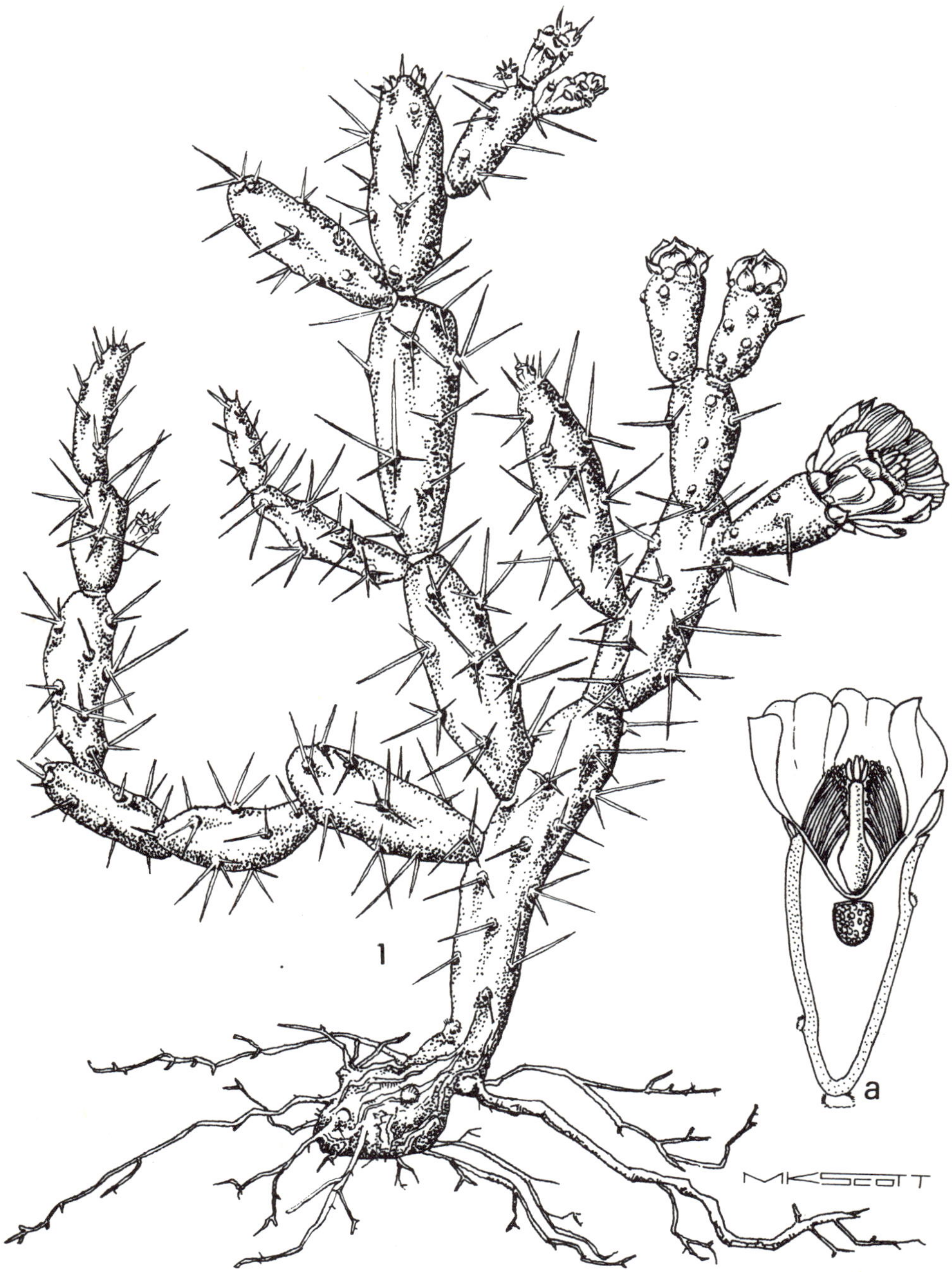

FIG. 49.—1, **Opuntia aurantiaca,** habit, × ½; a, longitudinal section of flower, × 1 (after Landbouweekblad, 22 Sept. 1970 and Henderson & Anderson, 1966).

4. **Opuntia aurantiaca** *Lindl.* in Bot. Reg. 19, t.1606 (1833); Britton & Rose, The Cactaceae 1: 107 (1919); Lansdell in J. Dept. Agric. Reprint 72 (1923); Burtt Davy, Fl. Transv. 1: 235 (1926); Phill. in Fmg. S. Afr. 13: 216 (1938); Archibald in S. Afr. J. Sci. 36: 195–211 (1939); Henderson & Anderson, Mem. Bot. Surv. S. Afr. 37: 220, f. 109 (1966). Type: Botanical Register, Vol. 19, t. 1606, from South America.

Low spreading and creeping shrublets up to 1 m high, rooting where contact is made with the soil. *Stems* with a tuberous subterranean portion, terete above ground, much branched. *Joints* brittle, subterete to linear, 6–20 cm long, 1–2,5 cm broad, light green; areoles close together, woolly, with glochidia and 2–7 sharp strong radiating spines, up to 3 cm long, minutely barbed apically. *Leaves* rarely present. *Flowers* many, c. 4 cm long; hypanthium tubular, usually spineless; petaloid segments lemon yellow to golden yellow. *Fruit* globose, 3–4 cm in diam., red, often bearing a few spines, inedible; able to form roots and reproduce vegetatively; seeds numerous, reniform, light brown, with low viability. FIG. 49.

Recorded as a pest in the eastern Cape; occasional in the Orange Free State, Natal and Transvaal. Originally from South America.

NATAL.—2830 (Dundee): Niekerkskraal, Wesselsnek, *Paterson sub PRE 32309.*

CAPE.—3326 (Grahamstown): near Grahamstown, *Leemann sub PRE 23290.*

According to Phillips (Fmg. in S. Afr. 13: 216, 1938) the species was introduced between 1850 and 1860 as an ornamental plant in a garden adjoining the old Mission Station at Hertzog, a village in the Stockenstroom district in the eastern Cape. (Grid 3226 DA). At present the infestation in the eastern Cape has become uncontrollable and eradication by means of biological control and chemical poisons, while proving extremely costly, have so far met with little success. The taxon appears to be of hybrid origin.

Common names: Katjie, Litjieskaktus, Litjiesturksvy, Platturksvy, Rankturksvy, Taaietjie, Jointed Cactus.

5. **Opuntia spinulifera** *Salm-Dyck,* Hort. Dyck. 364 (1834); Britton & Rose, The Cactaceae 1: 182 (1919); Phill. in Fmg. S. Afr. 15: 120 (1940). Type: Described from plants in cultivation in Europe.

Much-branched, forming large dense clumps up to 3 m tall (not arborescent). *Joints* flattened, orbicular, 20–40 cm in diam., glaucous-green, densely beset with areoles in a regular sub-spiral arrangment, areoles small, c. 4 mm in diam., sunken, with numerous glochidia and 3–6 spines of varying lengths mostly pointing downwards, the longest c. 2 cm long, the marginal spine clusters more strongly developed. *Flowers* with a cup-shaped hypanthium, 2,5 cm long wide, profusely covered by areoles containing bristles around the mouth; petaloid segments yellow, up to 4 cm long. *Fruit* not seen. (Britton & Rose do not describe it).

Recorded as a pest in the eastern Cape and Natal. Also reported as a pest in Australia. Originally from Mexico.

NATAL.—2930 (Pietermaritzburg): Pietermaritzburg, *Harding 5.*

CAPE.—3226 (Fort Beaufort): Stockenstroom district, *Oates s.n.;* Hertzog and Seymour, *Van der Merwe sub PRE 10054.*

Common names: Blouturksvy, Rondeblaar, Groot Rondeblaar, Saucepan cactus, Large Roundleaved Prickly Pear.

6. **Opuntia** sp. aff. **O. lindheimeri** *Engelm.* in Bost. J. Nat. Hist. 6: 207 (1850); Britton & Rose, The Cactaceae 1: 165 (1919); Phill. in Fmg. S. Afr. 15: 119, t.7 (1940); Lyman Benson in Fl. Texas 2: 246 (1969). Type from Texas.

Erect spreading shrubs 1–1,5 m tall. *Joints* flattened, orbicular to broadly obovate, c. 20 cm long and 16 cm broad, glaucous-green; areoles widely spaced, lower often atrophied, raised, c. 5 mm in diam., woolly and with glochidia; spines mainly towards the apex, 1–3, up to about 5 cm long, about equal in length, straight or mostly slightly curved, stout, flattened, pale grey to yellow. *Flowers* close together above, in enlarged bristly areoles often attended by a spine; hypanthium cylindrical, with bristly and woolly areoles mostly around the mouth; sepaloid segments broadly obtriangular, apiculate; petaloid segments yellow, retuse. *Fruit* obovoid, c. 7 cm long, purple, fairly smooth; seeds c. 5 mm in diam., white.

Recorded as a pest in the Fort Beaufort district in the eastern Cape.

CAPE.—3226 (Fort Beaufort): Seymour, Kat River Valley, *Van der Merwe sub PRE 10055;* Fort Beaufort, *Geyer sub PRE 31400.*

Specimens from an infestation, confined to an area in the Fort Beaufort district, cannot be placed with certainty in any of the described species. Mr. E. J. Alexander of New York Botanic Gardens in 1938 identified this taxon as *O. tardospina* Griffiths. In the Flora of Texas, Lyman Benson reduces this species to a synonym of *O. lindheimeri.* This species is described by Britton & Rose as extremely variable, composed of many races, differing in armament, etc. This suggested that its affinity could be with this species, even though the spines are not described as flattened and curved which is the case in the local plants.

Common names: Klein Rondeblaar, Small Roundleaved Prickly Pear.

7. **Opuntia dillenii** *(Ker-Gawl.) Haw.*, Suppl. Pl. Succ. 79 (1819); Britton & Rose, The Cactaceae 1: 162 (1919). Burtt Davy, Fl. Transv. 1: 236 (1926); Phill. in Fmg. S. Afr. 15: 119 (1940). Type: Based on an illustration by Dillenius.

Cactus dillenii Ker-Gawl. in Bot. Reg. 3: t.255 (1818).

Low spreading bushes forming clumps or more upright, up to 2 m tall. *Joints* flattened, pale green, narrowly elliptic, c. 20 cm long and 10 cm broad; areoles large, 10–12 mm in diam., densely fringed with bristles up to 8 mm long and with white curly wool in the centre; spines up to 8 in each areole, spreading, varying in length, the longest up to c. 5 cm long, straight or slightly curved, acicular or somewhat flattened, yellow, indistinctly banded when young, becoming brown with age, occasionally absent. *Flowers* with a slender curved hypanthium, narrowed in lower half; petaloid segments a dirty greenish yellow. *Fruit* globose, purple, c. 5 cm long, narrowed and curved towards the base, with few areoles; seed c. 5 mm in diam., brownish.

This species from Central America is now widely distributed through cultivation in warmer parts of the Old World. It is recorded as a pest in southern India and Australia. It is fairly common around Pietermaritzburg in Natal, where it is known as the Pipestem Prickly Pear or Pypsteel-turksvy, because of the curved fruit resembling a tobacco pipe.

NATAL.—2930 (Pietermaritzburg): Pentrich, near Pietermaritzburg, *Harding 3;* Nagle Dam, *Paterson sub PRE 32671.*

Common names: Geeldoringturksvy, Pypsteelturksvy, Pipestem Prickly Pear.

O. dillenii Ker-Gawl. is placed as a variety under *O. stricta* (Haw.) Haw. by Lyman Benson in Cact. Succ. J. 41: 126 (1969), and in Fl. Texas 2: 248 (1969). In South Africa the two taxa appear to be distinct.

8. **Opuntia stricta** *(Haw.) Haw.*, Syn. Pl. Succ. 191 (1812); Britton & Rose, The Cactaceae 1: 161 (1919); Burtt Davy, Fl. Transv. 1: 236 (1926); Lyman Benson in Fl. Texas 2: 248 (1969). Type: A cultivated plant of unknown origin.

Cactus strictus Haw., Misc. Nat. 188 (1803).

Spreading much-branched dwarf bushes 60–80 cm tall, forming thickets. *Joints* flattened, broadly to narrowly obovate, 8–15 cm long, glaucous-green, with several straight, spreading, acicular, yellow-brown spines up to 4 cm long, emerging from prominent areoles, or spines few, small or absent. *Flowers* fairly large, c. 7 cm long, yellow, the hypanthium narrowly tubular, narrowed towards the base, straight or slightly bent, with few areoles; stamens and style about half as long as light yellow petaloid segments. *Fruit* narrowly obovoid, 4–6 cm long, purple, smooth, with very few or without areoles, apex hollow; seeds c. 5 mm diam., yellowish.

Recorded as a pest in the eastern Cape, Natal and recently from northern South West Africa. It is the well known Pest Pear of Australia. It grows wild in western Cuba, Bahamas, Florida to Texas, etc.

S.W.A.—2016 (Otjiwarongo): Otjiwarongo, *Dreyer s.n.*

TRANSVAAL.—2528 (Pretoria): Pretoria, wasteland, *Codd 10760.*

NATAL.—2929 (Underberg): Estcourt, farm near town, *Bedford s.n.* 2930 (Pietermaritzburg): Pietermaritzburg, outskirts on road to Richmond, *Hilliard sub PRE 32306.*

CAPE.—3226 (Fort Beaufort): Fort Beaufort area, *Geyer sub PRE 32237.* 3326 (Grahamstown): around Grahamstown, *Du Toit sub PRE 24035.*

Common Names: The Pest Pear of Australia. (Possibly the names Engelseturksvy, Luisies- and Suurturksvy and Sour Prickly Pear also apply to this species; cf. *O. vulgaris*).

9. **Opuntia vulgaris** *Mill.*,* Gard. Dict. ed. 8, No. 1 (1768); Britton & Rose, The Cactaceae 1: 156 (1919); Burtt Davy, Fl. Transv. 1: 235 (1926); Hunt in F.T.E.A. Cactaceae: 3 (1968). Type: Based on an illustration by Bauhin, Hist. Pl. 1: 154 (1650) after L'Obel (Ic. 2: 241, 1591).

Cactus opuntia L., Sp. Pl. 1: 468 (1753) pro parte minore, quoad syn. Bauhin.

Spreading shrubs up to 5 m tall, forming sturdy, often branching trunks. *Joints* flattened, usually fairly thin, bright green when young, variable in shape and size, narrowly obovate to narrowly elliptic, 10–30 cm long, 5–10 cm broad, attenuate below; areoles small, woolly with brown glochidia, spines straight, acicular, 1–2 (–3) per areole (on the trunks up to 12), unequal in length, the longest 2–3 (–7) cm long, strong, greyish with a brown point. *Flowers* c. 7–9 cm long, with a tubular, narrow hypanthium attenuate at the base, spineless but with

*Known to entomologists as *O. monacantha* Haw., a synonym of *O. vulgaris*.

fairly many areoles; petaloid segments yellow or the outer maroon. *Fruit* edible, obovoid, c. 6 cm long, reddish purple with several persistent areoles; seed 2 mm in diam., brown with a white aril.

Origin unknown, probably from South America, now widespread. Fairly common in the Republic; planted as hedges around homesteads and kraals, also for fodder and fruit.

TRANSVAAL.—2528 (Pretoria); Tweefontein near Premier Mine, *Du Toit sub PRE 23033*. 2531 (Komatipoort): Komati River Poort, *Van der Schijff 3994*.

NATAL.—2632 (Bela Vista): Ndumu, near store, *Stephen & Van Graan 1338*. 2732 (Ubombo): Mkuzi Game Reserve, *Ward 4468*. 2832 (Mtubatuba): Hluhluwe Game Reserve, *Ward 4423*. 2930 (Pietermaritzburg): Pietermaritzburg, *Harding 6*. 2931 (Stanger): Durban, Salisbury Island, *Ward 6253;* Bluff, *Pickworth sub PRE 32669*.

CAPE.—3325 (Port Elizabeth): Uitenhage, *Petty sub PRE 23282*. 3326 (Grahamstown): Bushmans River Mouth, *Johnston 851;* Bathurst, *Comins 2033*.

Common names: Engelseturksvy, Luisiesturksvy, Suurturksvy, English or Sour Prickly Pear.

10. **Opuntia ficus-indica** *(L.) Mill.*, Gard. Dict. ed. 8, No. 2 (1768); Britton & Rose, The Cactaceae 1: 177 (1919); Burtt Davy, Fl. Transv. 1: 236 (1926); Phill. in Fmg. S. Afr. 15: 119, t.3 (1940); Hunt in F.T.E.A. Cactaceae: 2 (1968); Lyman Benson in Fl. Texas 2: 250 (1969). Type: Based on a plant in the Leiden Botanic Garden, of unknown origin.

Cactus ficus-indicus L., Sp. Pl. 1: 468 (1753). Type as above.

O. maxima Mill., Gard. Dict. ed. 8, No. 5 (1768); Britton & Rose, The Cactaceae 1: 180 (1919); Phill. in Fmg. S. Afr. 15: 119 t.2 (1940). Type: a cultivated plant of unknown origin. *O. megacantha* Salm-Dyck, Hort. Dyck. 363 (1834); Phill. in Fmg. S. Afr. 15: 119 (1940); Henderson & Anderson, Mem. Bot. Surv. S. Afr. 37: 225, f.111 (1966). Type: cultivated; probably from Mexico.

Shrubs or trees up to 5 m tall, forming sturdy trunks with age. *Joints* flattened, narrowly elliptic to ovate, varying in size, 30–60 cm long and 6–12 cm broad, attenuate below, often acute above, fairly thick, glaucous-green; areoles small to large, and then raised and woolly, with 3–6 radiating, unequally long, greyish white spines up to 3 (–10) cm long, straight or occasionally slightly curved, or spineless (in older plants and some cultivars). *Leaves,* if developed, minute, subulate, early deciduous. *Flowers* about 7 cm long; hypanthium broadly cylindrical, contracted below, with numerous raised areoles spirally arranged, densely woolly and filled with glochidia, occasionally also bearing small spines and minute leaves; petaloid segments yellow or orange. *Fruit* ellipsoid, c. 7 cm long, reddish, succulent, edible; seeds about 5 mm long.

Benson believes the species to be a native of Mexico. It was introduced to southern Europe, Africa and India very long ago and is used for hedges, fodder and edible fruit. It is the most common and widespread species in South Africa.

TRANSVAAL.—2429 (Zebediela): Sekukuniland, Magnet Heights, *Mogg 795; Barnard sub PRE 23276; 585; 586*. 2527 (Rustenburg): Jacksonstuin near Brits, foot of Magaliesberg, *Van Vuuren 335*. 2528 (Pretoria): Pretoria, near Prison Reserve, *De Winter s.n.*

NATAL.—2632 (Bela Vista): Ndumu Game Reserve, *Oatley sub NH 50818*. 2732 (Ubombo): Mkuzi Game Reserve, *Ward 4450*. 2930 (Pietermaritzburg): Pentrich Grange near Pietermaritzburg, *Harding 743N;* Ashburton, *Paterson sub PRE 32307*.

O.F.S.—2827 (Senekal): frequent on eastern slopes of Groot Doornkop, *Goossens 935; 936*.

CAPE.—3326 (Grahamstown): Farm Belmont near Grahamstown, *Du Toit sub PRE 23254; 23256*–23260.

Recorded as a pest in parts of the eastern Cape, being especially plentiful around Uitenhage; also abundant along the Mediterranean and the Red Sea. Thunberg, in his Flora Capensis ed. Schultes on pp. XV and XVIII (1823), enumerates it (as *Cactus Ficus*) in his list of introduced plants he found there during his travels in the Cape, 1772–1775, recording it as an escape, sometimes used for hedges.

Mr W. G. Barnard in 1937–1939 described the Prickly Pear groves found in Sekukuniland in the eastern Transvaal and their importance to the native tribes living there (cf. notes and photos at PRE). By pruning and grazing the plants are encouraged to develop trunks up to 3 m high and are then left to branch freely, in time forming a dense canopy 4–5 m high. At the base of the Lulu Mountains one of these groves, 30–60 years old, measured about 2 km by 100 m. "Such groves afford fine shelter against sun, wind and cold for farm livestock as well as pleasant play grounds for children. Natives depend upon the fruits for food (and beer) for two months in the year at the critical time between the current season's grain planting and the exhaustion of the previous season's stocks. Inside such groves there is a hive of industry—women gathering and packing fruits; girls and children feasting and dancing; the local chief holding an Indaba (Council); others milking cows and goats—in fact all the common activities of a tribe were being carried on beneath grateful shade on a scorching day in mid-summer". Barnard observed that old plants, 30–50 years old, became spineless, viz. 'Kaalblaar'. Broken-off joints of the Kaalblaar, left exposed to the elements, revert to the spiny condition and were placed around older Kaalblaar trees to protect them. They are also used for fences. Curiously, a joint planted in good soil and well cared for may remain spineless. The fruits of the Kaalblaar are favoured most, having a better taste. He was told that, according to tradition, a joint was introduced into Sekukuniland from Kimberley in 1875.

Common Names: Boer(e) turksvy, Kaalblaar, Doringblaar, Grootdoringturksvy, Mission Prickly Pear.

FIG. 50.—**Pereskia aculeata,** apical portion of branch showing leaves, flowers and fruits, × 1 (*Pickworth* s.n.).

5421

4. PERESKIA

Pereskia *Mill.*, Gard. Dict. Abr. ed. 4 (1754); Britton & Rose, The Cactaceae 1: 8 (1919); Vaupel in Pflanzenfam. ed. 2,21: 615 (1925); Hunt in Hutch., Gen. Flow. Pl. 2: 433 (1967). Type species: *P. aculeata* Mill. (*Cactus pereskia* L.).

Trees, shrubs or scramblers with normal stems and leaves, the raised pubescent areoles bearing spines in pairs or clusters. *Leaves* somewhat fleshy, deciduous, alternate, flat and broad. *Flowers* solitary or in terminal or axillary panicles or corymbs. *Perianth* rotate with the numerous segments entire or fringed. *Stamens* numerous. *Ovary* often with scales, spines or hairs, few- to many-seeded. *Berry* soft, globose or obovoid, unarmed or spiny; seeds black, shiny.

Distributed in Mexico, West Indies, Central and South America, mostly in the coastal regions. A few species cultivated in Natal, one apparently naturalized.

Pereskia aculeata *Mill.*, Gard. Dict. ed. 8 (1768). Type: West Indies.

Cactus pereskia L., Sp. Pl. 1: 469 (1753).

Pereskia pereskia (L.) Karsten, Deutsch. Flora 888 (1882); Britton & Rose, The Cactaceae 1: 10 (1919).

Shrubby to clambering with long whip-like branches; areoles with 1–3 straight spines near the base of the plant but towards the top with pairs of short hooked spines in the leaf axils. *Leaves* shortly petiolate, oblong to ovate, 5–7 cm long, 3–4 cm broad, dark glossy green. *Flowers* aggregated towards the top of the branches, white, cream, yellow or pinkish, rotate, 2,5–4,5 cm in diam., with numerous spreading segments and stamens, strongly scented. *Ovary* beset with leafy scales, often spiny. *Berry* yellow, about 2 cm in diam., quite smooth when ripe; seeds black, somewhat flattened. FIG. 50.

A garden escape in Natal and countries further north, forming impenetrable hedges. Planted over graves for protection by the Zulus. The fruits are edible and known under various names such as Barbados Gooseberry, Lemon Vine, etc.

TRANSVAAL.—2229 (Waterpoort): Waterpoort area, *Burger 109*.

NATAL.—2732 (Umbombo): Lake Sibayi, climber in forest, *Vahrmeijer 723*. 2831 (Nkandla): Empangeni, forestry plantation, *Venter 855*. 2931 (Stanger): S.E. of Stanger on fixed sand dunes, scrambling in a dense mass over Zulu graves, *Smith sub PRE 32236;* between Stanger and Stanger Beach, Ngongoni veld, fairly frequent in bush clumps, *Acocks 10396*. 3030 (Port Shepstone): Isipingo Beach, *Ward 4926*.

CAPE.—3326 (Grahamstown): Grahamstown, near 1820 Settler Museum, *Troughton 100*.

INDEX

*An asterisk signifies exotic species or genera which are not naturalized; synonyms are in italics.